Henry Charlton Bastian

A treatise on Aphasia and Other Speech Defects

Henry Charlton Bastian

A treatise on Aphasia and Other Speech Defects

ISBN/EAN: 9783744751209

Printed in Europe, USA, Canada, Australia, Japan

Cover: Foto ©berggeist007 / pixelio.de

More available books at **www.hansebooks.com**

A TREATISE ON

APHASIA AND OTHER SPEECH DEFECTS

BY

H. CHARLTON BASTIAN, M.A., M.D.Lond., F.R.S.

*Fellow of the Royal College of Physicians of London;
Hon. Fellow of the Royal College of Physicians of Ireland; Hon. M.D. Royal University
of Ireland; Corresponding Member of the Society for Physiological Psychology
of Paris, and of the Royal Academy of Medicine of Turin;
Emeritus Professor of the Principles and Practice of Medicine and of Clinical Medicine
in University College, London; Consulting Physician to University College
Hospital; and Physician to the National Hospital
for the Paralysed and Epileptic.*

WITH ILLUSTRATIONS

LONDON

H. K. LEWIS, 136, GOWER STREET, W.C.

MDCCCXCVIII.

(All Rights Reserved)

PREFACE.

IN this work I have reproduced, with a few additions, the Lumleian Lectures "On some Problems in connection with Aphasia and other Speech Defects," delivered before the Royal College of Physicians of London in 1897, and published in the columns of *The Lancet* during the months of April and May of that year. They form parts of Chapters i. and ii. and nearly the whole of Chapters vi., viii., and ix. The greater part of Chapter vii. was delivered as a Clinical Lecture at University College Hospital, and appeared in *The Lancet* for September, 1897.

With these exceptions the work is a new one. Although it treats the subject in a rather more complete way than has hitherto been attempted, I am fully conscious of its incompleteness and many shortcomings. The literature of the subject can only be described as enormous, and could not be dealt with in anything like a complete manner. Still I have striven to look at this very complicated subject as a whole, and while submitting many prevailing doctrines to a critical examination, I have interwoven therewith an exposition of the views which, after long study, I have myself been led to entertain.

As mere theoretical or speculative opinions on a subject of this kind are comparatively valueless unless they can be shown to be in accordance with facts, I have taken much pains to bring forward a large number of typical cases, some of which have come under my own observation, whilst the majority have been culled from various sources. My object has been to select simple, or comparatively simple, cases which have been completed by the records of a necropsy. This, of course, has not always been possible, so that a certain number of incomplete cases have also been recorded for the purpose of throwing light upon special points capable of elucidation by clinical evidence alone.

Conclusions drawn from very complex cases with diffuse lesions, and also from cases in which there has been no necropsy—and especially from the early symptoms presented by these latter cases—are comparatively devoid of value, and certainly should not be used as a basis for the establishment of new doctrines.

Some modern writers have, in my opinion, needlessly complicated a subject already too complex and obscure, by adopting merely speculative classifications devoid of any proved psychological and pathological bases; whilst others have, perhaps, shown a tendency to overburden it with a needlessly strange and new nomenclature. Both these tendencies I have tried as much as possible to avoid.

H. C. B.

MANCHESTER SQUARE, W.,
February, 1898.

CONTENTS.

CHAPTER I.
INTRODUCTION: PHYSIOLOGICAL AND PSYCHOLOGICAL DATA . . . 1

CHAPTER II.
THE RELATIONS EXISTING BETWEEN THOUGHT AND LANGUAGE . . . 37

CHAPTER III.
THE CLASSIFICATION OF SPEECH DEFECTS 51

CHAPTER IV.
DEFECTS OF SPEECH AND WRITING DUE TO STRUCTURAL OR FUNCTIONAL DEGRADATION OF MOTOR CENTRES 55

CHAPTER V.
DEFECTS OF SPEECH AND WRITING DUE TO LESIONS INVOLVING THE PYRAMIDAL FIBRES 61

CHAPTER VI.
STRUCTURAL DISEASE IN THE GLOSSO-KINÆSTHETIC AND CHEIRO-KINÆSTHETIC CENTRES 80

CHAPTER VII.
MERE FUNCTIONAL DISABILITIES IN THE GLOSSO-KINÆSTHETIC CENTRE . 114

CHAPTER VIII.
STRUCTURAL DISEASE IN THE AUDITORY AND IN THE VISUAL WORD CENTRES 140

CHAPTER IX.
STRUCTURAL DISEASE IN THE AUDITORY AND IN THE VISUAL WORD CENTRES (*Concluded*) 179

CHAPTER X.

INCOÖRDINATE AMNESIA: PARAPHASIA AND PARAGRAPHIA . 216

CHAPTER XI.

DEFECTS OF SPEECH AND WRITING DUE TO DAMAGE TO THE COMMISSURES BETWEEN THE DIFFERENT CORTICAL WORD CENTRES . . 234

CHAPTER XII.

MODES OF RECOVERY FROM APHEMIA AND APHASIA. MODES OF PRODUCTION OF RECURRING AND OCCASIONAL UTTERANCES. LICHTHEIM'S TYPES OF SPEECH DEFECT 262

CHAPTER XIII.

THE READING AND WRITING OF NUMERALS. PRESERVATION OR LOSS OF MUSICAL FACULTY. AMIMIA. MIRROR WRITING . . . 282

CHAPTER XIV.

THE ETIOLOGY OF APHASIA AND OTHER SPEECH DEFECTS . . 301

CHAPTER XV.

DIAGNOSIS IN SPEECH DEFECTS . 307

CHAPTER XVI.

PROGNOSIS IN SPEECH DEFECTS. CAPACITY FOR EXERCISING CIVIL RIGHTS 318

CHAPTER XVII.

THE TREATMENT OF SPEECH DEFECTS . , . . 338

A TREATISE
ON
APHASIA AND OTHER SPEECH DEFECTS.

CHAPTER I.

INTRODUCTION: PHYSIOLOGICAL AND PSYCHOLOGICAL DATA.

THE modern interest in, and development of knowledge concerning, aphasia and other speech defects dates from the publication of certain memoirs by Broca, some six and thirty years since, when he attempted to localise what he termed the "faculty of articulate language" in a limited convolutional region of the left cerebral hemisphere. The publication of his cases and conclusions formed the starting point for a whole new series of investigations, whose result has been a remarkable development in our knowledge of the localisation of functions in the cerebral cortex, while the discussions to which these investigations have given rise have materially helped to lead to a better understanding of the working of the complex cerebral mechanisms needed for the carrying on of speech and thought. We are thus at the present day capable of dealing with the subject of speech defects from a much broader basis of discovered facts, as well as with a greater critical insight than was at all possible at, or even long after, the time when Broca wrote his famous memoirs. Very much, however, remains to be discovered before the many differences of opinion that exist concerning obscure and complicated points in connection with the nature and exact mode of production of speech defects are likely to be set at rest.

The special tendency of aphasia and other forms of speech defect to be associated with disease of the left rather than of the right hemisphere is now a well recognised fact. This holds good for right-handed but not as a rule for left-handed persons. In these latter aphasic defects are produced by disease in similar parts of the right cerebral hemisphere. It is believed, therefore, that motor incitations for acts of speech are accustomed to pass off in the main from one cerebral hemisphere, and that whether this shall be from the left or from the right hemisphere, becomes determined principally by the increased dominance, or slightly earlier development of one of them, brought about more or less remotely in association with right- or with left- handedness.

No attempt will be made here to trace the development of our knowledge on this important subject in any systematic manner. This is all the less necessary because it has on several occasions been already done by others. I shall therefore, at once proceed to set forth the views at which I have arrived after a study from time to time of this subject during the last thirty years. The whole subject is, however, so complicated that a number of underlying problems will have to be considered before the different kinds of speech defect can be individually studied. To this task, therefore, we must now proceed.

SPEAKING, READING AND WRITING AND THE CEREBRAL PROCESSES ON WHICH THEY DEPEND.

A brief consideration of the mode in which the faculty of articulate speech, together with the superadded accomplishments of reading and writing are acquired, will facilitate our comprehension of the various defects in the power of intellectual expression with which we are concerned.

That thought in all its higher modes cannot be carried on without the aid of language is a proposition that will be almost universally admitted if we use the latter term in its broadest sense. For, as Thomson says,[1] "Language, in its most general acceptation, might be described as a mode of expressing our thoughts by means of motions of the body; it would thus include spoken words, cries, involuntary gestures that indicate the feelings, even painting and sculpture, together with those contrivances which replace speech in situations where it cannot be employed." Articulate speech, in one or other of its modes, is, however, the process which (for ordinary human beings) is found to be inseparably related with their thinking processes. Speech is, indeed, nothing else than "a system of articulate words adopted by convention to represent outwardly the internal process of thinking."

Taking the human race at the present stage of its history,

---

[1] "Laws of Thought," 1860, p. 27.

when most elaborate languages have long ago been acquired by different sections of it, we may now briefly set forth the principal step by which individual children learn to understand one of these languages; how afterwards they learn to speak, to read, and to write; and to what extent the symbols involved in these various processes recur to the mind as the framework of thought.

A brief sketch of the nature of the processes involved in these acquisitions was attempted by the writer in 1869 in an article[1] entitled the "Physiology of Thinking," and from this a few quotations may now be made.

"The young infant first begins to distinguish natural objects from one another by differences in shape, colour, touch, odour, etc., which these may present to its different senses; it is then taught (slowly and with difficulty) to associate some object possessing certain combined attributes by which it is remembered, with a certain articulate *sound* which has been often repeated whilst the object is pointed at, till by dint of continual repetition this sound (or word) becomes so identified with the various attributes of the object that, when heard, it invariably recalls to memory the object of which it may now be said to form a kind of additional attribute, just as the sight or touch of the object will in turn call up the memory of the *sound* which has been employed as its designation. At first these articulate sounds (or spoken words) are only connected with external objects, though soon certain adjectives, signifying approval or disapproval, are added as qualifying sounds. By degrees the number of nouns and of adjectives in use increases, and also other parts of speech are added. . . . . . The process of learning is the same in all cases, whether the spoken sound is to be associated with an external object, with an emotional condition, or with a conception of the mind: first, it is necessary that we should be able to recollect and identify, when again presented to consciousness, either the set of attributes belonging to the object, the peculiarities of the emotional state, or of the intellectual conception; and, secondly, that we should be able to recollect the particular vocal sounds which have been associated with these several modifications of consciousness when previously existing. . . . . . This is the first stage passed through in the acquirement of a language—it is the mere learning to associate particular *sounds* with particular mental impressions, which association at last becomes so strong as to be almost inseparable, the thing unfailingly recalling

[1] *Fortnightly Review*, January, 1869.

to memory the sound, and the articulate sound as surely conjuring up a more or less vivid *idea* of the thing . . . . . it would seem pretty obvious that so far as the infant thinks by means of language, it does so by means of the *remembered sounds* of words—these are its linguistic symbols of thought, which must, however, be mixed up inextricably in its mind with other sense-impressions, and more especially with those of sight. For it may fairly be said that the great majority of children can remember the names given to many external objects when they are four or five months old; their memory in this respect continually increasing through succeeding months, even whilst they still make no very distinct effort at articulating words for themselves."

The next step is the development or acquirement by the individual child of the power of articulating for himself the sounds which have hitherto been increasingly employed as mental symbols. The potentiality of attaining to such a power the child receives, in the main, as an inheritance from so many antecedent generations of men, that its actual manifestation—the acquisition, that is, of the power of speaking —can only be regarded as a motor achievement of an order similar to some of those which may be included among the instinctive acts of lower animals : the similarity being not so much with the instinctive acts that animals are born with the capacity of performing, but rather with those which manifest themselves a little later in life, and which (from their more gradual acquirement) might be thought not to be instinctive acts at all.[1]

A process of "learning" to speak intervenes in part in the former case, but it is whilst the inherited structures are undergoing development in the child's nervous system.

"A certain order of development is always observed in the various parts of the human body, and this holds good also with regard to the several parts of the nervous system. . . . . . Even though the child acquires the power of uttering articulate sounds slowly, still when we think of the delicacy of the muscular combinations necessary, and of the almost instinctive way in which they are brought about, we shall rather be impressed with the notion that this could not have been accomplished at all had not the infant been born with a nervous system tending to develop itself in certain special directions, and thus making the performance of the highly complex muscular acts necessary for articulate speech a possibility. Slowly elaborated

[1] "The Brain as an Organ of Mind," 1880, p. 561.

developments of the parts of the medulla oblongata and of the brain concerned in the acts of speech, we may presume, had taken place in remote individuals of the parent race, as they acquired additional powers in this respect; and the power of developing similar structural relations between nerve cells and nerve fibres, thus established, having been handed down and gradually rendered more perfect by hereditary transmission through countless succeeding generations, the infant of to-day is born, perchance, with the potentiality of developing a nervous system as complex and as perfect in this respect as any which may have preceded it in its own ancestral line." A slowly growing mechanism of this kind becomes perfected under the influence of suitable stimuli of a volitional order, which here, as in the case of the acquirement of new motor powers by an adult, have an unquestionable though an unexplained influence in bringing about the development of nerve-tissues in the centres to which they are directed.[1] "This impetus, we may presume, is given by the passage of nerve-currents downwards from those superficial portions of the cerebral hemispheres concerned in the acts of intellectual perception and of memory, to those parts which are the motor centres concerned in articulate speech."

"At first the child's articulatory capacity is confined to mimicking—that is to say, it repeats such words only as have just been spoken to it; but after a time, when the act of emitting this sound has become perfectly easy by constant repetition, the child gives utterance to it of its own accord, on the mere sight of the object with which the sound was originally associated in its mind. This then is the second stage in the acquirement of language; and the child only slowly attains to a more perfect performance of the mental and motor processes involved." After a time, however, thought and language become inseparably associated, so that words are voluntarily recalled by the renewal of previous nerve actions in the auditory and other word centres, and such nerve actions are followed by the complex combination of muscular actions concerned in the articulation of the several words as they arise in thought.

Since the foregoing views were expressed and published, the writer has met with an altogether unexpected confirmation of their truth. In the year 1877 he was consulted concerning the health of a boy, the son of a leading barrister, who was then twelve years old, and had been subject to "fits" at intervals. The first fits occurred in infancy, when the patient

[1] See "Brain as an Organ of Mind," 1880, p. 563.

was about nine months old. Towards the end of the second year these fits seemed to have ceased, and the child appeared sufficiently intelligent—to be well, in fact, in all respects except that he did not talk. When nearly five years old the little fellow still had not spoken a single word, and about this time two eminent physicians were consulted in regard to his "dumbness." But before the expiration of twelve months, as his mother reports, on the occasion of an accident happening to one of his favourite toys, he suddenly exclaimed, "What a pity!" though he had never previously spoken a single word. The same words could not be repeated, nor were others spoken, notwithstanding all entreaties, for a period of two weeks.[1] Thereafter the boy progressed rapidly, and speedily became most loquacious. When seen by the writer he spoke in an ordinary manner, without the least sign of impediment or defect.[2]

Although there seemed no room for doubt as to the credibility of the above narrative, still, on account of the extraordinary nature of the facts, it may be well to remark that it was completely confirmed by the governess who had previously had the care of the child, and who was present on the occasion of this first and untaught act of articulate speech. The above account has also been submitted to the father, who, in reply to my enquiry as to whether anything required to be altered in what is above stated, replied that "the statement was perfectly correct."

No explanation of such facts seems possible, except on the supposition that speech has now become a truly automatic act for human beings, and that if children do not speak at birth this is in the main due to the fact that their nervous systems are still too immature. But when, in the natural course of development, the parts concerned have become properly elaborated, the highly complex movements concerned

---

[1] An emotional is much stronger than a volitional stimulus—a thing of higher tension—so that it may occasionally force its way along channels and against resistance which the volitional stimulus alone has been unable to overcome. Illustrations of this are frequently to be met with among persons who, from the effects of disease, have temporarily lost the power of speaking. Such individuals occasionally utter some word or short phrase under the influence of emotion which they are afterwards quite unable to repeat.

[2] Mentioning this case some years since to Sir Richard Quain, he informed me of a closely related fact. His eldest daughter up to the age of two years had not walked a step, or even tried to walk, when one day he put her down in the standing position, and to his great surprise as well as that of the nurse, she walked from one side of the room to the other. This also was an untaught act, as there had been no previous trials and failures.

in speech may, under certain circumstances, be at once called into play, independently of previous trials and failures—just as the nervous mechanism concerned with the act of sucking may be called into play in the human infant at the time of birth, on the presentation of its proper stimulus. No such untaught acts of speech would however be possible, unless development had been taking place in the normal manner, and unless the auditory sense and intelligence were unaffected. The manifestation of attempts at speech are supposed in this case to have been merely retarded by some slight and quasi-accidental conditions, such as are occasionally operative in childhood—especially in those who suffer from epileptic or other convulsions.

Without an instance of this sort coming almost under one's own cognizance, neither the writer nor any one else might have been inclined to bestow much credence upon two very similar cases, the records of which have come down to us from writers of antiquity.[1]

The son of Crœsus who, according to Herodotus,[2] had never been known to speak, and whose cure had been in vain attempted, was, at the siege of Sardis, so overcome with astonishment and terror at seeing the king—his father—in danger of being killed by a Persian soldier, that he exclaimed aloud—Ἄηθρωπε μὴ κτεῖνε Κροῖσον—" Oh, man, do not kill Crœsus!" This was the first time he had ever articulated, though he is said thereafter to have retained the faculty of speech as long as he lived. Again, it appears that Aulus Gellius,[3] after repeating the above story from Herodotus, relates a similar fact in the following terms:—" Sed et quispiam Samius athleta, nomen illi fuit Αἴγλης, quum antea non loquens fuisset, ob similem dicitur causam loqui cœpisse. Nam quum in sacro certamine sortitio inter ipsos et adversarios non bona fide fieret, et sortem nominis falsam subjici animadvertisset, repente in eum, qui id faciebat, sese videre, quid faceret, magnum inclamavit. Atque in oris vinculo solutus, per omne inde vitæ tempus, non turbidè neque adhæsè locutus est."

Coming to more recent times, Wigan also in his work on

---

[1] The real import of these latter cases does not seem to have been apprehended, either by those originally recording them or by a modern writer who has lately referred to them (Bateman, "On Aphasia," 1st ed., p. 138). It need scarcely be pointed out that the sudden beginning to speak for the first time without previous prolonged trials and failures, is a matter vastly transcending in importance, the sudden resumption of speech, when it has been for a while suspended in consequence of brain-disease.

[2] "Herod.," Hist. i., 85.

[3] "Noctes Atticæ," lib. v., cap. ix.

"The Duality of the Mind," published in 1844, records (p. 377) a very similar case. Referring apparently to the son of Crœsus he says:—"There is a story of a youth whose tongue was suddenly unloosed at the sight of impending danger to his father, and doubt is often expressed as to the reliance to be placed on its truth," and then he gives the following details concerning a parallel case for the accuracy of which he was able to vouch.

"An export merchant in the present day whose immense establishment is one of the most conspicuous and remarkable in the city of London (and who consulted me professionally many years), had a son about eight years of age, perfectly dumb, and the family had abandoned the hope that he would ever be endowed with the gift of speech. There was no defect in intellect, nor lesion of any other faculty. In a water-party on the Thames, the father fell overboard, when the dumb boy called out aloud, 'Oh, save him! save him!' and from that moment spoke with almost as much ease as his brothers. Two of my intimate friends were present at the miracle, which was the subject of unbounded joy and congratulation. The young gentleman is now one of the most active and intelligent members of his father's firm."

The powers of *reading* and of *writing* are accomplishments superadded to that of articulate speech.

The child has already learned to associate certain objects, or visual states of consciousness, with definite sounds (or names); he has further gained the power of articulating these names for himself: so that when he begins to learn to read, he gradually builds up a still further "association," by which certain written or printed hieroglyphics, representing letters in definite combinations, are linked to the already known states of consciousness (perceptions, conceptions, etc.) and their sound representatives. The previous combinations are therefore supplemented by being correlated with new visual symbols; and it seems certain that in the act of reading the words which are primarily perceived in the visual centre would under ordinary circumstances almost simultaneously recall the corresponding sounds in the auditory centre, as part of the perceptive process involved in this act.[1] From the auditory centre the stimuli inciting to the articulation of the corresponding words would then pass to the glosso-kinæs-

---

[1] Where this cannot occur it must be more difficult for the person to understand what is read, and, as will be seen subsequently, it may be impossible for him to read aloud.

thetic centre, and thence on to the bulbar motor centres in precisely the same manner as in the case of ordinary speech.

"With reference to the process of writing, it almost invariably happens that this accomplishment is acquired after the individual has been taught to speak and to read more or less perfectly. During this course of instruction the pupil learns to associate the visual perceptions of the separate letters of words with certain muscular movements of the hands and fingers necessary to enable him to produce the written letters for himself, and afterwards to join them together so as to represent words. This involves a long and tedious process of education, and the muscular movements which are ultimately learned are in all probability more intimately associated with sight-perceptions than with sound-perceptions; though of course the word as a revived sound-perception may be said to exist also during the act of writing." As I have said elsewhere,[1] "although the individual thinks, previously to writing the actual words, in the ordinary way (the word being revived in thought as a sound-perception), when he does begin to write them, in all probability the acquired and now more or less automatic muscular acts needed for the writing of the words, are rather immediately incited by revival of visual impressions which have been called up by the previously revived sound impressions, than by the sound impressions themselves. So that whilst in speech the motor acts are immediately dependent upon revived sound impressions, in writing the motor acts are only mediately dependent upon these, and are the immediate results of the revived visual impressions of the letters, which the corresponding sound associations have called up. Thus, although the two processes of speaking and of writing are to a certain extent processes which have steps in common with one another, still they differ to such an extent that it is quite conceivable for the one power to be affected whilst the other may remain almost intact; and this is sometimes the case, although in the majority of instances, impairment in the power of speaking and of writing are met with in the same individual."

The muscles of the upper extremity being also to the fullest extent voluntary muscles, and therefore very different from those concerned in the acts of speech, the whole process of learning to write is one which comes much more within the ken of our consciousness than does the otherwise parallel process of learning to articulate words.

---

[1] "On the Various Forms of Loss of Speech in Cerebral Disease," *Brit. and For. Med. Chir. Rev.*, January and April, 1869.

We ought therefore to have much more power of recalling the kinæsthetic impressions concerned with writing movements, than those which are associated with speech-movements. But how difficult is the memorial recall of this order of sense-impressions, and how vague and blank a feeling is associated with the attempt, as compared with the recall of a visual or of an auditory impression, any one may easily convince himself who will make the following simple experiment. Let him close his eyes, and with pen in hand make movements in the air as though he were writing the word "London." He may thus assure himself that he has a set of sensations accompanying these movements. After an interval, say the next day, let him again close his eyes, and, without making any movement, attempt to recall "in idea" the muscular and other sensations he previously experienced when writing the above-mentioned word. Let him then contrast his comparative powerlessness in this direction, with his ability to recall in idea the visual appearance of this word when written or its corresponding sound.

And if from this standpoint we look to the relative definiteness and recoverability of the kinæsthetic impressions consequent upon speech-movements, we find that the impressions which accompany actual speech-movements for different words can only vaguely be realised as distinct from one another, and that they are certainly far less distinctive than the kinæsthetic impressions derived from the acts involved in writing different words. The general rule, that the vaguer the sensation the lower is its degree of recoverability in idea, certainly holds good here also—as any one may discover who will make the necessary comparative trials.

Thus slight as may be the power of recalling the kinæsthetic impressions derived from writing, the ability to recall those occasioned by speech is even less. But that there should be such a difference is no other than might have been expected, since a precisely similar difference obtains in regard to impressions from "automatic" and from "secondary automatic" movements generally, as compared with those of a more "voluntary" order.

The Various Kinds of Word Memory.

According to Sir William Hamilton, "Memory strictly so denominated is the power of retaining knowledge in the mind, but out of consciousness. I say retaining knowledge in the mind but out of consciousness, for to bring the *retentum* out of memory into consciousness is the function of a totally different faculty (recollection). . . . It is not enough that we possess the faculty of acquiring knowledge and of retaining it in the mind but out of consciousness; we must further be endowed with a power of recalling it out of consciousness into consciousness—in short, a reproductive power (recollection). This reproductive power is governed by the laws which govern the succession of our thoughts—the laws, as they are called, of mental association." This definition of memory implies the notion of an organic change taking place in definite nerve elements on the occurrence of each sensory or intellectual process—that is, the notion of a permanent nervous modification of some kind, plus the possibility of its renewal in more complete form from time to time. We may therefore suppose that on fitting occasions, by the intervention of associational activity, there will be revival in more complete form of something like the original molecular activity in the nervous elements concerned with the primary perceptional or intellectual process of this or that kind. It is not essential that the memorial revival of the sensory impression or of the intellectual process should after multitudinous repetitions be associated with any distinct conscious phasis. What Sir William Hamilton termed the *retentum* may, indeed, be revived as a mere unconscious nerve action—a link in a perceptive process or in a chain of thought represented merely (as John Stuart Mill put it) by "certain organic states of the nerves." This latter consideration is especially worthy of note from the point of view of the importance of revived kinæsthetic impressions for the guidance of movements, seeing that their revival may be unattended by any distinct conscious phasis; and much the same thing may often be said concerning that memorial recall of words in the auditory centre which immediately precedes speech.

From what has already been said it is evident that " loss of recollection " by no means implies nor is it to be taken as synonymous with " loss of memory." For instance, a patient may be unable to recollect words—that is, spontaneously revive them—for ordinary speech when his memory for such words, nevertheless, exists unimpaired, as may be shown by the fact that the patient is able at once to repeat the words in question when he hears them pronounced or sees them written. His defect, therefore, may consist in a mere lowered activity of the auditory word centre, in which words are primarily revived during thought. "Loss of recollection" may, in fact, depend upon one or other of two causes: either (*a*) upon some diminished functional activity—that is, diminished readiness to be roused—in the central nerve mechanisms in which the *retentum* is, so to speak, stored or rendered possible of revival; or else (*b*) upon some defect in this or that set of commissural fibres (associational channels). "Loss of memory," however, in the strict sense of the term implies disease of, or serious damage to, the central nerve units in which the particular *retentum* is stored up or registered.

Although it is important from a scientific point of view to make such a distinction as has been above pointed out, it will be found both convenient and practical to let the term "amnesia" stand for loss of recollection as well as loss of memory of this or that kind.

From what has been said it follows, that all cases of amnesia (in the broad sense of the term) ought, from an anatomical or localising point of view, to be divided into two generic groups: (*a*) cases in which there is centric defect (either structural or of marked functional type), with which there will often be loss of memory of words as well as loss of recollection; and (*b*) cases in which there is merely commissural defect (mostly structural), with which there may be loss of recollection of words, but no necessary loss of word-memory. It will be one of my objects in the present work to dwell upon the fact (which I pointed out some ten years ago) that it is often in the case of speech defects extremely difficult, if not impossible, from clinical evidence alone to decide whether the underlying lesion or default leading to a particular kind of amnesia be centric or commissural in seat. And yet from the point of view of the localisation of the lesion such a decision may be a matter of much importance, seeing that if the lesion were centric we should look for it in one part of the brain, while if it were commissural it might be found in a region comparatively remote therefrom.

In the case of words there are three distinct kinds or physiological types of memory to be considered—one of them existing in two forms, so as to make four varieties in all. These varieties of verbal memory are as follows:—(1) *Auditory memory*: the memory of the sounds of words—that is, of the auditory impressions representative of different words. (2) *Visual memory*: the memory of the visual appearances (printed or written) of words—that is, of the visual impressions corresponding with different words. (3) *Kinæsthetic memory*:[1] (*a*) the memory of the different groups of sensory

---

[1] These forms of word-memory were first definitely stated by me to be purely sensory in papers on the "Physiology of Thinking" (*Fortnightly Review*, 1869) and on the "Muscular Sense" (*Brit. Med. Jour.*, April, 1869);

impressions resulting from the mere movements of the vocal organs during the utterance of words (impressions from muscles, mucous membranes, and skin) — that is, of the kinæsthetic impressions corresponding with the articulation of different words, which for the sake of brevity I have proposed to speak of as "glosso-kinæsthetic" impressions; and (*b*) the memory of the different groups of sensory impressions emanating from muscles, joints, and skin, during the act of writing individual letters and words — that is of the kinæsthetic impressions corresponding with the writing of different letters and words, which I have for similar reasons proposed to speak of as "cheiro-kinæsthetic" impressions.[1]

The organic seat of each of these four different kinds of word-memory is in relation with its own set of afferent fibres; and the several centres must also be closely connected with one another by commissural or associational fibres, so that the memory of a word or the recollection of a word in one or other of these modes doubtless involves some amount of simultaneously revived activity in one or two of the other word centres.

The relative intensity or importance (in the process of recollection of words for ordinary speech) of the memorial revival in each of these centres is probably subject to more or less marked variations in different individuals. In the majority of persons, as I pointed out in 1869, the revival of words in the auditory centre is the most potential process, and that which occurs first in order of time. This seems to be now very generally admitted.

The Localisation of the Different Word Centres.

Although I am not a believer in the complete topographical distinctness of the several sensory centres in the cerebral hemisphere, I consider it clear that there must be certain sets of structurally related cell and fibre mechanisms in the cortex, whose activity is associated with one or with another of the

---

and the name "kinæsthetic" was subsequently ("The Brain as an Organ of Mind," 1880, p. 543) applied to the complex groups of impressions resulting from movements of this or that part of the body.

[1] Other French writers, as well as Ballet, attribute to Charcot the doctrine that "le mot n'est pas une unité, mais un complexus" ("Le Langage Intérieur," 1886, p. 13) because he also dwelt upon these four different kinds of word-memory. Charcot's lectures on Aphasia were delivered in 1883, but M. Ballet will find a full description of these four different kinds of word-memory in my book "The Brain as an Organ of Mind," 1880, p. 696, or in the French translation ("Le Cerveau et la Pensée," tome ii., p. 222), published in 1882.

several kinds of sensory endowment. Such diffuse but functionally unified nervous networks may differ altogether from the common conception of a neatly defined "centre," and yet for the sake of brevity it is convenient to retain this word and refer to such networks as so many "centres."

Looking to the extremely important part that words, either spoken or written, play in our intellectual life and to the manner in which they are interwoven with all our thought-processes, it becomes highly probable that most important sections of the auditory and the visual sensory centres are devoted to the reception, and consequently to the revival in thought, of impressions of words: and for convenience of reference it is permissible to speak of these portions as auditory and visual "word centres" respectively. Similarly, there must be what I have termed kinæsthetic word centres of two kinds (the one in relation with speech-movements, and the other with writing movements) holding a like all-important relation to the expression of our thought by speech and writing. It is possible that the particular parts of the general auditory and visual centres which are in relation with word-impressions may be more or less distinctly defined, like the analogous parts of the general kinæsthetic centre that are in relation with speech or with writing movements. Certain it is that there are some varieties of amnesia in which the part of the visual centre in relation with words seems to be specially at fault (as in "word-blindness"), just as there are other cases in which the part of the auditory centre in relation with words is either wholly or partially inactive (as where we have to do with different degrees of "word-deafness")—in each case without defect in other parts of the general visual or auditory centres, as evidenced by the fact that the persons so affected can still quite well see and recognise ordinary objects, or hear and recognise the nature of ordinary sounds.

In regard to the visual centre as a whole, it seems now to be established that it is more or less diffused through the convolutions on the inner aspect of the occipital lobe and probably even more widely throughout this lobe. The particular part of the visual centre that is most concerned with the appreciation and memorial recall of words—or, in other words, the destruction of which most certainly gives rise to word-blindness—is now fairly well settled. There is much evidence to show that this region corresponds with the angular gyrus either alone or in association with part of the supra-marginal lobule—and therefore that it is situated just beyond the confines of the occipital lobe, and in the region originally assigned by Ferrier, upon the basis of his experiments with monkeys, as the centre for vision as a whole.

In regard to the localisation of the general auditory centre considerable doubt now exists, since the researches of Schäfer and Sanger-Brown do not support Ferrier's allotment of this endowment to the upper temporal convolution. Curiously enough, however, it again happens that the localisation of the part of the general auditory centre most concerned with the appreciation of words (as based upon clinico-pathological evidence in man) must be regarded as being in the posterior half or two-thirds of the upper temporal convolution. The above-mentioned results of Schäfer and Sanger-Brown as to the localisation in monkeys of the general sense of hearing were of a negative character, and cannot be said to afford any definite evidence against this presumption as to the site of the auditory word centre in man.

The situation of one of the two kinæsthetic word centres can be rather more certainly localised. Having elsewhere stated very fully my reasons for believing that the so-called "motor centres" of Ferrier and others are really sensory centres of kinæsthetic type by means of which movements are guided, I shall not now attempt to set forth the evidence in detail in favour of this opinion, but will content myself by making a few quotations from my last communication in reference to this important subject, namely, an article on "The Neural Processes underlying Attention and Volition."[1]

"The act of willing any particular movement consists essentially in a consent (after the balancing of reasons that may exist for or against) to the occurrence of such a movement; the movement itself being at the time mentally prefigured by certain revived sensations—such revival representing, as James Mill said long ago, 'the last part of the mental operation.' The occurrence or not of the movement is to a certain extent an accident, and one which when it occurs lies altogether outside the mental process itself.

· "Let us look then now a little closer at these last links in the chain of association—that is at 'the last part of the mental operation' which leads on to the performance of a voluntary movement. It consists in a revival of the idea or conception of the movement to be executed. This idea has always at least a two-fold basis, though the actual production of the movement often requires a three-fold excitation of sensory centres in immediate succession, as I shall presently show.

"For limb and trunk movements the idea is composed of revived visual and kinæsthetic impressions which have previously been received during the execution of similar movements. The reawakened activity of these sensory centres affords the necessary stimulus and guidance for the reproduction of the movement—the molecular actions associated with their excitation evoke, that is, the related suitable activity of motor centres in the spinal cord. So that as W. James puts it, 'every representation of a motion awakens the actual motion which is its object, unless inhibited by some antagonistic representation simultaneously present to the mind.' The same kind of thing happens in regard to speech-movements, only here

[1] *Brain*, vol. xv., 1892, pp. 25-27, and 30-33.

we have the reawakened activity of auditory in conjunction with kinæsthetic centres starting the stimuli needful for calling into activity the proper motor centres for speech in the bulb.

"I would only add here that this view of James Mill is also the doctrine taught as to the mode of production of voluntary movements by Lotze[1] and by Herbert Spencer;[2] that it has been advocated independently by W. James and myself since 1880; and later by Münsterberg,[3] Horsley and others.

"These kinæsthetic centres are, I think, afferent centres in every way analogous to the afferent centres situated in the spinal cord. The so-called 'motor centres' of the cortex were not, of course, originally supposed to be termini for afferent impressions; when first discovered they were said, and they are still maintained by Ferrier to be, true motor centres. Now, however, in spite of the different interpretation which has been given of their functions, many still cling to the belief that the centres in the Rolandic area must be motor centres because internuncial fibres connect them with the real motor centres in the bulb and spinal cord—and because, therefore, 'motor incitations' must pass along such internuncial fibres. 'What initiates a motor process' they say,[4] 'is to all intents and purposes motor.' Or, as Ferrier[5] puts it, 'centres immediately concerned in effecting voluntary movements' are 'as such motor.' Both these statements I believe to be altogether erroneous. As I have said elsewhere,[6] 'The plan on which nervous centres generally are constructed, of whatsoever grade, makes it essential that the stimulus which awakens the activity of a "motor" ganglion or centre shall come to it through connecting fibres from a "sensory" ganglion, centre, or knot of cells—that is, from cells which stand in immediate relation with ingoing fibres.' Thus, we should not call a cortical centre for afferent impressions 'motor,' any more than we should call the group of ganglion cells on the afferent side of a spinal reflex arc 'motor.' In each case the nerve cells that receive the afferent impulses are in association with channels which convey 'motor' impulses; and in each case the stimulation of such internuncial fibres or of the centres from which they proceed would give birth to definite movements. The course of these internuncial fibres is for the most part horizontal in the spinal cord, though more rarely it may be an ascending one. But from the kinæsthetic centres in the brain the course of the internuncial fibres is downwards (in the pyramidal tract); hence the current is commonly spoken of, truly enough, as an 'outgoing current'; but with the effect, apparently, of fostering some confusion in the minds of not a few persons."

"From what has been above set forth, as well as from the facts and arguments detailed in a previous communication to this Society,[7] it seems to me quite clear that there is no reason for postulating the existence of cortical motor centres for the production of voluntary movements; that whatever the mode in which simple movements are produced, that is,

---

[1] "Medicinische Psychologie," 1852.

[2] "Princip. of Psychol.," 1st ed., 1855, p. 613; and 2nd ed., 1870, vol. i., p. 496.

[3] Münsterberg's theory (see *Mind*, 1888, p. 463) concerning the "muscular sense" and its relation to voluntary movements is identically the same as that published independently by Prof. James and myself in 1880.

[4] W. L. Mackenzie, in *Brain*, 1887, p. 433.

[5] "Functions of the Brain," 2nd ed., p. 348.

[6] "The Brain as an Organ of Mind," p. 585.

[7] The Neurological Society of London. The communication referred to is a paper on "The Muscular Sense: its Nature and Cortical Localisation," which appeared in *Brain*, April, 1887.

whether they are voluntary or reflex, only one set of motor centres is called into play, and that these motor centres are situated in the bulb and in the spinal cord; and, further, that the functioning of motor centres generally is attended by no psychical accompaniments."

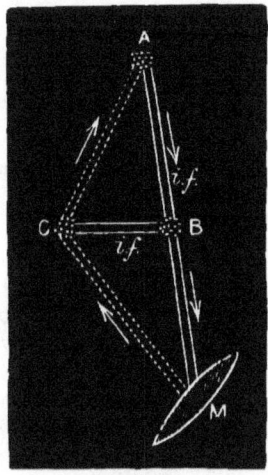 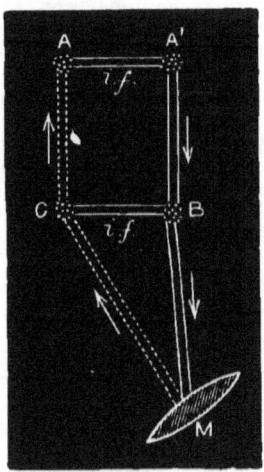

Fig. 1.   Fig. 2.

Fig. 1.—Diagram illustrating the origin of kinæsthetic impressions (so far as they come from muscle) and their relation to the production of voluntary movements in accordance with my views.

A Cerebral afferent (kinæsthetic) centre, receiving and registering ingoing impressions from muscle by way of an afferent spinal centre C; *i.f. i.f.*, two sets of internuncial fibres; B, spinal efferent or motor centre, which receives incitations from A, whence they are sent on to the muscle M.

(A is an afferent centre in the same sense that C is an afferent centre and each of them may convey "motor incitations" along internuncial fibres to the motor centre B.)

Fig. 2.—Diagram illustrating the origin of muscular sense impressions and their relation to the production of voluntary movements in accordance with the views of Dr. Ferrier.

A Cerebral afferent (tactile) centre receiving and registering ingoing impressions from muscle by way of an afferent spinal centre C; A' a supposed motor centre, which operates through commissural fibres upon the spinal motor centre B.

⁎ No psychical processes are believed by either of us to be associated with the functional activity of the tracks represented by unbroken lines and the centres lying in their course.

I say that the functions attributed by Ferrier to A' are really performed by A as in Fig. 1, and that if A, and A' existed as in Fig. 2, there ought to be two sets of excitable areas in each hemisphere.

The cerebral cortex is in my view, to be regarded as a continuous aggregation of interlaced "centres" and associational fibres, towards which ingoing impressions of all kinds converge from all parts of the body; and also of other related regions in which higher, or derivative, mental processes are in part carried on. Here all the various ingoing currents evoked by the organism's relations with its outside world, and by changes in its own non-neural tissues and organs, come into relation with one another in various ways, and conjointly give rise to nerve actions which have for their subjective correlatives all the sensations and perceptions, all the intellectual, and all the emotional processes which the individual is capable of experiencing. From these terminal and complexly related "end-stations" for ingoing currents, and from various annexes in connection therewith, above referred to, outgoing currents issue which rouse in definite ways the activity of the motor centres in the bulb and in the cord, so as to give rise to any movements that may be "desired," or which are accustomed to appear in response to particular ideas or sensations.[1]

The phenomena of volition are not the result of the working of any special faculty or mysterious entity, and are not carried on in motor centres; they are merely certain exemplifications of intellect in action; so that ample justification is found for the dictum of Spinoza, "Voluntas et intellectus unum et idem sunt."

I may now, therefore, state my belief that Broca's region —namely, the posterior part or foot of the third frontal convolution—is in reality the part of the brain to which I have been alluding as the "glosso-kinæsthetic" centre. The situation of the "cheiro-kinæsthetic" centre cannot be localised with nearly as much confidence. The tendency for some years has been to follow Exner, who believes it to be situated in the posterior part of the second frontal gyrus, though, as we shall see later, the evidence in favour of this localisation is at present almost non-existent. All that can be said on this point, therefore, is that we know approximately where to look for the cheiro-kinæsthetic centre.

For the purpose of our discussion, then, it may for the present be assumed that the two kinæsthetic word centres are situated as above stated: that the auditory word centre is situated in the posterior half or two-thirds of the upper temporal convolution; and the visual word centre in the angular and part of the supra-marginal convolutions. It must also be

[1] As a means of affording some sort of reconciliation between the views of Ferrier and myself the late Dr. Ross said (*Brain*, vol. x., p. 103):—" I think it, however, quite likely that the kinæsthetic and motor centres coincide, in so far as that the former are situated in the two outer and the latter in the third layer of the cortical cells of the parieto-frontal area of the cortex." But this view meets with no favour from me simply because, in my opinion, (1) the kinæsthetic centres do precisely the kind of work which cortical motor centres (or "psycho-motor" centres as they are sometimes termed) were originally supposed to perform; and (2) because there is no independent evidence whatever in favour of the existence of cortical motor centres that cannot be equally well explained in accordance with my views. The occurrence and distribution of "secondary degenerations" starting from the Rolandic area, as well as the mere size and histological characters of the third layer of cells in this region of the cortex have no independent value, as they are just as easily explicable in accordance with my views, as I have elsewhere endeavoured to show (*Brain*, April, 1887, pp. 82 and 127).

assumed, upon grounds subsequently to be set forth, that these last-named word centres are connected together by a double set of commissural fibres. We must likewise suppose that two other important sets of commissural fibres exist between the different word centres—namely, one set through which the auditory word centre acts upon the glosso-kinæsthetic centre for the production of speech-movements, and another by means of which the visual word centre acts upon

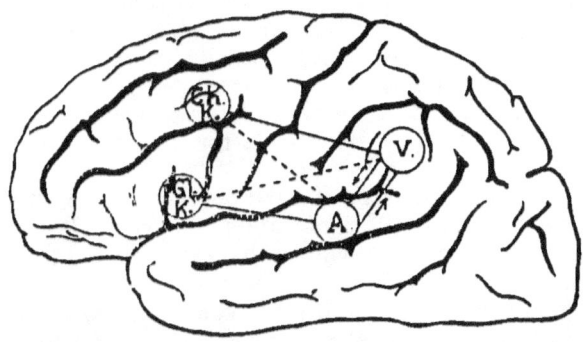

Fig. 3.—Diagram showing the approximate sites of the four Word Centres and their Commissures.

the cheiro-kinæsthetic centre for the production of writing movements.

In the study of speech defects it is therefore necessary to consider the effects of lesions in the following situations; (a) in the different kinds of word centres; (b) in the different commissures by means of which these centres are connected with one another; (c) in the internuncial fibres connecting the two kinæsthetic word centres with their related motor centres, in the bulb and in the cervical region of the spinal cord; and (d) in these motor centres themselves which are concerned with the actual production of speech and writing. But before dealing with any of these problems in detail a few other aspects of the questions relating to word-memory, as well as the modes of activity of the brain in perceptive and speech processes require to be considered.

THE PRIMARY SITE OF REVIVAL OF WORDS IN SILENT THOUGHT.

It seems clear that words are the symbols with which our thoughts are inextricably interwoven, and that the revived feelings, ideas, or "images" of words may enter into thought processes by a more or less simultaneous renewal of activity in different regions of the cerebral cortex. There may be a revival of sounds of words as we hear them in ordinary speech; there may be a revival of visual impressions of words as we have seen them in written or printed characters; and lastly, there may be a revival of the feelings of muscular contractions concerned in the pronunciation of words. Of these modes of "ideal" recall of words the two former are distinct and easily recoverable, while the latter is vague and difficult of conscious realisation. Let anyone, as I have said, contrast his idea of the sound of a word, or his idea of the appearance of the word when printed or written, with his idea of the muscular and other feelings associated with the articulation of the same word, and the inferiority in definiteness and recoverability of the latter will at once become obvious.

It is, however, a matter of extreme importance for the due understanding of the different kinds of speech defect and for the success of our endeavours to refer them to defective activity in this or that physiological region of the hemispheres that we should definitely know in what sensory region of the cortex words are principally recalled to mind during ordinary thought-processes. Two distinct views radically opposed to one another have been advocated on this subject.

There is the view (1) that words are revived as "motor processes"—that is, as faint excitations of the processes occurring in motor centres during the articulation of words; and (2) there is the view that words are revived in ordinary thought in the main as auditory ideas or images. The former view has for a long period been promulgated by Hughlings Jackson, and has more recently received the strong support of Stricker, both of them regarding Broca's region as a motor rather than as a sensory centre. The latter view is that which the writer has now for many years advocated. The fundamental nature of my opposition to the former view may be gathered from the fact that I have for many years expressed the opinion that motor centres, wherever they may be situated, are parts whose activity appears to be wholly free from subjective concomitants. No "ideal" reproductions seem ever to take place in such centres; they are roused into activity by outgoing currents, and, so far as we have any evidence, the induction in them of molecular movements which immediately afterwards issue through cranial and spinal motor nerves to muscles are, like those which they engender, simply physical phenomena.[1] It is true that the

[1] "The Brain as an Organ of Mind," 1880, p. 599.

altered condition of the muscles and of contiguous parts induced by these outgoing stimuli together engender a body of ingoing impressions, the terminus for which is the kinæsthetic centre. This latter is therefore a true sensory centre, and in it images of movements or "ideal movements" may be revived in a more or less vague manner as already indicated.

But, even if we adopt what appears to me to be the more legitimate view that articulatory movements as well as movements in general are represented in the cerebral cortex only by sensory centres, there are still, I think, good reasons for rejecting the notion that the "material of our recollection" in the use of words during silent thought is primarily revived as glosso-kinæsthetic impressions in Broca's centre. The principal reasons opposed to this view seem to me to be as follows:—

In the first place it must be evident from the mode in which speech is acquired by the child that during the few months in which words enter into the simple trains of thought, before he has acquired the power of articulating them for himself, they must be revived as auditory impressions. Secondly, there is, as we have seen, a much greater definiteness of impression and readiness of recall for auditory than for articulatory feelings; and so far, therefore, there is a greater fitness in the former for serving as the "material of our recollection" of words in ordinary thought-processes. Thirdly, there is reason to believe that revived auditory feelings continue after the acquirement of speech by the child to have the same relation to his thought-processes as they must have had before his acquirement of the power of speaking. If this were not so it would be impossible to understand why total deafness supervening in a child in full possession of speech as late as the fifth, sixth, or even the seventh year, will certainly entail dumbness unless the child be drilled in lip-reading, that is, unless the primary incitation to acts of speech be gradually transferred from the auditory to the visual centres. Fourthly, because, as we shall subsequently find, there is much evidence against this view to be derived from the study of speech defects, and none that I am aware of telling unmistakeably in its favour.

It seems to me to be an error to attempt to settle such a question by endeavouring to ascertain which form of word-memory reveals itself most in consciousness; and I venture to think that the recent principal advocate of the view that words are primarily revived during silent thought as "motor processes, relies too much upon what is in reality an untrustworthy method—that of introspection. By concentrating his attention upon the genesis of his own speech, Professor Stricker inevitably brings its expressive side into undue prominence. As Taine says:[1] "Plus on imagine nettement et fortement une action, plus on est sur le point de la faire. . . . quand l'image devient très lumineuse elle se change en impulsion motrice."[2]

The primary revival of words during thought takes place, I submit, in the great majority of persons by a sub-conscious process in the auditory centre, and tends to be immediately followed by correlated revivals in the glosso-kinæsthetic centre, and these again by incipient or complete activities in the bulbar motor centres. It may be perfectly true that kinæsthetic memories vary in intensity in different individuals, as do auditory and visual memories; and it might have been thought that Stricker and also Ballet (who adopts his view in a more discriminating manner) were to be counted as persons in whom the former kind of memory is highly developed, but for the fact that Ballet expressly states[3] his inability voluntarily to recall cheiro-kinæsthetic impressions. He believes, moreover, for reasons given, that the same holds good for Stricker. Thus they cannot voluntarily recall the kinæsthetic impressions associated with the act of writing a word, but they say they can recall the kinæsthetic impressions associated with the articulation of the same word, even although the movements in the latter case are much more of an automatic type.[4]

These apparently contradictory results are easily explicable in accordance with my views, when one considers how the thought of a word naturally runs on to an incipient articulation;[5] while, on the other hand, the mere thought of the same word has no appreciable tendency to evoke writing movements. The negative results of Ballet and Stricker with cheiro-kinæsthetic revivals of words would seem, therefore, to have all the force of a crucial experiment and to throw

---

[1] "De l'Intelligence," troisième édition, tome i., p. 482.

[2] See his work, "Le Langage et la Musique," Paris, 1884.

[3] "Le Langage Intérieur," 1886, p. 55.

[4] See what has been already said in reference to this point on p. 10.

[5] Ballet, in fact, says, *loc. cit.*, p. 52:—"Chez moi, comme chez la plupart des moteurs je pense, la parole intérieure devient souvent assez vive pour que j'arrive à prononcer à voix basse les mots que dit mon langage intérieur." What he says on the next page shows that he is not a strong "visual," but does not tell against his being a strong "auditive." His better memory of a lecture after delivering it might be explained by the help derived from auditory or kinæsthetic impressions of the words, or from the two in combination.

great doubt upon their supposed ability primarily to revive words in the glosso-kinæsthetic centre.

It seems pretty certain, therefore, that the real linguistic counters for thought are the auditory and visual memories of words. And in regard to these two, I believe that in the very large majority of persons it is the auditory word-memory which is first revived in silent thought. This view is now very generally adopted, as may be seen by reference to the works of Ross[1] and Bernard.[2] It appears also to be the view of Herbert Spencer, since he says:[3] "Our intellectual operations are, indeed, mostly confined to the auditory feelings (as integrated into words) and the visual feelings (as integrated into ideas of objects, their relations and their motions)." Although the first stage in the revival of words seems to occur, however, in the auditory centre, the molecular disturbance thus initiated is immediately transmitted to a varying extent in two directions. It is transmitted to the visual word centre on the one side and to the glosso-kinæsthetic centre on the other—to the latter strongly where the thought is to issue in speech, and to the former strongly where it is to issue in writing.

There can be no doubt that in some persons the visual centres as a whole are more highly developed and organised than the auditory centres, and *vice versâ*. The existence of such differences has been thoroughly shown by Francis Galton.[4] Thus he took steps to test the power of what he calls "visualising"—that is, the ability of persons to see with the mind's eye distinct images of objects in their natural grouping and colouring. The one hundred answers that he received as to the power possessed by the persons selected of visualising the breakfast table to which they had sat down in the morning disclosed an extraordinary range of variation in this respect. One of those who had the highest power of visual recall answers: "Thinking of the breakfast table this morning, all the objects in my mental picture are as bright as in the actual scene." One of those who had the lowest power answers: "No individual objects, only a general idea of a very uncertain kind." Galton adds: "There are a few persons in whom the visualising faculty is so low that they can mentally see neither numerals nor anything else; and again there are a few in whom it is so high as to give rise to hallucinations." In this general sense persons may well be classed as "visuals" and "auditives" respectively;

[1] "On Aphasia," 1887, p. 114.   [2] "De l'Aphasie," 1885, p. 49.
[3] "Principles of Psychology," vol. i., 1870, p. 187.
[4] "Inquiries into Human Faculty," 1883, p. 83.

and we may thus indicate for this or that person which is the more potent sensory endowment.[1]

Supposing, however, that a person is a "visual" in this general sense, it should not, in my opinion, be taken as necessarily implying that visual memories of words are for him the first to be revived in silent thought. If it were so we should expect to find speech greatly interfered with in many cases of simple word-blindness; but this seems comparatively rarely to be met with, the contrast in this respect being most striking between destruction of the visual and destruction of the auditory word centres respectively. The primary revival of words may, perhaps, still occur in a "visual" in the auditory centre, only in such a person these primary revivals may be strongly backed up and reinforced by visual images. So that where the auditory word centre is damaged in such a person its action, as a leader in the memorial recall of words, might to a certain extent be taken on by the visual word centre; while in a person who is not a "visual" but little compensation of this kind could occur for the loss of the auditory word centre. Thus the clinical effects in the form of speech disturbance would be different and altogether more marked in the latter case. It will be found that many recorded cases of speech defect lend support to these views, and that this, in fact, is the real practical outcome of all that has been said by Charcot and his followers as to the division of persons (in reference to the interpretation of speech defects) into four categories—"auditives," "visuals," "motors," and "indifferents"; a doctrine which has been fully set forth by Ballet.

I contend that this doctrine must not be taken to mean anything more than that the different kinds of word-memory referred to in the first three types may be met with in different degrees of excellence, and that in the fourth type there is no one kind of memory which is developed to a preponderating extent. In the latter class there would be no ground for supposing that the ordinary rule as to the primary revival of words taking place in the auditory centre is ever altered. As to the existence of persons representative of the third class I have already expressed my doubts—that is to say, I can find no good reasons or evidence to show that words ever arise in silent thought primarily in the kinæsthetic centres. I doubt whether these memories are capable of separate voluntary recall, and do not believe there is any evidence to show, as Bernard says,[2] that "the motor centre for speech may become independent of the sensorial centre which had presided over its education."

[1] On this subject many interesting details have been given by Ballet, *loc. cit.*, pp. 17-45.
[2] "De l'Aphasie," 1885, p. 48.

The case is altogether different in regard to individuals of the second class—the "visuals"—because not only does the power of memorial recall in the visual centre vary immensely in different persons, as we have seen, but there is also good reason to believe that in a small minority of persons the primary revival of words during a process of silent thought may take place therein, just as it does in the act of reading. Thus Galton says :[1] "Some few persons see mentally in print every word that is uttered; they attend to the visual equivalent and not to the sound of the words, and they read them off usually as from a long imaginary strip of paper." It is possible, of course, that a person possessed of this high visualising power, whilst he may be thus aided in the delivery of a speech previously written, may nevertheless, in a process of silent thought, conform to the ordinary rule and revive words voluntarily in his auditory centre. But it may be otherwise with some few persons, so that, as Ribot supposes, they may habitually think and represent objects "by visual typographic images." Something of this kind we are bound to suppose must occur in a word-deaf person whose auditory word centre is destroyed and who is yet able to speak—a rare conjunction that has occasionally been met with; just as we are bound to suppose that when a congenitally deaf and dumb child is taught to speak by the lip-reading process he brings this about by means of a primary revival of visual images, which act directly upon the glosso-kinæsthetic centre and thence upon the motor centres in the bulb.[2]

Bernard has cited[3] a remarkable instance of a patient who possessed a visual memory exceptionally strong, whilst his auditory memory was proportionally weak. In this patient the visual images of words and figures were evidently predominant during thought-processes. "If he recited a lesson whilst at college, or an extract from a favourite author later

[1] *Loc. cit.*, p. 96.

[2] In the case of a deaf and dumb child taught to speak and understand others by means of hand and finger movements, their thought-counters would be combined visual and kinæsthetic impressions, or in the case of one who had been taught to read much from early life they might be "visual typographic images." On the other hand, in the case of a congenitally blind person who has been taught to read aloud by means of raised letters and words excitation of the centres for touch and kinæsthesis would constitute the initial processes in the act of reading aloud, although revived auditory impressions might constitute, as in ordinary persons, the habitual thought-counters. A patient of the former class whose history has been reported by E. Fournier ("Essai de Psychologie," part ii., chap. v.) says : "Je sens quand je pense que mes doigts agissent, bien qu'ils soient immobiles. Je vois intérieurement l'image que produit le mouvement de mes doigts."

[3] *Le Progres Médical*, July 21, 1883.

on two or three readings were sufficient to fix in his memory the page with its lines and its letters, and he recited by mentally reading the desired passage, which, at the first summons, presented itself with the greatest distinctness." Ballet[1] also quotes a letter from a librarian at Geneva which is extremely interesting. He writes: " When I think of a word or a phrase I see very distinctly this word or this phrase printed in ordinary characters, or written in my own writing or that of another person; the letters of a word are distinct from one another, and the intervals between each word written in black appear to me also. I see them in white. All my reproductions of words are visual." The same author mentions an orator, Hérault de Séchelles, who was accustomed, as it were, to read his speeches; and cites an epigram of Charma, also a " visual," who, speaking of the class to which he belongs, said, "Nous pensons notre écriture, comme nous écrivons notre pensée."

Undoubtedly, then, it would seem that there are certain exceptional persons who, as it were, read rather than hear their thoughts; and in whom, as Ballet says, the visual images of words acquire such an importance that they alone constitute the medium of their internal speech. Still, in spite of these exceptions, the general rule is that auditory images constitute the most potent representations of words, whilst visual images form the most potent representations of ordinary external objects.

THE REVIVAL OF WORDS FOR SPEECH IS A COMPLEX PROCESS.

The views just expressed refer, as I have said, to the seat of the primary revival of words in memory, because there is strong reason for believing that the activity when once initiated does not limit itself to a single centre. It must be borne in mind that the structural relations existing between the different seats of word-memory and the modes in which these are functionally related is in accordance with what occurs in ordinary processes of perception. In regard to such a process I have elsewhere said:[2] "When any one constituent of a natural cluster of sensations comes within the range of the corresponding sense organs of an animal the other possible impressions composing the cluster (and representing the organism's knowledge of the external object) become simultaneously nascent in memory, so that the object is perceived

---

[1] *Loc. cit.*, p. 44.
[2] "The Brain as an Organ of Mind," p. 176.

or recognised. If in a dark room my hand comes upon an orange or a book, either of these sensations of touch will immediately fuse with nascent ideas of other possible sensations from the same object (whichever it may be), so that this object is perceived as a present external reality. . . . . . Thus it happens that an object is recognised immediately or intuitively, not so much by the mere single or double impression present, as by the blending of this with more or less fully revived memories of other impressions which have at various times been associated with the same object. Truly enough, as Bain says, 'When we see, hear, touch, or move, what comes before us is really more contributed by the mind itself than by the present object.' . . . . . It is therefore by the simultaneous consciousness and fusion, as it were, of the subjective side of various new and old impressions that a present object is perceived and recognised. It could only be by the previous establishment of structural communications between the severally related sensory cells (in different centres) that the excitation of those of any one order would suffice to revive more or less strongly in other groups just such molecular changes as like objects had on previous occasions excited. And it may be easily understood that the molecular movements initiated by any one or two sense impressions may start from such groups of cells and thence flow over into all communicating channels between them and the cells of other related groups."

In a similar manner it may be supposed that the revival of activity in the auditory centre during the voluntary recall of words does not remain limited to that centre, but gives rise, as I have already said, to molecular movements in two directions—that is, forwards to related portions of the kinæsthetic centre and backwards and upwards to related portions of the visual centre—though with different degrees of intensity in different persons.

It is of importance to remember that for ordinary persons (that is, for those who are neither congenitally blind nor congenitally deaf) the four memories of words seem to be mainly called into play in definite couples—namely, the auditory and the glosso-kinæsthetic revivals taking place during articulate speech, and the visual and cheiro-kinæsthetic revivals taking place during ordinary writing. By this I mean that in expressing one's self in spoken words memories of such words are first principally revived in the auditory centre, and that the nerve units thus called into activity immediately rouse through commissural fibres (see Fig. 4, *c*) the corresponding glosso-kinæsthetic elements before the pronunciation of the word can be effected through the aid of the motor centres in the bulb. Probably in a healthy brain there is also some amount of concurrent activity during speech in related portions of the visual word centre.

Again, when we are expressing our thoughts by writing, though the memories of the words are probably first revived in the auditory word centre, corresponding memories are almost simultaneously revived (through the intervention of the audito-visual commissure, *b*) in related parts of the visual word centre, and from this region stimuli pass by another commissure (*d*) to corresponding cheiro-kinæsthetic elements before the actual writing of the words can be effected through the instrumentality of motor centres in the cervical and upper dorsal regions of the spinal cord.

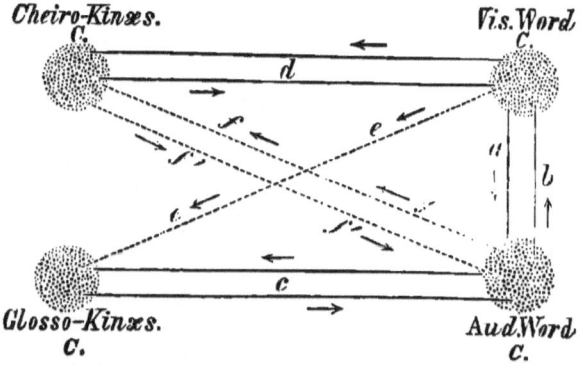

Fig. 4.—Diagram illustrating the relative positions of the different Word Centres and the mode in which they are connected by Commissures.

The connexions represented by dotted lines indicate possible but less habitual routes for the passage of stimuli.

There can be no doubt, in fact, that the functional association existing between the auditory and the glosso-kinæsthetic centres, as well as that between the visual and the cheiro-kinæsthetic centres, is of the closest kind. No less intimate, however, is the bond of association between the activity of the auditory and the visual word centres themselves. These latter centres are often necessarily called into associated activity in immediately successive units of time. This happens, for instance, in two such common processes as reading aloud and writing from dictation.

In reading aloud the primarily excited visual word centre

must arouse, through (*a*) the visuo-auditory commissure, related parts of the auditory word centre, since this is the part which ordinarily calls the glosso-kinæsthetic centre into activity, whence properly coördinated incitations issue from the cortex, in order to call into play the motor centres in the bulb.

Again, in writing from dictation the sounds of the words reach the auditory word centre, and the activity thus aroused becomes transmitted through (*b*) the audito-visual commissure to related parts of the visual word centre, this being the part that usually rouses the cheiro-kinæsthetic centre into activity, whence properly coördinated incitations issue from the cortex in order to call into play the motor centres in the cord concerned with the act of writing.

In exceptional cases it seems that the ordinary functional coupling of the auditory with the glosso-kinæsthetic and of the visual with the cheiro-kinæsthetic word centres is not adhered to. Thus the deaf-mute thinks in the main with revived visual symbols (either of hand or of lip movements), and it is from the organic seats of these that incitations pass to related parts of the glosso-kinæsthetic centre (*e*, *e*). A similar stimulation of these centres direct from the visual word centre seems occasionally to obtain in persons who are not deaf-mutes. We have already assumed it to be possible for some few "visuals"; and subsequently for the interpretation of certain cases of disease in which the auditory word centre has been damaged we shall find ourselves obliged to have resort to such a mode of action.

Again, in those children who have been born blind, but have nevertheless ultimately learned to write, a direct association must become established between the tactile and the auditory as well as between the auditory and the cheiro-kinæsthetic word centres (*f*, *f*). The same kind of associated activity between these two centres must exist in certain other persons not born blind, seeing that some patients suffering from word-blindness owing to destruction of the visual word centre are able, nevertheless, to write, either spontaneously or from dictation.

This functional unification of the word centres of which I have been speaking does not, however, remain in a state of isolation. Words that are heard are first of all associated in the mind of the child with external objects, so that such auditory impressions become linked, by means of associational fibres, with the organic seats of the several sensory impressions —also freely connected with one another—that the child has been able to derive from this or that object. Thus the name subsequently becomes revived as part of the perceptive process whenever the object is again presented. The sight of the mother recalls the name "Mamma," just as the sound of this word would revive the corresponding visual, tactile, and other images. After a time the auditory representative of the name becomes reinforced by glosso-kinæsthetic impressions as soon as the child learns to utter the word; later by visual impressions, when it learns to read, and by cheiro-kinæsthetic impressions when it has learned to write.

The several components of word-percepts are thus brought, not only into relation with one another, but become no less intimately associated with the several sensory components of object-percepts. The result is that the hearing or the sight of the name of any object immediately calls up in the older child or in the adult an idea of the object, just as its presentation to sight or other sense produces a nascent activity in the several centres in which its name is registered, though mainly in the auditory and the visual centres.

Words soon become to a very large extent the symbols whereby we carry on our thoughts, and this thinking by means of words becomes all the more thorough as thoughts become more complex. We can think of a particular person or of a particular plant as well by recalling the visual image as by recalling the name. But when in thinking we have to recall general names, such as "animal" or "tree," or still more abstract names such as "virtue" or "vice," it becomes certain that in silent thought we use the words as symbols for the more or less complex ideas which they represent. And if we bear in mind that the seats in which these words are registered (our word centres) are in organic and functional relation with the seats of registration of the corresponding percepts, concepts, etc., we are enabled in a measure to understand how the revival of the word in the mind (as the thought-counter) is associated with an almost simultaneous activity in the seats of registration of the corresponding percepts and concepts.

We may thus recognise how it happens that in simple acts of perception, and still more in simple thought-processes, we have no limitation of cerebral activity to narrowly localised centres, but rather widespread processes of activity in very varied regions of the cortex, and that, too, in both hemispheres of the brain. In our talk about individual centres and their functions we are apt to forget how much the brain acts as a whole even in operations that seem comparatively simple, and many of those who have written on aphasia have, as I venture to think, not sufficiently taken into account the fact that the name constitutes an integral element of the percept or of the concept.

On the Different Modes of Excitation of the Word centres, and on the Reason of their Functional Predominance in the Left Hemisphere.

It is important to recollect that word centres are naturally called into activity in states of health in three modes, but that failure in the possibility of their excitation by one or other of these modes occurs in various morbid conditions, leading to this or that kind of speech defect. The three modes of excitation are these: (1) By means of "sensory" impressions coming from without; (2) by "association"— that is, by impulses communicated from another centre during some act of perception or thought-process; and (3) by "voluntary" recall of past impressions, as in an act of recollection. It is convenient for practical purposes to separate these second and third modes of calling into activity the perceptive centres; still they are probably closely related to one another and scarcely separable in nature, although the associational processes in the case of "volitional" excitation may be so difficult to trace that the recall assumes a quasi-spontaneous character.

The excitability of the centres—that is, the molecular mobility of their nervous elements—may vary with age, state of health, or the existence of different morbid conditions. Their excitability may be so much lowered that they are only capable of responding to powerful stimuli; so that while volitional recall (or " recollection ") may be impossible within their province, such centres may still be capable of acting in " association " with others—that is, in an automatic manner during an ordinary process of thought; or at least under the " sensory " stimulus which initiates a perceptive process. At other times the excitability of the perceptive centres may be unduly exalted, so as to lead to hallucinations, illusions, and a wholly different class of defects such as are often met with among the insane.

From what has been said in a previous section it will be seen that I entertain doubts whether kinæsthetic word-memories of either kind are capable of being revived alone in a voluntary manner.

As to the causes which have determined the greater or almost exclusive influence of the left hemisphere in inciting speech movements, and therefore in acting upon the bulbar motor centres for words, only conjectures can be offered. It is, however, now pretty generally agreed that the immediate or proximate cause is to be found in the fact of the predominant use of the right hand, which entails a greater functional activity upon the left cerebral hemisphere. This view rests principally upon the now ascertained fact that in the great majority of cases in which aphasia has occurred as a result of brain lesions in the right hemisphere (with or without the association of left hemiplegia) those so affected have been left-handed persons.[1] It would thus seem to follow that the predominant use of the right hand or of the left hand carries with it, as one of its associated effects, the leading activity in the production of speech by the left or the right hemisphere respectively; and that we must consequently push our question further back, and inquire as to the causes that have led to this predominant use of the right hand.

This problem has been considered by Pye-Smith,[2] and also in a more exhaustive manner by William Ogle,[3] but can only be briefly alluded to here. It has been thought to depend largely upon tribal or social customs and early tuition of young children by their mothers, though doubt has been thrown upon these views as giving any adequate explanation of the problem. William Ogle looks rather to certain peculiarities in the mode of origin of the left and right carotids respectively as calculated to favour a slightly freer blood supply to the left cerebral hemisphere. If this were really so, it is conceivable that it might after a time have led to a slightly earlier development of the convolutions on the left side of the brain (which has been affirmed to occur by Gratiolet, though denied by Carl Vogt and Ecker), and thus perhaps have led to a preferential use of the right hand in the early and more definite movements of the child. Such preferential use being once started, and maintained through several generations, it is only to be expected that it would continue; for if the effects of use and disuse are inherited as, with Herbert Spencer, I believe to be the case (in spite of all that Weissmann says to

---

[1] According to Seguin, out of 266 cases of aphasia he found 243 with right and 17 with left hemiplegia; that is, somewhere about 6 per cent. W. Ogle, on the other hand, from inquiries among 2,000 hospital patients, found that 85 of them were left-handed; that is, about 4½ per cent. Left-handedness, curiously enough, was found to be "twice as common in men as in women." Ogle shows, too, how very frequently there is hereditary transmission of this peculiarity.

[2] *Guy's Hospital Reports*, vol. xvi., 1871.

[3] *Transactions of the Royal Medical and Chirurgical Society*, 1871.

the contrary) heredity would coöperate with the original cause in confirming and making general this preferential use of the right hand.

But whatever may have been the original cause of this now very general attribute of right-handedness—and the chief difficulties of the problem with which we are concerned lie here—it would easily follow that with its continuance in process of time the left hemisphere should become slightly heavier than the right (as it has been found to be), and that the left carotid might become very slightly larger than the right carotid (as it is said to be). It is practically certain that the great preponderance of right-hand movements in ordinary individuals must tend to produce a more complete organisation of the left than of the right hemisphere. We may confidently look for the existence in it of the organic basis of a vastly greater and more complex tactile experience; and as movements of the right hand and arm are more frequent, both as associated factors of this experience and in other ways, we have also a right to suppose that the kinæsthetic centres will be similarly developed to a notably greater degree in the left hemisphere.

Many years ago I ascertained a fact which then seemed very difficult to understand—namely, that the specific gravity of the cortical grey matter of the brain in the left frontal, parietal, and occipital regions is often distinctly, though slightly, higher than that from corresponding regions of the right hemisphere. Full details as to this will be found in a paper "On the Specific Gravity of the Human Brain," published in 1866.[1] Now such an increase in specific gravity might easily be produced by the greater number of cells and associational fibres which the extra-sensory and derivative functions above referred to would necessarily entail.

Thus it may, in brief, be said that this preponderating activity of the left hemisphere in regard to tactile and kinæsthetic impressions (about which there is no room for doubt) may have much to do with the fact that the left hemisphere is the more potent and seems to take the lead in some thought-processes, as well as in giving rise to the voluntary excitations which determine the muscular acts involved in articulate speech. A greater convolutional complexity of the lower part of the left frontal lobe as compared with the right, which has been found to exist, is thought to be a result of this preponderant activity of the left hemisphere in the production of articulate speech. Some evidence in

[1] *Journal of Mental Science*, pp. 28, 32.

favour of this view is to be found in the fact that Broadbent, in his examination of two brains taken from left-handed persons that were submitted to him by William Ogle, found in them the reverse condition—that is, that the greater convolutional complexity was on the right rather than on the left anterior lobe.[1]

There still remains the question why one cerebral hemisphere only should be efficiently educated for the perception of speech and for the production of speech-movements, when, seeing that the bulbar motor centres are bilateral, it might seem that such centres would be likely to receive their stimuli from both hemispheres. In partial reply to this question Moxon, in a very able paper,[2] long ago suggested that the call upon attention was so great for the production of the extremely elaborate movements concerned in speech that the necessary concentration of attention would be much facilitated if this process of education were limited to one hemisphere. Whatever may be thought of this ingenious suggestion, nothing better has hitherto been forthcoming.[3]

What has just been said as to the action of the left hemisphere rather than the right in the production of speech must be understood with certain reservations. It must not be supposed that the right hemisphere remains entirely uneducated, either in regard to the comprehension or to the production of speech. Little is said by writers generally on this subject, and what has been said has reference principally to the third frontal convolution. A hypothesis as to the functions of this part of the brain has been enunciated by Hughlings Jackson,[4] and another view, as to what is figuratively termed the "overflow of education into the opposite hemisphere," has been expressed by Wyllie.[5] But neither

[1] *Transactions of the Royal Medical and Chirurgical Society*, 1871, p. 294.

[2] *British and Foreign Medico-Chirurgical Review*, 1866, p. 481.

[3] Moxon's paper is, indeed, in other respects remarkable, looking to the date at which it was written. Thus he speaks of spoken words as represented in Broca's region by "educated associations of movements"; and the meaning that he attaches to this phrase is rendered evident further on when he dwells upon the "inconceivable pitch of education which is given to these supra-motory departments, if I may so call them, of the brain, which hold ready for use the memorial forms of outgoing words," or still better where he says, "The situation of the ideas of *associated motions* which form the faculty of speech is supra-motory, whilst the situation of the ideas of *associated sensations* which form the faculty of language comprehension is supra-sensory."

[4] *Brain*, vol. ii., 1879, p. 327.

[5] "The Disorders of Speech," 1895, p. 262.

of these hypotheses appears to me to be in accord with existing knowledge.

The most important points to be borne in mind in reference to this subject would seem to be these. All the movements concerned in speech are movements produced by symmetrically placed muscles on the two sides of the body; there is just as much reason, therefore, for the registration of kinæsthetic impressions resulting from speech-movements in the right as in the left third frontal convolution. Again, it cannot be supposed that auditory impressions from spoken words do not pass from each ear to similar regions in the opposite cerebral hemisphere; and our present knowledge makes it equally improbable that the visual impressions of words are not registered in a visual word centre in each hemisphere. For each of these three sets of afferent impressions, therefore, there is, as I believe, no question of an "overflow" from one to the opposite hemisphere; on the contrary, each hemisphere receives its own proper share of ingoing impressions, and doubtless registers them in a more or less similar fashion. That this should be so seems an obvious truth, though it is one that hitherto appears to have been very imperfectly realised. Again, it may be taken as established, for reasons which will subsequently appear, that the separate centres composing each of these pairs are brought into functional union by means of commissural fibres forming part of the corpus callosum. So far, therefore, we seem bound to admit that each hemisphere has the chance of being equally educated, so far as the mere *reception* of speech-impressions is concerned.

It is, however, when we come to the executive side of speech functions that the great difference begins to appear between the relative activity of the two hemispheres of the brain. For reasons which we have previously been dimly endeavouring to indicate, the executive speech functions become relegated to one hemisphere, and this leading hemisphere, generally the left, becomes for this and for the other reasons which we have adduced more highly organised. The fact that the volitional impulses for the incitation of speech pass off from this hemisphere leads of necessity to the gradual perfecting of the associational channels between the auditory and the glosso-kinæsthetic centres, as well as between the auditory and the visual word centres of the left hemisphere, and also to the perfecting of the outgoing channels of communication between these sensory centres and the true motor centres in the bulb. The corresponding associational and outgoing channels in the opposite hemisphere, however, being but little used, may remain comparatively undeveloped.

This difference in the degree of activity of the two cerebral hemispheres during speech would tend to become more and more accentuated in future years, and would in all probability lead to a much higher grade of functional activity, even on the receptive side, in the three kinds of word centres pertaining to the leading hemisphere.

Views of this kind as to the partial education of the word centres of the opposite hemisphere will be found to be of considerable importance when we come to speak of the modes in which destruction of one or other of the word centres in the leading hemisphere may be compensated—or, in other words, of the way in which a cure may be brought about in this or that form of speech defect.

CHAPTER II.

THE RELATIONS EXISTING BETWEEN THOUGHT AND LANGUAGE.

WHILE it is universally admitted that thought in all its higher modes cannot be carried on without the aid of language of some kind, it seems to many no less certain that the simpler modes of thought can be performed without any such aid.

All degrees of complexity exist between the lowest and the highest modes of thinking, but roughly speaking we may describe two grades or degrees of complexity.

There is the simplest mode (a) in which thinking is carried on only in trains of feelings and perceptions. On this low level our thoughts perhaps correspond pretty closely with those that may be conjectured to pass through the minds of many of the lower animals. There is much truth in the statements of Ross[1] when he says: "Now a great part of our thinking consists of perceptions experienced at the moment, and of reminiscences of previously experienced perceptions. Suppose one takes a walk in a green field, he perceives the greensward, the trees, and the foliage, from which rays of light are reflected to his eyes, and the cattle and birds, whose lowing and singing fall upon his ear, and the more vivid of these perceptions start a train of reminiscent perceptions, probably from scenes in his childhood, while there is with him an ever present feeling of the great conditions of all our sensuous perceptions—time and space. It is possible that in addition to these thoughts he may carry on a train of reasoning in words, but whether that be so or not, there can be no question that the greater part of his thinking has consisted of perceptions actually experienced, and trains of perceptions ideally revived. Thinking by means of percepts is, therefore, common to man with the lower animals, and although the thoughts of the former, even on this level, are much more varied and complex than those of the latter, yet there is no essential difference in principle between the two."

[1] "On Aphasia," 1887, p. 121.

(b) At a higher level thought is largely accomplished by means of words, and chiefly by the aid of general or abstract names—the names of concepts. Thinking of this kind is that which is carried on by the great mass of mankind, though of course it comprises innumerable grades of complexity, answering to the different degrees of knowledge and intellectual development in those who are the subjects of it. Starting from the comparatively simple modes of thought that are common to young children and primitive races of mankind, it becomes more and more complex till it merges into that highest mode of abstract thought, which forms almost a class apart—that which is often carried on by philosophers and advanced mathematicians largely by means of abstract terms and mathematical symbols.

It is obvious, therefore, that just as all gradations exist between the thought-processes of lower animals and those of children and the lowest races of men, so are all grades of complexity to be met with between the thought-processes of these latter and those that constitute the highest manifestations of human intelligence.

That this should be so may be the more easily understood if we bear in mind (1) that sensorial activity, which is common to animals and men, contains the germs of all mental processes; and (2) that language becomes, when once initiated, a most powerful factor in the development of thought—an instrument, indeed, by which alone the most complex processes of thinking are rendered possible.

(1) Professor Bain says,—"Some sensations are mere pleasures or pains, and little else; such are the feelings of organic life, and the sweet and bitter tastes and odours. Others stretch away into the regions of pure intellect, and are nothing as regards enjoyment or suffering, as, for example, a great number of those of the three higher senses."

Sensation is, in fact, a complex rather than a simple mental process. It is invariably compounded of cognition and feeling. That there is a discriminative or intellectual side to even the most subjective of our sensations has been strongly insisted upon by Sir Wm. Hamilton and others. Any sensation, however simple, can only be recognised as such—can only become an element of our consciousness, (a) by the simultaneous memory or revival of some past impressions, (b) by the intuitive recognition of their likeness or unlikeness to the present impression, and (c) by the simultaneous recognition that this is felt as in a certain place. This holds good even for touches and tastes, which are habitually referred to some part of that inner circle of the non-ego, represented by the organism's own body. And in reference to such odours, sounds, and sights as are referred to the outside world beyond the organism, it becomes plainly impossible to attempt to preserve any real distinction between sensations and perceptions—since precisely similar mental processes are involved in both.

It seems plain, therefore, that a gradual transition may be traced between simple sensations and the most elaborate perceptions; that there is a difference in degree rather than in kind, between these two processes; and that James Mill, in his "Analysis of the Human Mind," was not without justification in making no use of the latter term, and in speaking merely of "simple" and of "complex" sensations. Moreover, it must be steadfastly borne in mind, that in every complex sensation (or perception) of an external object, there occurs an embodied cluster of judgments and inferences, similar in kind to those which compose the basis of all intellectual action. Thus the notion that an intellectual element enters into the very groundwork of all sensations is so well founded as to make it not at all surprising that such an opinion should have been held alike by ancient and by more modern philosophers.

Sir Wm. Hamilton says, "It is manifestly impossible to discriminate with any rigour, sense from intellect." And again Max Müller writes:—"though sensations, percepts, and concepts may be distinguished, they are within our mind one and indivisible. We can never know sensation except as percepts, we can never know of percepts except as incipient concepts."

This double or two-sided nature of sensation, and the necessary development from it of the germs of intellect on the one side, and of emotion on the other, as from a common root, is a fact of the greatest interest from a physiological as well as from a philosophical point of view.

There is, however, a third aspect of sensation or perception, which has not yet been mentioned, though it seems to be one of great importance in helping to determine the development of nervous structures, and the correlative increasing complexity of mental phenomena. This is to be found in that exercise of volition or "Will" which enters into every perception under the form of attention. Nor must it be here forgotten that in still another way are sensations related to volitions. The pleasures and pains of sense, either actually present or represented in idea, seem unquestionably to constitute in large measure the subjective sides of those neural processes which most frequently issue in the production of so-called volitional movements.

It is of great importance, therefore, here to note that intelligence, sensation, emotion, and volition are mental processes, the primary stages of which are dependent upon, and inseparably connected with, different modes or aspects of the functional activity of the sensory centres in the cortex of the brain.

(2) Although lower kinds of thinking may, as we have seen, be carried on without the aid of words by means of feelings and percepts, associated with images of percepts, it is, as I have said, admitted by all that words are essential for complex thought, and become most important instruments in the development of the power of thinking. Thus Sir Wm. Hamilton says:[1] "A word or sign is necessary to give stability to our intellectual progress, to establish each step in our advance as a new starting-point for our advance to another beyond. A country may be overrun by an armed host, but it is only conquered by the establishment of fortresses. Words are fortresses of thought. They enable us to realize our dominion over what we have already overrun in thought; to make every intellectual conquest the basis of operations for others still beyond. . . . . Though, therefore, we allow that every movement forward in language must be determined by an antecedent movement forward in thought; still, unless thought be accompanied at each point of its evolution by a corresponding evolution of language, its further development is arrested."

The power of language in aiding cerebral development and thinking processes, although it must have been great from the first, and ever tending to increase, did not reveal itself so forcibly till means had been adopted for the preservation and communication of human experience and thought from generation to generation, by the use either of "picture-language," hieroglyphics or more modern forms of writing and printing. When these latter came into common use and when books began to circulate, then at last language began to exercise its full influence as an aid to and developer of thought.

As Thomson says:[2]—"Language becomes more analytic as literature and refinement increase. This property indicates as we should expect, corresponding changes in the state of thinking in different nations, or in the same at different times. With increasing cultivation, finer distinctions are seen between the relations of objects, and corresponding expressions are sought for to denote them; because ambiguity and confusion would result from allowing the same word, or form of words, to continue as the expression of two different things or facts. . . . . . A discovery can hardly be said to be secured, until it has been marked by a name which shall serve to recall it to those who have once mastered its nature, and to challenge the attention of those to whom it is still strange. Such words as

---

[1] Lectures, vol. iii., pp. 138-140.   [2] "Laws of Thought," p. 28.

inertia, affinity, polarisation, gravitation, are summaries of so many laws of nature. . . . . . Names then are the means of fixing and recording the results of trains of thought, which without them must be repeated frequently, with all the pain of the first effort."

What has been said hitherto will enable us to realise the importance of two problems which now require a brief consideration. They are, (1) to what extent are thoughts and words inseparable? (2) Is there a difference in this respect between different classes of words?

(1) Upon this question as to the extent to which thoughts and words are inseparable two quite different views have been expressed. On the one hand Kussmaul and others hold such views as these. He says,[1] "Let us now consider for a moment the most essential points in the process of talking. In the first place a *thought* is indispensable which we have *conceived*, and then an *impulse of feeling* urging us to express it. Next we *choose* and *say* the *words* which the acquired language in our memory places at our disposal. Finally, the *reflex apparatuses* are called into play which give outward utterance to the words . . . . . although we arrive at our ideas through speech, these, as soon as they are formed, possess an independence of words."

This seems to be practically making thought independent of language, in adults at least, though a careful study of the next chapter in which Kussmaul deals with this subject, does not seem to me in any way adequate to substantiate this position.

On the other hand Max Müller is a still more determined adherent of the opposite view that "thought and language are inseparable." But this is, as I also believe, altogether too sweeping a statement. It can only be made by denying the existence of what I have referred to as the lowest mode of thinking—that which is carried on by trains of actual and revived feelings and percepts—that is denying that this process is worthy of the name of thinking.

His experimental proof that "thought is impossible without words" is not quite so convincing as he imagines. He triumphantly says,[2] let any reader try, for at least five minutes, to think the saying of Descartes, *Cogito ergo sum*, without allowing either these Latin words or their equivalents in some other language to pass through his mind. The impossibility

---

[1] *Ziemssen's Cyclopædia*, pp. 595 and 603.
[2] "The Science of Thought," 1887, p. 56.

of this may be frankly conceded; but then this is just one of those abstract reasonings which many would regard to be impossible without the aid of language. And of course the fact that this is so does not at all prove that in simpler modes of thinking visual images at least may not to some extent take the place of words.

(2) The question whether all kinds of words are equally inseparable from thought is also one of much difficulty. Thus, it would seem that the separability of words from thought is most marked in the case of names of persons or things,— visual percepts of which may take the place of words. It is worthy of note, therefore, that in cases of amnesia these are always the words that first fail to be revived. On the other hand, no such substitution would seem possible in the case of such words as "virtue," "charity," "gravitation," "polarisation," and similarly for other parts of speech.

Again, a partial independence of thoughts and words, in the direction suggested by Kussmaul, certainly seems to be indicated by the very fact that we "select" our expressions. Thus, according to the different shades of meaning sought to be conveyed in our propositions, we often deliberately weigh or select the abstract nouns, the adjectives and verbs that we deem most expedient for the complete communication of our thoughts to others. The process by which such words are associated with our thoughts, seems to be a little less automatic than that by which external objects or the images of such objects associate themselves with words—that is, in the latter case the name, in a state of health, forms an integral part of the percept or of its revived image, whilst our use of abstract nouns, adjectives, verbs, prepositions, etc., is not thus strictly tied down—they become associated with our thoughts in a less formal manner, so that the thought is able to couple itself with this or that word out of two, three, or four that may simultaneously become nascent with the thought itself.

Does Conception take place in a Centre altogether apart from Perception?

The first authoritative writer on defects of speech who started the notion of the existence of an altogether separate centre for conception or ideation was Broadbent. This he did in a very important memoir, "On the Cerebral Mechanism of Speech and Thought," published in 1872.[1] Whilst there adopting some views which I had previously published on the

---

[1] *Transactions of the Royal Medical and Chirurgical Society*, 1872, pp. 180, 181, and 191.

subject of the cerebral processes that occur in perception,[1] he made a departure from them in the direction above indicated. The functions that I spoke of as being carried on in the perceptive centres he divided into two stages. He said: "There is a primary or rudimentary perceptive act in which the external cause of a given set of sensations is recognised as such, and in which the simple attributes, as of form, colour, hardness, etc., are perceived. And there is a higher degree of elaboration in which, by the combination or fusion of perceptions derived from the various organs of sense, a conception or idea of an object as a whole is obtained. This is a new and distinct process, and is usually accompanied by the affixing of a name to the object. To the 'perceptive centres' I relegate simply the translation of sensations into rudimentary or primary perceptions, and these centres must lie somewhere in the marginal convolutions which receive radiating fibres from the crus and central ganglia, upon which therefore impressions will first impinge, and which are symmetrically combined by the fibres of the corpus callosum. The higher elaboration, the fusion of various perceptions together, and the evolution of an idea out of them, will be accomplished not by radiation from one perceptive centre to all the others, but by convergence of impressions from the various perceptive centres upon a common intermediate cell area, in which a process analogous to the translation of an impression into a sensation and of a sensation into a primary perception will take place. This intermediate cell area will form a part of the supreme centre, and will be situated in the superadded convolutions which receive no radiating fibres."

In a subsequent communication some years later these views were still further developed by Broadbent.[2] The centre for concepts was then termed the "naming centre," whilst a related higher motor centre was postulated as a "propositionising centre," in which words other than nouns were supposed to be registered, and where sentences were formulated preparatory to their utterance through the instrumentality of Broca's centre.

This supposed convergence of impressions from the various perceptive centres into a new region in which "conception" and "naming" take place is a view totally different from my own, which nearly thirty years ago was thus expressed:[3] "In the perceptive centres the primary impressions made upon the organs of sense are converted into 'perceptions proper'—that

[1] "On the Muscular Sense and the Physiology of Thinking," *Brit. Med. Jour.*, May, 1869.
[2] *Brain*, vol. i., 1878.   [3] *Brit. Med. Jour.*, May, 1869.

is to say, they receive their intellectual elaboration, and this elaboration implies an intimate cell and fibre communication between each perceptive centre and every other perceptive centre, since one of the principal features of a perceptive act is that it tends to associate, as it were, into one state of consciousness much of the knowledge which has been derived at different times and in different ways concerning any particular object of perception. An impression of an object, therefore, made on any single sense centre on reaching the cerebral hemispheres, though it strikes first upon the perceptive centre corresponding, immediately radiates to other perceptive centres, there to strike upon functionally related cells, all this taking place with such rapidity that the several excitations are practically simultaneous so that the combined effects are fused into one single perceptive act."[1] Thus the work which Broadbent supposes to be done in his "naming" or "concept" centre is according to my view carried out by the simultaneous activity of the different perceptive centres, and certain annexes that are derivative developments therefrom, of which I shall subsequently speak. And, again, I have postulated, instead of a single separate "naming centre," the existence of four "word centres" as important and intimately correlated parts of the more general auditory, visual, and kinæsthetic centres.

Consequently I never seek to explain certain cases of speech defect, as he and others whom I am about to mention not unfrequently do, by supposing the existence of lesions in a localised centre for "concepts," or of lesions involving the commissural fibres proceeding to or issuing from such a centre.

With variations in detail, views similar to those of Broadbent as to the existence of a special centre for concepts were subsequently published by Kussmaul[2] and Charcot,[3] whilst the views of the latter were adopted and further promulgated by Bernard[4] and Ballet.[5] The existence of a separate centre for concepts is also postulated by Grasset.[6]

The views of Kussmaul were embodied in an elaborate and

---

[1] From this it is evident that there can in reality be only one perceptive centre, although its different component parts were formerly by myself (as well as by Broadbent) spoken of as "perceptive centres" because their functional activities were integral components of a perceptive act. I recognise now that it would have been better not to have used this term as a name for these centres—that they are, in fact, more appropriately spoken of as sensory centres.

[2] *Ziemssen's Cyclopædia*, vol. xiv., p. 779.

[3] *Le Progrès Médical*, 1883.

[4] " De l'Aphasie," 1885, p. 45.   [5] " Le Langage Intérieur," 1886.

[6] " Leçons de Clinique Médicale," 1896 (Montpellier).

complicated diagram which it is unnecessary to describe. A more simple and easily to be comprehended diagram was also produced by Charcot (and is to be found in the work of Bernard), which is based upon the same principles as the scheme issued by Broadbent, with the exception that he omits the "propositionising centre" and also adopts my view as to the existence of separate word centres. He, however, makes the several sensory impressions converge to an "ideational centre" in order that their meaning should be realised, and indicates that the speech centre (that of Broca) and the centre for writing movements (both of them as "motor" centres) are called into activity directly from this ideational centre, as well as from the auditory and the visual centres respectively.

Some diagrams that have attracted much attention were also published by Lichtheim in illustration of his well-known paper on aphasia.[1] They likewise show a centre for concepts altogether apart from the sensory centres, and in his endeavours to explain the different kinds of speech defect he refers two of his types to a destruction of supposed afferent and efferent fibres proceeding to and from this conceptual centre. While, therefore, his diagrams and his language in many parts of his paper would make one think that he adopted in full the view as to the existence of a wholly separate centre for concepts, he says towards the end of his paper[2] "this has been done for simplicity's sake," and that he does not "consider the function to be localised in one spot of the brain, but rather to result from the combined action of the whole sensory sphere." This statement, though it is quite in accordance with my own view, seems to me to invalidate much of his exposition and to make it almost impossible for him legitimately to suppose, as he does, that two of his types of speech defect are to be explained by the supposition of the existence of a lesion involving either the afferent or the efferent fibres pertaining to such a widely diffused centre. In addition there is the serious defect that his diagrams are at variance with his views on this important subject.[3]

My dissent from these particular views of Broadbent was expressed in 1880,[4] and again more strongly from them as

[1] *Brain*, 1885. The reader will find several of the diagrams previously referred to in an article on "The Sensory Side of Aphasia," by E. A. Shaw, in *Brain*, 1893, p. 492.

[2] *Loc. cit.*, p. 477.

[3] I shall later (in chap. xii.) state what I believe to be the interpretation of his seven types of speech defect, when it will be seen that they may be explained without any necessity for supposing the existence of lesions in any such parts.

[4] "The Brain as an Organ of Mind," p. 636.

well as from the allied views of Kussmaul, Charcot, and Lichtheim, in 1887.[1] I am glad to say that this dissent from these doctrines has been followed by that of A. de Watteville,[2] Ross,[3] Allen Starr,[4] Wyllie,[5] and Byrom Bramwell,[6] all of whom have likewise decided against the propriety of postulating the existence of a separate centre for ideas or concepts.

My opposition to this postulation of a separate "centre for concepts" was based originally upon psychological considerations. It seemed to me wholly unnecessary and at variance with what appeared to be the real nature of the process of perception.

Then, again, I am unable to find any clear evidence from clinical data tending to prove the existence of a separate "centre for concepts"—or, in other words, any existing forms of speech defect that can only be explained by supposing the existence of a lesion in such a centre or in the course of its afferent or efferent fibres. I am convinced that the supposed necessity for assuming the existence of a "centre for concepts," when seeking to interpret different forms of speech defect, may in many cases be obviated by a fuller recognition of the different degrees of functional excitability that may obtain in the auditory and the visual word centres respectively. We shall see that their molecular mobility may be so much lowered that they are only capable of responding to powerful stimuli. Thus, whilst volitional recall, or recollection, may be impossible within their province, they may still be capable of acting in association with other centres—as when reading may be fluent though voluntary speech is greatly impaired; and still more easily under a direct sensory stimulus—as when a word is repeated which the patient has just heard pronounced.[7]

[1] *Brit. Med. Jour.*, Nov., 1887.
[2] *Le Progrès Médical*, March 21, 1885.
[3] "On Aphasia," 1887, p. 129.
[4] *Brain*, 1889, p. 97, where Starr says: "The results of pathological observation fail to give any support to the hypothesis of an 'ideational centre' which Broadbent and Kussmaul have introduced into their diagrams."
[5] "The Disorders of Speech," 1894, pp. 230 and 235. Wyllie, however, I think, tends to convey an erroneous impression by representing in a diagram, on p. 276 of his work, the seats of word revival in a lower plane and of percept revival in a higher plane, and assuming that the stimulation of Broca's centre takes place from the seat of percept revival rather than from the auditory word centre. This seems to be also the position of Lichtheim, judging from his Fig. 7 and what he says in reference thereto, *loc. cit.*, p. 477.
[6] *Lancet*, vol. i., 1897.
[7] Examples will be given later when speaking of the defects due to disabilities in the auditory word centre.

Another cause that has led to this postulation of a "centre for concepts" is traceable, I think, to an inadequate realisation of the nature of a perceptive process and of the fact that a name may and does normally constitute an integral part of the complexus of revived sensory impressions which go to constitute such a process.

Again, as I have previously endeavoured to show, perceptive processes vary greatly in complexity and merge by insensible gradations into processes of conception. It seems thoroughly legitimate, therefore, to suppose that these latter more specialised modes of mental activity, whilst having their roots in perceptive centres, may be completed in outgrowths therefrom—that is, in parts of the brain which are in close relation structurally and functionally with the several sensory centres. I have commonly spoken of such regions as "annexes" of the perceptive centres.

Of late Flechsig has called special attention to four areas of the cortex that differ from the sensory areas, since they are neither in relation with afferent nor with efferent fibres. He assumes[1] that these regions subserve higher mental functions than those carried on in the sensory centres, and terms such regions "association areas." These regions seem, therefore, to correspond with what I have referred to above as "annexes of the perceptive centres." They occupy a large proportion of the cerebral cortex, and are thus located by Flechsig: (1) in parts of the pre-frontal lobes; (2) in a large portion of the temporal lobes; (3) in a considerable area of the posterior parietal region; and (4) in the island of Reil. These four fairly well-defined areas are, as above stated, not directly connected with afferent or with efferent fibres, and in addition to this two other reasons are given for supposing them to be concerned with the carrying on of higher functions. Flechsig points out, in the first place, that these regions remain immature and completely devoid of myeline for several months after birth, though the sensorial centres have arrived at comparative maturity; and, secondly, that these association centres are the parts which are especially developed in the brain of man as compared with that of the lower animals.[2]

It is only fair to Broadbent to point out that more than twenty years previously he had cited almost exactly these regions of the cerebral cortex as parts that were neither in

[1] *Neurologisches Centralblatt*, 1894, pp. 674 and 809; and *Gehirn und Seele*, Leipzig, 1895.

[2] Since the above was written I have received from Déjerine the reprint of a communication made by him to the Société de Biologie, on Feb. 20, 1897, in which he contests these views of Flechsig.

direct relation with peduncular fibres nor with those of the corpus callosum, and that he had attributed to these regions (for almost exactly similar reasons) just the same functions as those now assigned to them by Flechsig.[1] He adds: "Now the convolutions which I have enumerated as having no direct communication with the crus, central ganglia, or corpus callosum are, in the first place, those which are latest in order of development, and on this ground alone might be supposed to be concerned in the more strictly mental faculties which are latest in their manifestations. They are those which constitute the difference between the human cerebrum and the cerebrum of the quadrumana; and it would, moreover, seem to accord with the general plan of construction of the nervous system and with what we know of the mental operations that these convolutions, which are withdrawn, so to speak, from direct relation with the outer world, should be the seat of the more purely intellectual operations."

It is, I think, perfectly legitimate to suppose that the annexes of the sensory centres, to which I have previously referred, tend to be developed in the directions above indicated by Broadbent and Flechsig, though how much of these territories they occupy must remain altogether uncertain. It seems also probable that there is no sharp line of demarcation between these annexes and the several sensory areas, and that *the combined sensory areas together with the annexes are accustomed to be thrown into functional activity more or less simultaneously.* Thus the processes of perception and conception, together with revival of linguistic symbols, are probably almost as inseparable in their localisation as they are in their nature and modes of occurrence; and their anatomical substrata must be supposed to occupy a very considerable extent of the cortex of both hemispheres.

A final question now remains for consideration related to this other which we have just been considering. It is this— Where are we to look for the registration and revival of words in the cerebral cortex? It may be said that this question has been already answered. And so it has tentatively and in a general sense. We have laid stress upon the existence of four different kinds of memorial registration of words and the probable sites of such word centres in the hemispheres. We have indicated also that the glosso- and the cheiro- kinæsthetic centres constitute definite parts of the general kinæsthetic

[1] *Transactions of the Royal Medical and Chirurgical Society,* 1872, p. 178.

centres, and that the auditory and the visual word centres probably also constitute more or less separate parts of the general auditory and visual centres.

Something additional, however, may now be said concerning the sites of these word centres which could not well have been said at an earlier period, and that is that each of them is probably to be found partly on the confines of its percept centre and partly on that of its related annexe. This supposition is made because some words (especially names of things, persons, and places) are in closest relation with sensory centres; whilst others, such as verbs, adjectives, prepositions, and other parts of speech constituting the framework of language, are in closer relation with conceptual processes. These two modes of functional activity are, as I have said, absolutely inseparable from one another, and therefore the several word centres must be in most intimate relation both with the sensory centres and with their annexes.

If the views above expressed be anything like an approximation to the truth, it may be judged how vain it would be to attempt to base our explanation of any of the different kinds of speech defect upon the supposed existence of some one separate centre for "ideation," "conception," or "naming," and to imagine that it is connected by means of commissures (long enough and separate enough to permit of isolated damage) with sensory centres on the one side and with motor centres on the other.

I am pleased to find that the views at which I have arrived on this important subject—while they differ so much from those held by some of the most authoritative writers on speech defects—are nevertheless in very close accord with opinions expressed by a very distinguished writer and thinker on these subjects—viz., the late Dr. James Ross. Thus, in his work "On Aphasia," he gave expression to the following views (page 125) :—" And on passing from thinking by percepts to thinking by concepts, and from that to thinking by abstracts, there are no new centres introduced, but only a complication upon complication of the one perceptive centre. All that can be said is that the correlative of perceptive thinking is excitation of that portion of the cortex of the brain which is directly connected with the sensory inlets; of conceptive thinking, excitation of portions of the cortex which are indirectly connected with them; and of abstract thinking, excitation of portions which are still more remotely connected with them. It must, however, be remembered, that the effective working of the portions of the cortex which are remotely connected with the sensory inlets will in a great measure depend upon the integrity of those that are in direct relation with them.

Let us now attend to the effects of dissolution of this structure. A destructive lesion of the portions of the cortex which are most remotely connected with the sensory inlets would destroy the capacity of the patient for highly abstract reasoning, and would no doubt inflict considerable damage on the language in which abstract thought is embodied, but this condition would not be recognised as an aphasia; and even the intermediate portions of the cortex in which conceptive thought is carried on might be seriously damaged without giving rise to a special speech disorder, inasmuch as any impairment of speech which might be present would only be regarded as a part of a general decay of the reasoning faculties. When, however, the lesion is situated in or near to the sensory inlets a disorder of language results which is out of all proportion to the general impairment of the reasoning faculties."

All that has been set forth above is thoroughly in harmony with the dictum of Max Müller, who says: "Though sensations, percepts, and concepts may be distinguished, they are within our mind one and indivisible. We can never know sensations except as percepts, we can never know of percepts except as incipient concepts."

CHAPTER III.

THE CLASSIFICATION OF SPEECH DEFECTS.

Although in this work we shall deal with what are known as "Acquired Defects of Speech"—that is, such defects as may have supervened after the power of speaking has been attained—still, it seems desirable that a very brief reference should be made, as a preliminary, to the principal defects of speech resulting either from congenital disease or from cerebral diseases occurring at some date prior to the manifestation of articulate speech.

The most important of these congenital defects is deafness, which of itself entails mutism, the individuals thus afflicted being known as "deaf-mutes." It must be borne in mind, however, that this mutism or dumbness may also be brought about by absolute deafness occurring from any cause after birth, but before the child has begun to talk, or even after it has learned to talk, up to the fifth or the seventh year. In cases of the latter type the child soon (when thus left without the customary guidance in articulatory acts derived through the sense of hearing) forgets how to speak and becomes dumb.

In addition to this class of cases there are also those of congenital idiotcy without deafness, in which the child never learns to talk. There are also other cases, allied to the last, in which, owing to some intracranial lesion occurring either before, during, or soon after birth, the brain becomes so damaged as to arrest the development of the child's mental as well as of its motor powers. In some of the less severe examples of this type (now commonly spoken of as "birth palsies") that have come under my observation speech has been merely deferred—perhaps till the fourth, fifth, or even the sixth year—and then after a time has become established in a natural manner. In other cases the occurrence of epileptic fits in infancy seems to be the retarding cause. One of the most remarkable cases known to me of deferred speech in connexion with the occurrence of epileptic attacks happened in one of my own patients, to whom reference has already been made (p. 6).

Lastly, some curious cases of congenital speech defect were described by Hadden[1] to which the term "idioglossia" has been applied. These children have to a certain extent a language of their own, so that when asked to repeat phrases they make use of different, though definite, sounds instead of those proper to the words that should be employed. The sounds which they substitute are said to be always the same for the same words. Some of these patients have been capable of writing correctly from dictation, and they have also shown a fair amount of general intelligence.

Turning now to our own special subject of "Acquired Defects of Speech" (under which defects of writing are also to be included) it may be said that loss of speech is associated with affections of the brain of all degrees of severity. Frequently speech disorders are owing to mere functional defects entailing failure of memory for words, caused perhaps by over-exhaustion from excessive mental or even physical exertion, or by the occurrence of some febrile illness; at other times loss of speech may be only one symptom of some grave structural disease of the brain which has also produced hemiplegia or other well-marked paralytic conditions. Again, even when due to structural diseases, loss of speech in all its grades up to complete inability to utter any distinct articulatory sounds may exist without the coexistence of any motor paralysis whatever of the hemiplegic type.

The classification of speech defects is best made to depend upon the regions of the brain that are damaged. A study of the peculiarities of the several kinds of defect due to lesions in this or that region of the brain, will thus, by their subsequent recognition in individual cases, afford valuable data for the establishment of a regional diagnosis.

In enumerating the different kinds of speech defect we may begin with the simpler varieties of disorder dependent upon defects in bulbar and spinal motor centres; we may then pass on to those forms caused by lesions in different regions of the cerebral hemisphere below the cortex—that is, to lesions in some part of the course of the pyramidal fibres. Taken together these constitute the different kinds of speech defect due to *sub-cortical* lesions.

On the other hand we have another very much larger group of defects due to lesions or disabilities in different regions of the *cortex* itself. These latter defects may be situated either in one or more of the word centres, or else in

---

[1] *Journal of Mental Science*, January, 1891.

the course of one or other of the commissures uniting these word centres with one another.

From this basis we shall be led to such a classification as that which I now subjoin, if we use the term speech in its larger acceptation, so that it may also include defects of writing.

It must always be borne in mind, however, that two or three different forms of speech defect often coexist in the same patient, so that the nature of the particular combination presented can only be ascertained by submitting the patient to a systematic method of examination, by aid of some such scheme as will be given further on in the chapter on diagnosis.

CLASSIFICATION OF SPEECH DEFECTS.

I. Defects due to Sub-Cortical Lesions.

A. Disabilities in Bulbar and Spinal Motor Centres.
B. Lesions in the course of the Pyramidal System of Fibres.

II. Defects due to Cortical Lesions.

(a) Paralytic Amnesia.

C. Disabilities in Cortical Word Centres.
   (1) In the Glosso-Kinæsthetic and the Cheiro-Kinæsthetic Centres.
   (2) In the Auditory Word Centre.
   (3) In the Visual Word Centre.

(b) Incoördinate Amnesia.

Showing itself in Speech (Paraphasia).
Showing itself in Writing (Paragraphia).

D. Lesions in the course of different Commissures.
   (1) In that between the Auditory and the Glosso-Kinæsthetic Centres.
   (2) In that between the Visual and the Cheiro-Kinæsthetic Centres.
   (3) In those between the Auditory and the Visual Word Centres.
   (4) In those between the general Visual Centre and the Auditory Word Centre.

This classification is by no means exhaustive—it is, in fact, rather the outline of a classification—as will be found when the several varieties of speech defect are considered in detail. This is especially the case for the defects due to cortical lesions included under the sections (2) and (3). In these sections various sub-varieties of defect will have to be considered. Thus we shall have to take into account whether we have to do with single or with conjoint lesions in the auditory and the visual word centres, and whether such lesions exist only in one or in both hemispheres. Differences may arise also according as the lesions occur in "auditives" or in "visuals." Then again, it will be found that the division is not absolutely sharp between cortical and sub-cortical lesions, since a class of defects will have to be considered, in which, though the symptoms may be similar to those producible by lesions of the auditory or of the visual word centres in some persons, they may also be actually produced by a lesion of the afferent fibres going to one or other of these centres combined with some damage to the commissure in the corpus callosum that connects it with its fellow in the opposite hemisphere.

My object will be to simplify a naturally very complex subject as far as it may be possible; and also to eliminate as much as possible from the general treatment of the subject the bewildering nomenclature and the crowd of special terms which have been so freely introduced from time to time by many different writers.

Wernicke, some years since, starting from Lichtheim's classification, and influenced by what I have endeavoured to show to be the erroneous supposition of the existence of a separate "centre for concepts" introduced a rather complicated classification including real with imaginary speech defects.[1] The actual existence of several of these varieties has been held in some doubt by Wyllie, who has nevertheless proposed new names for them;[2] and these different nomenclatures again will be found duly set out, and in part adopted, by Elder in his recent work.[3] I believe, however, after the most careful consideration of the subject that, as I have already stated, no cases exist which need for their interpretation the postulation of a separate centre for concepts, and that, consequently, no varieties of speech defect are only explicable on the supposition of damage to certain imaginary commissures which are supposed to bring this imaginary centre into relation with the auditory word centre on the one side, or with Broca's centre on the other.

---

[1] *Fortschr. der Med.*, 1885, p. 825.
[2] "Disorders of Speech," 1895, pp. 363-368.
[3] "Aphasia and the Cerebral Speech Mechanism," 1897, p. 66.

CHAPTER IV.

DEFECTS OF SPEECH AND WRITING DUE TO STRUCTURAL OR FUNCTIONAL DEGRADATION OF MOTOR CENTRES.

Loss or extreme difficulty in speech due to structural disease in the bulbar speech centres is now commonly known by the names Anarthria or Dysarthria. In the different forms of this trouble we have to do with various grades of difficulty in utterance, while the power of communicating thoughts by writing, as well as the mental power of the patient, may be unimpaired. It is, at all events, unnecessary that there should be any accompanying defect of this kind. Both spoken and written speech may also be comprehended as well as ever.

There are two general diseases of the nervous system, namely Chorea and Insular Sclerosis, in which slight anarthric defects are commonly met with. And there are other localised diseases of the bulb, in which the structural lesions are better known and of a grosser type, where there is a more uniform association of profound disabilities in regard to speech. A brief reference must now be made to these several conditions, in order that the kind of speech defect with which they are associated may be pointed out.

The alteration of speech met with in *Chorea* is generally comparatively slight and represented mostly by a mere lalling or indistinct utterance, caused by incoördinate action on the part of some of the groups of muscles concerned in the articulation of words. It is by no means certain that such disordered activity is due to morbid conditions of the bulbar motor centres, although it seems probable that this is so. Still it must be borne in mind that great uncertainty prevails as to the localisation of the defects causing the various clinical phenomena met with in chorea.

The speech in *Insular* or *Disseminated Sclerosis* is also variable in its character and degree of indistinctness. Although the speech in this affection is commonly spoken of as being of a "scanning type," this particular form of alteration is by no means uniformly met with. There may be merely thickness of speech—an indistinct or slurring utterance. The scanning speech which is occasionally encountered has to be distinguished from the "blundering enunciation of syllables" that commonly characterises General Paralysis of the Insane, a defect which is supposed (and probably with justice) by Kussmaul to be due rather to degeneration with perverted activity in the cortical centre (kinæsthetic) concerned with speech. He says:[1]—"The scanning utterance of syllables must not be confounded with blundering enunciation. In the former, articulate sounds are correctly combined to form syllables and words: the syllables are merely separated from one another by brief pauses. In the latter, both articulate sounds and syllables are misplaced and thrown into confusion. To borrow Westphal's admirable illustration, the general paralytic in trying to say 'artillery,' calls it 'artrillerary,' the patient with insular sclerosis, on the other hand, pronounces it 'ar-til-le-ry.'"

A still more marked, progressively developed, failure in power of articulation is met with in *Labio-glosso-laryngeal Paralysis*—a disease which is caused by degenerative changes gradually advancing in the nerve cells entering into the composition of the bulbar nuclei concerned with speech. This entails a gradually increasing paralysis of the muscles in the parts included in the name of the disease, as well as in the palate. The kind of difficulty first experienced in articulation varies a good deal in different cases, according as the lips, the tongue or the palate are the parts that first show signs of paralysis; ultimately, however, the patient's utterance becomes so gravely affected as to be almost unintelligible. The act of deglutition also becomes more and more gravely affected. In addition to paralysis of the tongue there is generally more or less well-marked atrophy of this organ. The disease is likewise generally either preceded or followed by the affection known as "progressive muscular atrophy," which is a disease of the same order affecting some of the motor nuclei in the anterior horns of the spinal cord.

In *Acute Labio-glosso-laryngeal Paralysis*, or "acute bulbar disease," as it is now frequently termed, the attack may commence with more or less distinct apoplectic symptoms, and be subsequently associated with profound anarthria and dysphagia, together with some amount of bilateral paresis of limbs and trunk, due to a small focal lesion occurring in the bulb. The lesion is either of the nature of a softening caused by thrombosis of one of the vertebral arteries, or of some

---

[1] *Loc. cit.*, p. 663.

one or more of the median branches of the anterior spinal artery; or else, more rarely, it is due to a small hæmorrhage into the substance of the bulb, owing to the rupture of one of the small vessels just mentioned.

Occasionally a rapidly growing tumour in the bulb may produce a very similar set of symptoms, though developing more slowly and without the initial apoplectic attack.

More frequently a lesion in the pons varolii may produce a very similar set of symptoms to those met with in cases of "acute bulbar disease," either by extending into the bulb or else owing to the very close proximity of the lesion to the bulbar speech centres. I say more frequently because both hæmorrhages and softenings are distinctly more common in the pons than they are in the bulb.

Strictly speaking, many of these cases, in which the bulbar centres are not actually damaged, would belong rather to the next category of speech defects; yet an example may be cited here as these cases lead to disabilities in speech and deglutition of exactly the same order as if the lesions were situated in the bulb itself.

A briefly related case of this kind may be quoted from Wilks,[1] which will, therefore, serve as a transition to the aphemic defects of the next category.

CASE I.—A lady fell in a so-called fit during dinner. She was taken up speechless and put to bed. She lay with her mouth open and with the saliva running from it, and she was unable to swallow or to speak. There appeared to be no paralysis of her limbs, and from her gestures and expression there was every reason to believe that she was perfectly sensible. She was soon able to leave her bed, and recovered her usual health, but she never lost the paralysis of the tongue and palate. She wrote down all her wants on a slate. She swallowed with difficulty, and the saliva was continually flowing from her mouth; but she was able to walk three or four miles a day, and was accustomed to join in a game of cards. About two years after the first attack she had another apoplectic fit, in which she died.

On *post-mortem* examination there was found to be a great amount of disease of the cerebral vessels; much blood, which had escaped from the pons, was effused at the base. Within the pons there was an old brownish cyst. The central ganglia were healthy.

Other speech defects of the anarthric type are due to functional perversions rather than to structural causes. Three varieties will be very briefly referred to: namely, (*a*) Lalling: (*b*) Stammering: and (*c*) Aphthongia.

(*a*) *Lalling* is, as Kussmaul says (*loc. cit.*, p. 844):—"The term used to characterise the speech of children before they have learned to pronounce their words so as to be intelligible

[1] *Guy's Hospital Reports*, 1867.

to all persons." It is, therefore, a defect due to a want of precision in the oral articulatory mechanism. In an ordinarily healthy child who is properly taught and who shows an adequate amount of attention, this want of precision soon disappears: but where the attention and training are defective it may linger long beyond the usual age. On the other hand, in those children whose brains and intelligence are defective this imperfect articulation may never wholly disappear. Thus, as Wyllie truly remarks:[1]—"Apart from those cases in which defective articulation is displayed only in connection with one consonant, and those other cases where local paralysis or deformity is sufficient to account for it, lalling articulation, in the great majority of cases, means imbecility of mind. It is very different with stammering (Stottern). Stammering associated with lalling is indeed often enough found in imbeciles; but in the great majority of cases stammering exists by itself, and is in no way indicative of deficient intelligence."

Lalling in its most typical form may, however, be met with temporarily in persons who are partly intoxicated.

(b) *Stammering* is a hesitating, spasmodic speech defect often beginning about the fourth or fifth year, and much more common in boys than girls, which is due not so much to a want of precision or coördination in the action of the several groups of muscular elements constituting the oral articulatory mechanism, but rather to a want of accord between the action of the laryngeal and the oral speech mechanisms. There is mostly a lagging action of the former.

Some such view has been held by many observers, but this doctrine has lately been strongly advocated and strengthened by Wyllie, who has given much attention to this particular question in his book on "The Disorders of Speech." This subject is too large to be entered upon here, and too important to be briefly dealt with. It lies, moreover, outside the immediate subject of this work, which is rather that of Aphasia and its allied states. These conditions are here merely mentioned from a classificatory point of view, and for the purpose of showing how they are related to what, in the broad sense of the term, are commonly regarded as the aphasic group of speech defects.

It may be well to point out, however, that what we in this country call stammering is described by Kussmaul and other German writers as stuttering (*Stottern*); whilst he applies the word stammering (*Stammeln*) in much the same sense that lalling is here employed—in fact to a minor degree of lalling.

[1] "The Disorders of Speech," 1895, p. 138.

In this country the terms stammering and stuttering are regarded as interchangeable.

(c) *Aphthongia* is a very rare affection of which only a few cases have been recorded. The condition was first named and described by Fleury.[1] He spoke of it also as a "reflex aphasia," and regarded its essential nature to consist of "cramps in the territory of distribution of the hypoglossi, which set in whenever an attempt to speak was made, and which render articulate expression impossible."

The affection has shown itself in two or three cases after periods of great emotional excitement, and once after an operation for excision of the tonsil. According to Kussmaul the first case was recorded in 1855 by Panthel of Ems, and on account of its remarkable nature I quote from him the following details.

CASE II.—A peasant boy, aged 12, was greatly excited by the sudden death of his father. He fainted at the funeral and remained unconscious for a quarter of an hour. After recovery from the syncope he was sound in body and mind, but for three days he was unable to speak, although he could move his tongue and lips freely, and could swallow without difficulty. When he attempted to speak his mouth, jaw, and tongue remained immovable, but the large laryngeal muscles supplied by the hypoglossi (the sterno-thyroid, thyro-hyoid and sterno-hyoid muscles) fell into visible and violent vibrating movements. The spasms ceased when the efforts to speak were discontinued. Pressure on the muscles relieved the spasms and enabled the patient to speak; this effect persisted as long as the pressure was kept up. Fourteen days after recovery a relapse set in, which was occasioned by fright, and lasted two days. A second relapse of a few hours' duration was brought on by violent emotion a few weeks afterwards.

In Fleury's case the patient was a man who had undergone the operation of tonsillotomy. The operation was followed by marked disturbances of sensation, loss of taste, aphonia, cerebral congestion, and epileptiform attacks. On every attempt to speak the tongue became immovably fastened to the hard palate. The intelligence was perfect, and the patient was able to write and to calculate.

Another case has been more recently recorded by Professor Ball of Paris, of which the following is an abstract.[2]

CASE III.—A man aged 45, in a fit of passion, was suddenly struck dumb. Although speechless he experienced no difficulty in expressing his thoughts in writing. There was no paralysis of motion or of sensibility, the tongue was easily protruded and moved without difficulty from one side to the other; but when the patient attempted to speak there was a spasmodic contraction of the muscles of the tongue, causing that organ to assume the form of a convex dome, and to be closely applied to the roof of the

[1] *Gaz. Hebdom.*, 1865, No. 15.
[2] Quoted from Bateman " On Aphasia," 2nd ed., 1890, p. 145.

palate. When he thus made useless attempts to speak, the tongue became hard and stiff, but when at rest it was as soft and flexible as in the normal state. At the end of a few days, and without treatment of any kind, speech was restored as quickly as it was suppressed.

These are three very interesting and unusual cases, and it will be observed that in the last two the spasm seems to have been limited to the intrinsic muscles of the tongue, whilst in Panthel's case the extrinsic muscles in the neck were obviously if not principally involved.

A very few words will suffice for all that need be said concerning defects in writing due to structural or functional degradation of the motor centres in the spinal cord upon which these acts depend.

It is obvious that structural disease involving the motor centres in the spinal cord concerned with the act of writing will lead, through different degrees of paralysis of the hand and arm, to defects in this power comparable with different degrees of anarthria caused by lesions in the bulbar motor centres. Disease of the peripheral nerves leading from the spinal centres would produce very similar disabilities in regard to writing—they are, in fact, sufficiently common in cases of peripheral neuritis—though well-marked defects of speech caused by disease of peripheral nerves are very much less common.

Again in Writer's Cramp we have a spasmodic and disordered action of the muscles concerned in writing, due to a functionally perverted activity of these spinal motor centres, leading to disabilities in connection with the act of writing which are very analagous to Stammering and Aphthongia as speech defects.

CHAPTER V.

DEFECT OF SPEECH AND WRITING DUE TO LESIONS INVOLVING THE PYRAMIDAL FIBRES.

THE speech defects belonging to this category I have for some years past thought it best to group under the term Aphemia. They vary much in degree in different cases—that is, from mere "thickness" or indistinctness of articulation (incomplete aphemia) to actual loss of speech or dumbness (complete aphemia).

The cases of incomplete aphemia cannot, so far as the mere form of the speech defect is concerned, be strictly differentiated clinically from forms of anarthria of similar severity. The distinction between aphemia and anarthria must, therefore, depend upon the nature of the collateral associated conditions; one of the most important distinctions being the fact that the impaired utterance in bulbar diseases is much more apt to be associated with more or less difficulty in deglutition than is that due to damage to pyramidal fibres (except where this damage is situated in the pons varolii); and also apt to be associated with more severe paralysis of the sixth, seventh, eighth, or twelfth cranial nerves. There is likewise a much greater tendency to the existence of some amount of bilateral paralysis of limbs and trunk muscles in the speech defects due to bulbar lesions.

It is, however, useful to keep up a difference in name, founded upon the different site of the lesion in the two cases. This course contributes to clearness in the classification of speech defects.

In all the cases of aphemia, whether slight or severe, the essential characters of the speech defect taken alone are, as I have said, just like those of anarthria—that is, there may be no trace of mental impairment; where the right hand does not chance to be paralysed, the patient's power of expressing his thoughts by writing is in no way diminished; and he is perfectly able to understand both spoken and written speech.

So far as we know at present the internuncial fibres with

which we are now concerned pass downwards on each side through the "internal" capsule in the situation of the *genu*, forming a slender band named by Brissaud the "geniculate fasciculus,"[1] which descends in the foot of the peduncle between the internal and the middle third of this body, on its way to the nuclei in the bulb that are concerned with the movements of the lips, tongue, soft palate, and larynx in the production of articulate speech.

Similar amounts of damage to these fibres alone ought always to produce similar degrees of aphemia, in whatever part of their course the damage may occur.

The varying degrees of aphemia met with in different cases will depend, therefore, not so much upon the site of the lesion and the particular part of the longitudinal extent of the pyramidal fibres that is damaged, as upon the degree to which the functions of the fibres (all or some) are interfered with in some part of their course.

Aphemia is clearly not a sensory defect—it is not a form of amnesia—because the subjects of it can revive words in all possible modes, and are, therefore, able to think and express their thoughts with an unimpaired freedom by writing. If the aphemia be in any way incomplete, moreover, such a case can be easily discriminated from a case of aphasia by the fact that the aphemic patient will always at once make an attempt, when bidden, to pronounce some simple word or syllable (however poor the attempt may be), while the typical aphasic patient is unable to make any such attempt—he will not try, or at all events will not succeed in repeating even the simplest vowel sound.

The reason of this important distinction seems to me to lie entirely in the situation of the lesion in the two cases. In aphasia the most important word centre for the expression of thought is affected; whilst in aphemia, as I understand it, all the centres in which the memory of words can be revived are intact, the damage occurs beyond these, and there is consequently nothing to interfere with the flow of thought and, in incomplete cases, nothing whatever to prevent attempts at articulation being made, just as similar attempts can always be made by patients suffering from disease which involves the bulbar articulatory centres.

Then, again, the degree of severity of the aphemia that would be occasioned by a given amount of damage to these fibres, when it occurs in the right hemisphere of the brain, is

---

[1] And by Pitres "le faisceau pédiculo-frontal inférieur," which seems altogether too unwieldy.

altogether less than that which would be produced by a similar lesion in the left hemisphere, so long as the person is right-handed. This is now fully recognised, and the reason for it was long ago clearly stated by Kussmaul.[1] He says:—"The familiar fact that centro-hemispheric hemiplegia of the right side is often associated with defects of articulation, frequently amounting to total loss of speech, which are at once more lasting and more severe than the slighter and more transient dysarthries occurring with left hemiplegia, proves that the main current of the centrifugal impulses of speech passes downwards through the left cerebral hemisphere. But since dysarthric troubles—though usually of a trifling kind—are noticed, as a rule, in connection with left hemiplegia likewise, we must conclude that, besides the main current just referred to, a weaker accessory current is transmitted through the right hemisphere."

It is, indeed, a matter of common observation that aphemic speech defects when they are of any severity, are, as with aphasic defects, commonly associated with right-sided and only very rarely with left-sided hemiplegia, in right-handed persons.

Numbers of cases of lesion in the course of the pyramidal speech fibres on the left side will be found recorded as cases of "aphasia," or, in more modern literature, as cases of "motor aphasia." Several of them are referred to by Pitres[2] and Boyer,[3] and others have been more recently cited by Ross;[4] but very few of these cases, unfortunately, are capable of affording absolutely conclusive evidence in proof of the view here maintained as to the precise nature of the speech defect occasioned by damage of these internuncial fibres. Some of them are doubtful cases where there was more than one lesion; others are cases in which the lesion was too extensive to be at all conclusive; and others, again, are cases in which no adequate clinical details are given to enable us to say whether the defect of speech from which the patient suffered was aphasic or aphemic in character.[5]

Many years ago, however, a fairly typical case was recorded by Dieulafoy[6] as an instance of "aphasia" due to

[1] *Loc. cit.*, p. 676.
[2] "Lésions du Centre Ovale," Paris, 1877.
[3] "Etudes Topographiques sur les Lésions Corticales," 1879.
[4] "On Aphasia," 1887, pp. 59 and 60.
[5] In only one of these cases (Pitres, No. xix.) have I found clinical details which look more like aphasia than aphemia. An explanation of this apparent discrepancy will be found later on in chapter xi., when dealing with defects in the *Audito-Kinæsthetic Commissure*. (See Case lxxxvi.)
[6] *Gaz. des Hôpitaux*, May 16, 1867, p. 229.

a lesion bordering upon, but not actually involving, the third left frontal convolution. The essential details of this case are as follows :—

CASE IV.—A man, 44 years of age, was admitted on January 8, 1867, into the Hospital of St. Antoine suffering from Bright's disease. There was general dropsy; also a very large quantity of albumen, together with fatty and hyaline casts, in his urine. There was likewise an apical systolic bruit.

On January 22, without previous premonitory signs, he found one morning that he was speechless; and at the hour of the visit he was seen to be gloomy and restless, pointing with his finger to his lips and tongue to indicate that he wished but was unable to speak.

There was slight right facial paralysis but no weakness of limbs. He recognised and took up, when bidden, such common objects as his glass or his fork, though he could not name them. " When one pronounced before him a clearly articulated word, he watched carefully the movements of the lips, and then uttered such monosyllables as '*bien*,' '*non*,' but that was all."

" A pencil was put into his hand, and several times he attempted to write his name, but he could never get beyond the first letters." Apart from these different troubles, " his intelligence was only slightly enfeebled"; so that his cerebral symptoms were thought to be due to a small embolism.

The speechless condition was of short duration, as by January 30 the patient could distinctly articulate several words, " especially when one was careful to pronounce them first of all"; and before the end of February he had almost completely recovered his speech.

The renal affection increased, and he died on April 22.

At the necropsy two small foci of softening were found in the left anterior lobe. One, the size of a pea, was situated to the right of the other, which was three times as large and located in the white substance just beneath the third frontal convolution, but in no way involving the convolution itself. No lesion was found in any other part of the brain.

It seems tolerably clear that this was a case of temporary aphemia and not of aphasia; though in cases where the lesion exists just beneath the third frontal convolution it is always possible that the functions of Broca's centre may be for a time deranged. The fact that the patient's speech was restored in the course of four or five weeks would point to the probability that the foci of softening only involved some of the fibres of the geniculate fasciculus. With regard to the patient's inability to write, this observation seems to have been made on the first day, and we are not told that it persisted for days or even weeks like the aphemia. It may therefore have been a quite temporary disability due to mental confusion, and persisting for some hours only after the occurrence of the lesion.

Other cases recorded by different observers, and collected by Pitres in his work " Sur les Lésions du Centre Ovale," seem to be of the same kind, though the clinical evidence as to the nature of the speech defect is scanty and very incomplete. A few brief details are subjoined concerning four of these cases, Nos. xxv., xxxviii., xxii. and xxi.

CASE V.—A woman, aged 56, became paralysed one day in the right side of the face and the right arm (the leg escaping), together with loss of speech. "She could only reply to questions by signs as she had completely lost her speech. She was only able to utter unintelligible sounds, and was very impatient because she could not be understood." On the second day paralysis of the right leg also supervened.

She died after three weeks, and the only lesion of the brain found at the autopsy was a patch of softening in the white substance of the left hemisphere "above the lateral ventricle," about an inch in diameter (*loc. cit.*, p. 68).

CASE VI.—A woman, aged 66, became completely paralysed on the right side and lost her speech. There was no loss of consciousness. Five days afterwards she was admitted to hospital, when her condition is described as follows, "The patient hears and comprehends what is said to her; she notices what goes on around her, but she cannot pronounce a single intelligible articulate sound." Death occurred on the thirteenth day after the onset of her illness, and during this time she had only been heard to pronounce the words "aie, aie, aie" and "Oh! mon Dieu."

At the autopsy, "On the lobule of the angular gyrus, and on the left superior parietal lobule, two minute yellow patches were found, each having a diameter of four millimetres. Everywhere else the cortical grey substance was completely healthy." On section of the hemispheres the only lesion found in the brain was a large ill-defined patch of softening in the white substance of the left hemisphere, extending from near the foot of the third frontal convolution back beyond the posterior extremity of the thalamus. The grey matter of the convolutions above the softening "presented no appreciable histological alteration" (*loc. cit.*, p. 94).

CASE VII.—A woman, aged 65, became paralysed on the right side with loss of speech, and six days afterwards was admitted to hospital. "The tongue could be protruded from the mouth, and did not deviate: the patient understood quite well what she was told to do, but she could not pronounce a single word; she replied by gestures to the questions put to her. Death occurred on the tenth day after the onset of the illness.

"At the autopsy the cortex of the left hemisphere was found to be perfectly healthy; but on section a hæmorrhagic focus of the size of a nut was exposed in the white substance of the fronto-parietal lobe, and extending in front to the summit of the *faisceau pédiculo-frontal inférieur*" (*loc. cit.*, p. 65).

CASE VIII.—A merchant, aged 79, became partly paralysed in the right arm and lost his speech on May 15, 1829. "After the expiration of several days his intelligence seemed perfectly restored; ideas, judgment, will, as expressed by signs, appeared intact, whilst speech remained absolutely lost. Gradually the right arm recovered its power, he became able to stand and walk, his various functions became regular, except that speech alone was not restored—no word could be articulated. Nevertheless after many attempts some monosyllables were after a long time pronounced, but always with difficulty." He died from troubles of another kind on Oct. 6, 1829— that is, in about five months after the onset of his illness.

At the autopsy a cavity 1 to 2 centimetres in diameter was found "in the middle and posterior part of the left anterior lobe, outside and in front of the corpus striatum." A few much smaller foci were found also in its immediate neighbourhood, and around these cavities for a depth of about two lines the brain tissue was softened. Other parts of the brain showed no lesion of any kind (*loc. cit.*, p. 64).

The details of these cases of incomplete aphemia are, as I have said, extremely meagre, but as a set off I may quote here some particulars given by Dr. Samuel Johnson, concerning what appears to have been a temporary attack of the same kind from which he suffered when in his 74th year. It will be seen that on June 17, 1783, he almost completely lost his speech for about twenty hours, though his understanding was not impaired, and he was able freely to express himself in writing, as is shown by two letters written on that day. Two days afterwards, in a letter to Mrs. Thrale, he thus quaintly describes his attack.

CASE IX.—" On Monday, the 16th, I sat for my picture, and walked a considerable way with little inconvenience. In the afternoon and evening I felt myself light and easy, and began to plan schemes of life. Thus I went to bed, and in a short time waked and sat up, as has been long my custom, when I felt a confusion and indistinctness in my head, which lasted I suppose about half a minute. I was alarmed and prayed God that, however He might afflict my body, He would spare my understanding. This prayer, that I might try the integrity of my faculties, I made in Latin verse. The lines were not very good, but I knew them not to be very good: I made them easily and concluded myself to be unimpaired in my faculties.

" Soon after I perceived that I had suffered a paralytic stroke, and that my speech was taken from me. I had no pain, and so little dejection in this dreadful state, that I wondered at my own apathy, and considered that perhaps death itself, when it should come, would excite less horror than seems now to attend it.

" In order to rouse the vocal organs, I took two drams. Wine has been celebrated for the production of eloquence. I put myself into violent motion, and I think repeated it; but all was vain. I then went to bed; and strange as it may seem, I think slept. When I saw light it was time to contrive what I should do. Though God stopped my speech He left my hand. . . . . My first note was necessarily to my servant, who came in talking, and could not immediately comprehend why he should read what I put into his hands.

" I then wrote a card to Mr. Allen that I might have a discreet friend at hand, to act as occasion should require. In penning this note I had some difficulty: my hand, I knew not how or why, made wrong letters. I then wrote to Dr. Taylor to come to me, and bring Dr. Heberden. . . . . I have so far recovered my vocal powers, as to repeat the Lord's Prayer with no imperfect articulation."

It appears that the loss of speech was of comparatively brief duration, for writing on June 25 to Mrs. Porter, Johnson says:—" Before night [on the 17th] I began to speak with some freedom, which has been increasing ever since, so that I have now very little impediment in my utterance." Also it appears that the loss of speech was not complete, as in a letter to Boswell dated July 3, he says: " I perceived myself almost totally deprived of speech. I had no pain. My organs were so obstructed that I could say *no*, but could scarcely say *yes*."

This attack was probably due to thrombosis in some small vessel supplying the geniculate fasciculus, the circulation through which may have been after a time restored. This interpretation is supported by the fact of the transient nature of the speech trouble, as well as by the fact that the limbs were not involved in the paralysis.

In regard to the duration of aphemia generally, this will depend upon two or three conditions. First, upon whether the geniculate fibres are actually destroyed or only temporarily disabled. When they are only pressed upon by extravasated blood or by an abscess, or temporarily cut off from their proper blood supply, the subsequent partial absorption of the clot or the opening of the abscess may relieve the fibres from pressure and so diminish the aphemia; whilst no such speedy disappearance would be possible when these fibres are actually torn across. Similarly the establishment of a collateral circulation may, after a time, restore the circulation to the peripheral parts of a region supplied by some vessel that has been plugged, and so lead to a partial restoration of function.

In cases where the internuncial speech fibres have been torn across or otherwise hopelessly damaged, however, there is still a possibility of recovery, though the facility with which it occurs seems to differ much in different individuals.[1] I allude to the possibility of the right hemisphere taking on compensatory functions in the manner first suggested by Broadbent—that is, through incitations passing across by fibres of the corpus callosum from the left to the right third frontal convolution, and thence downwards through the internuncial speech fibre of the right hemisphere to the motor centres in the bulb. This possibility will be more fully discussed further on (chap. xii.), where I shall point out some of the difficulties that stand in the way of such a mode of recovery from aphemic speech defects. Still it does seem to occur in some persons, though not in others, even where the damage on the left side of the brain has occurred in early life, and when the chance of the establishment of such a compensatory activity of the right hemisphere might be regarded as greatest.

An excellent example of recovery apparently taking place in this fashion has been recorded by Déjerine,[2] of which the following is an abstract.

CASE X.—A young woman, aged 23, was admitted into the Hôtel Dieu, on December 10, 1878. Three years previously whilst apparently in perfect health she lost her speech and became paralysed on the right side. She first felt tinglings in the right foot, then lost her speech; two hours after the right arm began to tremble and this lasted for nearly half an hour. It was only the next day, after she was taken to a hospital, that the right

---

[1] On this subject see chap. xvi., on Prognosis.
[2] *Bullet. de la Soc. Anatomique*, Jan., 1879, p. 16.

leg became paralysed. The face was also paralysed on the right side. No anæsthesia; no interference with special senses.

The aphasia was complete; to all questions the patient replied "*papa, papa*," and so she continued for nine months. At the expiration of this time she began to be able to pronounce some words [when bidden] but replied almost invariably to all questions by the words "*ma sœur Louise*." For the last twelve months she has recovered her power of speaking, and now it is difficult to believe when speaking to her that one has to do with an old aphasic, since she pronounces easily and without any trouble all the words that she employs.

She also began to be able to walk a little about twelve months ago, but still can only do so slowly and with help. The right leg is rigidly extended; and the deep reflexes are exaggerated. The arm is absolutely paralysed and rigid, the forearm being flexed upon the arm, and the fingers bent in upon the palm.

She is extremely emaciated, and has had a severe cough for twelve months, this being associated several times with hæmoptysis. There were well marked signs of phthisis at both apices, together with pleurisy on the right side. The patient died on December 29, 1878.

At the necropsy the cortex of both hemispheres was healthy, but the anterior one third of the left was slightly smaller than the corresponding portion of the right hemisphere. The third frontal convolution showed no sign of change in any part; its volume and form were similar to that of the opposite side.

The pedunculo-frontal section of Pitres having been made, an old focus of softening obliquely situated and of linear form was seen completely cutting across the fibres from the third frontal convolution, destroying also a part of the lenticular nucleus as well as the anterior part of the internal capsule. The focus was about two and a half centimetres long, by one broad and two in the antero-posterior direction. No other lesion was found in the left hemisphere, and the right was quite healthy.

There was well marked secondary degeneration, with atrophy of the left side of the pons and the left anterior pyramid, together with well marked sclerosis throughout the posterior part of the right lateral column of the cord.

In the large majority of cases of aphemia there is only more or less difficulty of articulation, as in those which I have hitherto been considering; these are cases of what may be termed *incomplete aphemia*. But at other times, though very rarely, we may have to do with absolute dumbness or mutism, in association with a more or less pronounced right hemiplegia. In these cases there is what may be termed *complete aphemia*. I formerly regarded these latter cases as being of functional or of structural origin; but the functional cases I now believe to be capable of quite a different explanation, and to be instances of what is known as Hysterical Mutism.

The cases due to structural disease are undoubtedly very rare; their existence is, in fact, denied by some.[1] Two typical cases with autopsies have, however, been recorded by Déjerine, and I have also seen what I take to be such a case occurring

---

[1] See chap. xii.

in a young gentleman who, after remaining completely dumb for nearly six months, ultimately very slowly but completely regained his power of speaking. Some particulars concerning these cases will now be given, commencing with the two recorded by Déjerine.[1]

CASE XI.—G., aged 66, a wood-carver, entered the Bicêtre in 1887 on account of a right hemiplegia without speech defect. Slight paralysis of the face, and the arm more paralysed than the leg. Sensibility unaffected. Deep reflexes exaggerated.

April 12, 1888. Three days ago on awaking he found himself unable to speak, and this disability has continued since. The loss of speech is absolute, he cannot utter a single word either spontaneously, by reading, or by repeating words that he hears. He is very intelligent. When shown an object and asked its name, though he cannot utter it he can indicate the number of syllables of which it is composed by making the corresponding number of expiratory movements, or by squeezing the hand of his questioner so many times. There is no deafness. He understands what he reads. He writes fluently with the left hand, either spontaneously or from dictation. The movements of the tongue and palate are unimpaired.

May 20, 1890. A new careful examination after an interval of two years showed the right hemiplegia to be slightly worse, as he could no longer walk. He still could not utter a single word, but indicated the number of syllables in the names of objects shown to him as before. A laryngoscopic examination showed a paralysis of the right vocal cord, its movements of abduction and adduction being extremely feeble. During deglutition the epiglottis remained motionless and the larynx gaping, so that violent paroxysms of coughing were produced, especially when swallowing fluids. His intelligence remained keen. He still understood all that was said or written, and himself wrote freely as before.

During the next six weeks he was in the Infirmary, and was frequently examined with similar results. He died suddenly on July 11, 1890.

At the autopsy the cortex of the left hemisphere showed nothing unnatural, and a microscopical examination of the third frontal gyrus and of the insula showed them to be healthy. On section of the hemisphere a focus of softening three centimetres long by two centimetres broad was found in the white substance just beneath the third frontal and the lower extremities of the ascending frontal and the ascending parietal convolutions. Two minute foci the size of a millet-seed, were also found in the putamen; also a tract of secondary degeneration in the internal capsule, occupying the genu and the anterior third of the posterior segment. This tract of degeneration was also traced through the crus, through the left side of the pons (which was smaller than the right half), and through the left anterior pyramid. The only other lesion found in these parts were two lacunæ, each the size of a millet-seed, situated in the pons, on either side of the middle line, "at the posterior limit of the motor region."

In the right hemisphere two minute lacunæ were found, each the size of a pin's head, one being in the putamen and one in the thalamus; while another, the size of a millet-seed, existed in the posterior segment of the internal capsule at the junction of its two anterior with its posterior third.

The other case recorded by Déjerine will be found to be very similar in all essential respects.

[1] *Compt. Rend. de la Soc. de Biologie*, 28 Fev., 1891, p. 155.

CASE XII.—Bourg., aged 67, an accountant, had been in the Bicêtre since 1879, on account of a speechless condition with right hemiplegia, following an apoplectic attack which occurred in the same year.

The right hemiplegia was only partial, associated with rigidity and exaggeration of deep reflexes, but no loss of sensibility. No facial paralysis. Intelligence normal.

He understands readily all questions addressed to him, either orally or by writing. He recognises all external objects. Cannot utter aloud a single word either spontaneously, by reading, repeating, or singing. But he can indicate the number of syllables in the words he wishes to utter, by so many expiratory efforts or pressures with the fingers. By bringing one's ear quite close to his mouth one can distinguish "certain words pronounced in an excessively low voice—less than a whisper." Some of these words are not pronounced correctly, the articulation of the labials especially being interfered with—the *b* being pronounced like an *m*. And the same mistakes occur when he attempts to read or repeat a phrase as when attempting to utter them spontaneously.

There is no trace of paralysis of the tongue or palate. Deglutition is in no way interfered with. On the other hand examination by the laryngoscope shows the existence of paralysis of the right vocal cord. It is in the cadaveric position but not atrophied.

The right hemiplegia being slight the patient can make use of a pen, and write either spontaneously, from dictation, or in copying. This he does fairly legibly and with no mistakes.

Death occurred on December 25, 1890, and at the autopsy the cortex of the left hemisphere was found to present a normal appearance throughout its whole extent; but three small foci of softening were present in the white substance of the hemisphere, just beneath the cortex, in the following situations:—(*a*) beneath the lower extremity of the fissure of Rolando; (*b*) beneath the posterior extremity of the third frontal convolution; and (*c*) beneath the lowest anterior portion of the ascending frontal convolution.

To these three subcortical lesions Déjerine ascribes (*a*) the slight right hemiplegia; (*b*) the speechless condition; and (*c*) the paralysis of the right vocal cord. On microscopical examination the third frontal convolution was found to be healthy, and so also were the pons and the bulb, except for slight tracts of secondary degeneration.

These are undoubtedly two interesting and important cases. In the first of them the speechless condition amounted to complete mutism; and in the second it was only a little short of this. In the first of the cases alone was there any lesion in the pons or bulb other than a tract of secondary degeneration, and these were two minute lacunæ in the pons, at the boundary of the motor region on each side of that body, whilst the bulb presented no sign of change. Déjerine contends, rightly enough I think, that no importance is to be attached, as regards the production of the symptoms present, to these minute lesions in the pons—more especially as in the second very similar case no such lesions existed. The unilateral laryngeal paralysis was almost similar in each of them, and this is a symptom rather difficult to explain in the present state of our knowledge, seeing that experimental evidence obtained with monkeys goes to show that there is a bilateral representation of the laryngeal muscles in each hemisphere;[1] and also that in man a unilateral paralysis of these muscles occurs only with extreme rarity as a result of lesions in the brain. Still a few instances of this latter kind are on record.[2]

The so-called bilateral representation of the laryngeal muscles in each hemisphere is based upon the fact that excitation of the cortex in monkeys on one side gives rise to contractions in these muscles (as in other muscles habitually associated in their action) on both sides of the larynx, and the simplest explanation of this appears to be that originally given by Broadbent, to the effect that the bulbar or spinal nuclei on each side, in relation with such bilaterally acting muscles, are freely united by commissures, so as practically to constitute only a single nucleus, which may be stimulated from either hemisphere. In man, however, this does not always obtain, as clinical observation shows. Thus, last year Wilfred Harris recorded[3] two cases of Jacksonian epilepsy in which, during the fit, what are usually bilaterally acting muscles were thrown into spasm on one side of the body only. In one case it was one half of the palate, and in the other only the abdominal muscles on one side, that were called into activity. While at the time of writing this I had under my care in the National Hospital for the Paralysed and Epileptic a man (J. K.), who was admitted on October 25, 1897, suffering from left hemiplegia, in whom the palate was paralysed only on the left side. The left trapezius and sterno-mastoid were also very weak, while the erector spinæ and the abdominal muscles were normal. Again, the movement of the left side of the thorax was markedly deficient, though there were no signs in the chest to account for such deficiency.

Another difficulty in connection with these cases is to understand why absolute mutism occurs as a result of a unilateral lesion in the course of the internuncial speech fibres of the left hemisphere, when a lesion of the left third frontal convolution (from which these fibres emanate) would have been quite consistent with the patients being able to say "Yes," "No," "Oh dear," and such-like recurring and occasional utterances. In the latter case the presumption at present generally entertained is that these thoroughly familiar expressions are produced through the instrumentality of the

---

[1] See chap. vii., p. 135.
[2] For references to such cases see Wyllie in "The Disorders of Speech," 1895, p. 310.
[3] *Lancet*, Oct. 24, 1896.

right hemisphere. Why, therefore, when a lesion occurs in the course of the left internuncial speech fibres should not the right hemisphere come into play in the same way? This question is, I think, just as pertinent in regard to incomplete cases of aphemia as for those which are complete—that is, in which mutism exists. But, as a matter of fact, in cases of incomplete aphemia the patient's utterance is consistently and uniformly bad. In the midst of such a patient's blurred speech we do not hear cases of clearly articulated words. The right hemisphere does not seem to be called into action in these cases; and it appears to remain similarly inactive in the cases where the aphemia is complete and mutism exists. The only explanation for this would seem to be that when the left third frontal gyrus is uninjured and the speech stimulus can pass therefrom through a varying length of the internuncial path, the right hemisphere is not called into play for speech purposes, or, at all events, only in certain cases, and mostly after a long time has elapsed. The speech stimulus continues to be sent along the old obstructed or blocked path. But where the left third frontal gyrus is itself destroyed the case is altogether different; then any speech that can be produced must come through the intervention of the right hemisphere, and when this is called upon, the speech for long can only be such simple well-worn utterances as "Yes" and "No," or common occasional expressions such as "Nurse," "Oh dear," or even oaths when the patient is angry or excited.

In the communication from which these cases are quoted Dejerine wrongly attributes to Lichtheim the recognition of the fact that the clinical characters of this kind of speech defect must be dependent upon a subcortical lesion, seeing that this recognition was fully made by me five years previously.[1] Again, following Lichtheim, he lays stress upon the point that his two patients had preserved intact their motor ideas of articulation (" leurs images motrices d'articulation ") because they could indicate the number of syllables in the name of any object shown to them by hand-pressures or by making the requisite number of expiratory efforts. I, of course, agree that such patients can revive their kinæsthetic impressions of words as much or as little as before the lesion; but I also hold that in a case where Broca's centre only had been destroyed some patients (and perhaps the majority of them) would be able to indicate the number of syllables in a word just as well, and in the same manner, so long as the

---

[1] "Brain as an Organ of Mind," 1880, p. 664 *et seq.*, or in the translation "Le Cerveau et la Pensée," 1882, t. ii., p. 258.

auditory and the visual word centres continued to remain intact. (See Cases xv. and xxii.) The test in question is, therefore, in my opinion, not adequate, as they suppose, to enable us to decide with certainty whether we have to do with an uncomplicated aphasia or with a case of aphemia.

The case of complete aphemia which came under my own observation, and in which recovery ultimately occurred was, I believe, one caused by a small lesion and was not a case of hysterical mutism. The slowness of the recovery after improvement commenced also seems to me suggestive of the gradual opening up of new channels in the nervous system, such as would occur if recovery took place in the manner suggested by Broadbent. Still, in the absence of an autopsy, I am free to admit that there is some room for doubt as to the exact nature of this case, of which the following are the most important particulars.

CASE XIII.—W. D., aged 20, was seen by me in September, 1885, in consultation with Dr. W. A. Phillips.

He left England in good health in October, 1884, for a temporary residence in Calcutta. After his arrival he soon commenced some light duties in a merchant's office and continued well till the beginning of the month of May, 1885. During the first three weeks of that month he suffered from a general eruption of boils. On May 25, a very hot day, he went to see a military review, and early in the evening did not feel well. It was thought he had been slightly affected by the heat, the next morning he complained of pain in the back and left side of his head. He had to leave the office about mid-day, not feeling well, and the same thing occurred on the following day. There was no actual sickness and fever, but he remained rather unwell, keeping either to his bed or sofa till June 3, when he went for a short sea voyage to Madras and back.

He returned to Calcutta on June 21, and while at dinner on that day suddenly became very excited and boisterous, and had a convulsive attack of some kind (no details as to its nature could be ascertained). For about a week after this he was at times odd in manner, sometimes muttering to himself at others taciturn, with occasional twitchings of the muscles about the face and shoulders. Then one day he suddenly lost his speech, though his intelligence was unaffected. He understood what was spoken or written, and could himself write freely to express his wishes or in reply to questions.

Nine weeks after this, having been completely dumb in the interval, he was first brought to me, on September 4, 1885, and he was then in the following condition. He was perfectly intelligent, understood readily all questions that were put to him, and wrote his answers freely and without any hesitation or mistake. He could move his tongue and lips in all directions, but could not utter a sound. The tongue came out straight. During the previous ten days (since his return to this country) he had suffered a good deal from pains in the left parietal and occipital region, and on two occasions had twitchings on the left side of the face. When seen, on tapping the head over the left posterior parietal region, there seemed to be tenderness. There was no lack of symmetry or mobility about the face. The pupils were equal, of medium size, and sensitive to light. The optic discs presented nothing distinctly unnatural. There was some paresis of both upper extremities, though this was most marked on the right side ; his grip, as measured by the dynamometer, being right 35, and left 47 pounds. Some distinct tremors of the right arm were noticed while the instrument was being pressed with the left hand. The knee-jerk on the right side was distinctly exaggerated, both actually and as compared with that of the left side. There was no lack of sensibility on either side of the face, trunk, or limbs. Pulse 100, regular; no cardiac *bruit*.

September 9. A small blister to the nucha having been ordered for the relief of the pains in the head, just after its removal, on September 7, the patient had a convulsive attack. He became rigid, made a guttural sound, had some slight convulsions of limbs, and remained unconscious five to ten minutes. Pulse to-day 104, regular. Tongue protruded straight, but covered with white fur on the right side, tremulous.

September 14. Pain in head now gone; no tenderness on left side; bears tapping there without flinching. Right knee-jerk still exaggerated.

The patient after this date went to his home in the country for six weeks, where he took six grains of iodide of potassium with three minims of liquor arsenicalis three times a day, and also a draught containing twenty grains of bromide of potassium every night.

November 26. During the first three weeks after his return the patient had seven fits, but none since that period. The total duration of each fit with subsequent stupor was said to be thirty to ninety minutes. As far as I could learn, the attacks were bilateral, associated with rigidity of limbs, or rigidity and tremors, rather than with actual convulsions, though sometimes these supervened towards the close. No headache now; this disappeared soon after return. No twitchings of face. Pupils equal, rather sluggish to light. No deviation of tongue or lack of symmetry about face. No tenderness to percussion anywhere over head. Knee-jerks now equal, no exaggeration on right. Can walk ten miles without fatigue. Grip much improved, but still weaker on right; right 70, left 93 pounds. Optic discs healthy. Pulse 84, regular. There was no improvement, however, in regard to speech ; I tried in vain to make him utter simple sounds, even after faradisation of throat and assuring him that he would then probably be able to speak.

December 2. Writes that yesterday he repeated to himself the whole of the vowel sounds, and also about twenty monosyllabic words of three letters.

December 4. Had a fit on the 2nd whilst at the dentist's, just as he was beginning to inhale laughing gas. I tried to induce him to read to me from a "school primer." He sat gazing at the book for several minutes, some tremors and slight twitchings of the facial muscles occurring while the efforts were being made. At last he uttered two or three monosyllables in an explosive fashion, at first very indistinctly in a sort of loud whisper, but afterwards others more plainly and at short intervals, not in quick succession. Thus, he uttered "cup," "boy," "hat," "hog," and afterwards read more currently these words: "Let us go to the cow."

December 8. In the interval he has been practising reading aloud, mostly when alone.[1] I now made him try again to read to me, and made this note: "He sits gazing at the book, with his hands between me and the upper part of his face; but I can see his mouth plainly. His lips move, his breathing is irregular, and he seems to be making efforts to pronounce the words he sees ; but no sound comes till the expiration of four and three-quarter minutes, when he said two words in a quick explosive manner, followed at

---

[1] About this time, when he told me he had been repeating the vowel sounds to himself, I said he should repeat them to some one else, and he at once wrote: "That is the difficulty. It seems so stupid. I can do it when I am alone, but not to anyone."

intervals of about a quarter of a minute by two or three more words, and so on through a page of Bell and Sons' "School Primer" composed of short monosyllabic words. Afterwards he tried to read another page more continuously. This was done rather better, but was accompanied by much working of facial muscles and apparent effort. His voice was cracked and squeaking in character.

December 15. Previously I had never been able to get him to repeat any words after me, nor to utter any words except what he read in the book as above described; but this morning he repeated after me the following phrases: "Good morning;" "It is a foggy day;" "If I go on like this, I shall soon go home." Each of these phrases was uttered after a moment or two of delay, with facial quiverings, and then sudden commencement after the fashion of a stutterer.

December 30. He has not spoken to anyone at home, but he reads better and with less delay. Has been reading to his sister and his mother. Read more distinctly to me also.

January 15, 1886. Hesitates and makes abortive stuttering efforts for three seconds before he can say "Good morning," and finally utters it very imperfectly. When set to read from the "School Primer," he began after one minute to read in a very weak, cracky voice. He read a page in an indistinct and very hesitating manner. Writes that he reads to himself daily, and speaks to himself when he is alone. When told to utter his own name, "William," he only pronounces it, after several abortive trials, in a very indistinct manner. Has no pain; sleeps well; appetite good. Grip, right 77, left 77 lbs.

February 15. He went on fairly well till February 4, though for some days previously, his sister informed me, he had not "seemed quite himself" —in fact, from the date when his father (thinking he was so much better) gave him, in accordance with his own wishes, some work to do, in the form of accounts and writing, in connection with his country estate. About 7.15 P.M. on February 4, he was found by his sister in a room alone and unconscious, in an armchair. He had not been in this room more than about fifteen minutes when he was thus found. He remained unconscious rather over an hour, having from time to time tremors over the whole of the body, the right leg being extended, stiff, and with the toes turned inwards. The right arm was also stiff, but in a flexed condition. His face was dusky. Five days afterwards he had another slight attack, a "sort of fainting fit." When over-tired since, some twitchings have been noticed in the right limbs. He has had no practice in reading or speaking since the fit, and has been more listless in manner. Previously he had been making some progress. He told me in writing that, after commencing the work for his father, he began to have pains in the left side of his head again, though these pains were not worse on the day of the fit, nor had he been overfatigued on that day. He also said he had not been sleeping well for two or three nights before the fit, having a few days previously left off a bromide draught which he had hitherto been taking every night.

I did not see him again till October 9, when he was brought to me by his mother. I learned that from early in June to the middle of August he went away alone to a village in Derbyshire, and lived at an inn there. One of his sisters then went to stay with him, and found him somewhat better. He has been at home for three weeks, and during this time has been communicating with his mother orally, not having occasion to resort to writing once. He has also been reading aloud daily.

On examination, I entered the following particulars in my case book:—
He complained of no pains in his head, and has no local tenderness. Appetite good; sleeps well. Pulse 88, regular; pupils equal and fairly

sensitive; face quite symmetrical; tongue protruded straight; grip, right 90, left 117. Knee-jerks slight, equal; no ankle clonus on either side. He reads aloud much better than he did; begins without delay, but reads with a weak, rather cracky voice, and with much apparent effort and facial contortion, a page from Humphry's essay on "Old Age," the monosyllables of the school primer being discarded. He speaks, too, in reply to questions, though much more slowly and indistinctly than he should do. His voice is husky; he articulates with much effort and facial contortion, and after an explosive fashion, somewhat like that of a bad stutterer. He was directed to practice reading several times daily, uttering each word as distinctly as possible, and also to resume doing some work for his father in connection with his farm.

After this I heard nothing till February 28, 1887, when I received a letter from Dr. Phillips, in which he says in reference to our patient, " he has now quite recovered his lost faculty, and is occupied in business in London. This result was not in any way sudden, but came about slowly and by continued effort and tuition." In short, he went on slowly but steadily improving from the time I saw him last, till he was sufficiently well to enter his father's office in London.

There are certain features about this case that may suggest its functional origin; but it seems to me that the balance of evidence is more in favour of its being a case of complete aphemia due to some small lesion in the course of the internuncial fibres in the left hemisphere. The patient's general health anterior to the occurrence of the mutism was obviously deranged, and might be said to favour the occurrence of thrombosis in some of the cerebral vessels. Then there was the pain on the left side of the head; the greater weakness of the grip on the right side (the patient not being left-handed); the exaggeration of the right knee-jerk; and the absence of hemianæsthesia. The extremely slow way also in which recovery was brought about, after its commencement, was certainly unlike what generally occurs in cases of hysterical mutism; and was distinctly more suggestive of an opening up of new cerebral paths and a progressive development of a previously almost unused cortical centre.

A case is briefly cited by Pitres [1] in which it would appear that something of the same kind had occurred. It is a case in which there was mutism at first, and after a long time complete restoration of speech, and where subsequently the only lesion found was a softening in the left centrum ovale. Clinical details as to the precise nature of the speech defect and the mode in which recovery occurred are unfortunately either meagre or wanting.

CASE XIV.—" Borda, aged 65 years, went to bed without experiencing anything unnatural on May 24, 1873. The next morning when she awoke at seven o'clock she was stretched out at the foot of the bed, paralysed on

---

[1] *Loc. cit.*, p. 96, case xxxix.

the right side and completely deprived of speech. During a period of fifteen days she could not pronounce a single word. During the following six months she could only articulate certain monosyllables. At last, at the expiration of this time her speech little by little returned."

"Admitted to the Salpêtrière in 1875 (under the care of M. Charcot) she was found to be in the following condition: right hemiplegia (face and limbs), with very strong secondary contracture, sensibility preserved. She could express her thoughts by speech, but she often hesitated before pronouncing certain words, and was obliged to make a real effort in order to do so. Death on December 2, 1876."

At the autopsy the right hemisphere was found perfectly healthy; the surface of the left hemisphere presented a normal appearance except for two very superficial yellow patches (one centimetre in diameter) not occupying the whole thickness of the grey matter, and situated the one upon the median portion of the first temporal convolution, and the other at the base of the second and third digitations of the island of Reil. The grey matter of the third frontal convolution presented a normal appearance. Transverse sections of the left hemisphere showed an old focus of yellow softening in the white substance outside the opto-striate bodies which involved the geniculate fasciculus, or what Pitres terms "le faisceau pédiculo-frontal inférieur." Well-defined secondary degenerations were present also in the peduncle, the bulb, and the spinal cord.

The lesion here must, of course, have been far larger than it was in my case, seeing that it involved complete paralysis of the right limbs as well as loss of speech, but it would appear that the restoration of speech must have taken place in the mode already suggested, and also in the same gradual manner as in my case.[1]

The evidence, therefore, supplied by the admitted great frequency of aphemic speech defects with lesions in the substance of the left hemisphere, and their rarity with similar lesions in the right hemisphere; the evidence afforded by Diculafoy's case of almost complete aphemia; and the similarly conclusive evidence afforded by the two cases above quoted from Déjerine, all go to prove that aphemia—incomplete or complete, according to the extent of the lesion—results from a damage to the left internuncial speech fibres.

It is, of course, self-evident that if these sets of internuncial fibres are damaged on both sides of the brain, there must necessarily be either incomplete or complete aphemia, according to the severity of the lesion. There are, however, notable differences between the effects produced by single and by bilateral lesions respectively; and one of much importance is this, that while recovery is possible with the former, severe and bilateral lesions, on the contrary, render recovery impossible.

[1] Case xl. of Pitres has an apparent similarity, but it is really different both clinically and pathologically. It is recorded in chap. xi. as Case lxxxvii.

A similar difference holds good for cases of aphasia due to a single lesion in the left third frontal convolution (from which the internuncial speech fibres issue), and those in which the corresponding convolution is also affected on the right side of the brain. Recovery is possible in the one case, impossible in the other.

In instances of the bilateral lesion, whether of the kinæsthetic speech centres or of their outgoing fibres, the condition of the patient comes to resemble so closely in many respects that caused by bulbar disease that all these cases have been grouped together under the name of "Pseudo-Bulbar Paralysis." An excellent example of this has been recorded by Barlow,[1] of which the following is an abstract.

CASE XV.—A boy aged 10, the subject of aortic disease, of which he ultimately died, was seized with right hemiplegia, chiefly brachio-facial, and aphasia. From this attack he had apparently recovered at the end of a month, the power of speaking having returned on the tenth day.

Three months later he was seized with left brachio-facial monoplegia. This time there was not only complete loss of speech, but paralysis of all voluntary movements of the face and tongue. Reflex deglutition, however, was unimpaired. There was no affection of sensation in the paralysed parts, either in the skin, or mucous membranes of the palate, etc., and the muscles reacted normally to the electric current. There appeared to be loss of voluntary motor power over the muscles concerned in deglutition and articulation. The only sound he was able to utter was "Ah." He could "understand all that was said to him. When asked his age he counted ten on the questioner's fingers; counted four, when asked how many brothers he had, and so forth."

"From the first he was able to write his name when asked, and after a few weeks could answer in writing any question that was put to him." He remained in the hospital till his death—a period of nearly two months—and during this time no change took place in his power of speaking or swallowing.

On *post-mortem* examination, a lesion about the size of a shilling was found in each hemisphere, in exactly corresponding situations. The region involved by each lesion—which was yellow softening—was "the lower end of the ascending frontal, and the hinder end of the middle and the inferior frontal convolutions" (as shown in a figure). The areas were less than one quarter of an inch deep; they involved the cortical grey matter and a little of the white substance.

The symptoms ultimately caused by a double lesion of the internuncial fibres emanating from these cortical centres would be almost exactly similar; and the same may be said of combined lesions in the third frontal convolution of one side and of the internuncial fibres of the opposite hemisphere in any part of their course.[2] Such cases are distinguishable from those of real bulbar paralysis from the fact that there is a

---

[1] *Brit. Med. Jour.*, July 28, 1877.

[2] Cases of this kind have been recorded by Ross in *Brain*, vol. v., 1882, p. 143, and also in his work on "Aphasia," p. 105.

history of two separate attacks, and that it was after the second that the paralysis of the lips, tongue, and pharynx occurred. Then, in addition, in pseudo-bulbar paralysis there is no wasting of the tongue or alteration in its electrical excitability ; and there is also no loss of the palatal or pharyngeal reflex. Still in many cases, especially where the history is imperfect, there may be great difficulty in arriving at a positive diagnosis between these two conditions.

No isolated defect of writing corresponding with aphemia as a speech defect seems to exist in a recognisable form. This is due to the fact that a lesion involving the internuncial fibres between the cheiro-kinæsthetic centres and the motor centres in the cervical region of the cord, in any part of their extent, would almost certainly cause paralysis of the hand also for movements other than those concerned with writing. Thus the mere agraphic defect would be merged in and concealed by a wider form of paralysis. The reason why this same kind of result does not occur in the aphemic class of cases is because here, even though the internuncial channels may be blocked in the left hemisphere, by means of which the specially combined movements needed for articulate speech are called into play, the right hemisphere is still capable of calling into action the bilateral bulbar nuclei concerned with other less specialised movements of lips, tongue, and palate ; consequently, it is only the articulatory movements of these organs that are paralysed in a case of aphemia. The special paralysis of speech is not merged in a wider defect simply because, for the actuation of movements other than those of speech, the bilateral motor centres of the medulla may still be called into play by the undamaged right hemisphere.

If the result is different in the case of writing with the right hand, this is simply due to the fact that here we have to do with unilateral motor centres, and that these unilateral centres in the right cervical region of the cord can only be roused into activity by fibres proceeding from the left cheirokinæsthetic centre, and that the internuncial fibres concerned with writing movements are perhaps so mixed up in their course from the cortex downwards with the internuncial fibres concerned with other movements of the right arm and hand, that agraphia as an isolated defect cannot be produced—that is, it is almost certain to be associated with paralysis of other less specialised movements, and thus to be unrecognisable as a distinct defect parallel with aphemia.

CHAPTER VI.

STRUCTURAL DISEASE IN THE GLOSSO-KINÆSTHETIC AND CHEIRO-KINÆSTHETIC CENTRES.

THESE kinæsthetic centres are, as I have already observed, concerned more with the expression of thought than with the thinking process. Their activity is in the main roused as thought is about to translate itself into action. These centres form the last outposts on the side of ingoing currents, and constitute at the same time the starting points for outgoing currents. They are situated at what psychologists have spoken of as "the bend of the stream." Although with lesions limited to these regions the power of thinking may not be very greatly interfered with, still it is nearly always interfered with to some extent, so that patients having such lesions do not usually exhibit anything like the same amount of mental clearness as that shown by patients suffering from aphemia. It is true that the latter very frequently preserve their power of communicating their thoughts by writing, whilst aphasic patients do so only rarely. Any mental disability in the latter would therefore tend to appear greater than it really is. But this only very partially explains the apparent difference in mental power that is commonly met with. The fact that speech and writing are so frequently involved together in typical cases of aphasia is due partly to the proximity of the glosso- and the cheiro- kinæsthetic centres, and, as we shall subsequently see, perhaps not less to the proximity of the two sets of commissural fibres connecting these centres with the auditory and the visual word centres respectively. Damage to these commissures may, in fact, be a cause of typical cases of aphasia and of agraphia, as I shall subsequently endeavour to show.

These modes of accounting for the co-existence of agraphia with aphasia undoubtedly hold good for certain cases, and especially for those in which, while there is no paralysis of the right hand and arm, the agraphia is complete—copying (except laboriously and slowly) being no more possible than writing spontaneously or from dictation. The destruction either of the cheiro-kinæsthetic centre or of the commissure connecting it with the visual word centre ought to produce such decided results as this. On the other hand, if no other disturbing cause were in operation, there is no reason why such a patient should not be able to write well with separate block letters (*écriture typographique*) either spontaneously or from dictation. And in cases where such a patient is paralysed in the right hand there would be no reason why he should not, with practice, become able to write in all modes with his left hand.

Whether the combination of agraphia with aphasia is mostly due, as I think, to this simultaneous implication or functional derangement of the cheiro-kinæsthetic centre or the commissure uniting it with the visual word centre, or to some other cause, must for the present be regarded as a moot point. Formerly a different view was taken. Thus, some of the earlier observers of aphasic cases, such as Trousseau, Hughlings Jackson, and Gairdner, seemed to consider that the inability to write was as much a result of a lesion in Broca's centre as inability to speak, and contended, indeed, that there was almost always a parity between these defects. Trousseau,[1] for instance, says: "The inability to write is proportionate to the inability to speak"; and again: "Aphasics write as badly as they speak, and those who do not speak at all are absolutely incapable of writing." Gairdner expresses himself in almost similar terms. Hughlings Jackson is even more decided in his opinion. Thus he says:[2] "If a patient does not talk because his brain is diseased, he cannot write (express himself in writing). . . . . I submit that the facts that the patients do not talk, and do write, and do swallow, are enough to show that there is no disease at all."

One of the first to dwell upon the marked inequality that may be met with between these two disabilities was W. Ogle,[3] to whom we owe the introduction of the term "agraphia." More than ten years previously, however, Marcé[4] had insisted upon the independence of these two defects. Thus, speaking of inability to write, he said: "It is independent of the faculty of expression [by speech], since twice the patients could write freely when they were quite unable to speak. It is independent of the motility of the hand, since even when

[1] Lectures (Translation by Bazire), 1866, p. 261.
[2] *Brain*, vol. i., 1878, pp. 320 and 329.
[3] *St. George's Hospital Reports*, vol. ii., 1867.
[4] *Mémoires de la Société de Biologie*, 1856.

the hand has preserved all its power, writing has been found to be impossible."

The dependence of ability to write upon the integrity of a definite centre altogether distinct from that by which articulate speech is effected is, I think, practically certain; this, however, is but another aspect of the question previously put, and will have to be considered more fully a little later. For the present it is assumed that there are two such separate but contiguous centres; and if this be true, that they are frequently affected simultaneously (either structurally or functionally) may be fully recognised. Still in some cases either the glosso-kinæsthetic or the cheiro-kinæsthetic centre seems to be in the main involved, so that in one case speech may be principally interfered with, while in another it is more especially writing that is made difficult or impossible. This latter kind of defect is much less frequent, and usually less noticeable, than the former, because the patient is also often more or less paralysed in the right hand and arm. In such cases attempts at writing would only be possible where the left hand and the right side of the brain have been more or less educated, and frequently no serious trials have been made in this direction.

Though aphasic patients are unable to give voluntary and preconsidered expression to their thoughts, words, or even short phrases and oaths, may occasionally be uttered under the influence of strong emotion. We often find these patients able to make use of short familiar words like "yes" or "no" in response to questions addressed to them, though they may be quite inappropriately employed. The articulation of such words, or "recurring utterances" as they are now commonly termed, is generally supposed to be brought about through the intervention of the comparatively uneducated right third frontal convolution, and this subject will later on be referred to more fully. As Hughlings Jackson originally pointed out, such a patient is quite unable to repeat one of these words which he is continually bringing out, or, indeed, any other simple vowel sound, when he is asked to do so. He cannot utter it, that is, in a purely voluntary manner, in response to a request or command. Another interesting peculiarity is also often seen when resident foreigners become aphasic. During recovery it is found that they are at first only able to express themselves in that language in which they are most thoroughly versed—namely, in their own native tongue. I have seen this in several patients. Two were Germans who had been long resident in this country; yet after an attack of right hemiplegia and aphasia each of them was for a long time unable to utter a word of English. When they began to speak they used German words only; and after they had further recovered, if occasionally in want of a word while speaking English, it was always a German equivalent that first presented itself.

There are many instances on record in which, though the aphasic condition itself has been complete and associated with more or less agraphia, the mental powers of the patients have been fairly well preserved. Many of such individuals are able to read intelligently to themselves, and play games, like draughts or cribbage, perhaps better than their neighbours.

The powers of articulation and of speech remaining to aphasic patients are very various, but may generally be ranged under one or other of the following heads: (*a*) Cannot utter a sound, or only mutters inarticulately; (*b*) constantly repeats some meaningless sound or sounds such as "tan-tan" or "cousisi"; (*c*) uses some one or two single words such as "yes" and "no"; and also mere sounds such as "ba-ba" or "poi-boi-ba"; (*d*) uses some short phrase habitually, such as "list complete," or "I want protection"; and (*e*) uses three or four words or expressions, though in themselves meaningless or irrelevant, in a constant and definite manner, as in Broca's case (Lelong) quoted below, in which the words "oui," "non," "trois," and "toujours" were so employed.

An examination of the recorded cases in which aphasia has been produced by lesions more or less completely localised to Broca's region—or to it and the foot of the second frontal convolution (the supposed seat of the centre for writing)—by no means suffices, as some imagine, to establish the view that agraphia is to be regarded as one of the effects of a lesion occurring in Broca's convolution. It will rather be seen, on the whole, to lend considerable support to the opposite notion that agraphia co-existing with aphasia is to be regarded as a result of a coincident damage to, or functional perturbation in, the foot of the second frontal convolution or thereabouts. The evidence afforded by such cases, however, is much less conclusive than it should be, owing to the fact that in so few of them have careful observations been made as to the ability of the patient to write some days after the occurrence of the lesion, when the mere perturbation of brain functions that it may have caused has had time to subside.

Of the first two cases of aphasia that were recorded by Broca,[1] in one the lesion was old and widespread, though the left third frontal convolution was one of the parts that were

[1] *Bulletin de la Société Anatomique* 1861.

destroyed; in the other the lesion was limited to the posterior parts of the left third and second frontal convolutions, and there was as a consequence loss of speech and writing, without any paralysis of limbs. This was, therefore, altogether the most memorable case, since by means of it more especially the first attempt was made to limit the situation of the region in the left frontal lobe, damage to which gave rise to aphasia. The following is an abstract of this classical case:—

CASE XVI.—Lelong, aged 84, was admitted into the surgical ward of the Bicêtre, October 27, 1861, under the care of Broca. In the month of April, 1860, he had a slight apoplectic attack, from which in a few days he was convalescent, no paralysis of limbs having been observed, though his speech was lost except for four or five words. His intelligence seemed unimpaired. He understood all that was said to him; and apart from his very scanty vocabulary he made use of expressive gestures. He remained in much the same condition for eighteen months, that is up to the time of his coming under Broca's care. To his questions he only replied by signs, accompanied by one or two syllables pronounced hastily and with visible effort.

Broca says:—"These syllables had a meaning. They were French words, viz., *oui, non, tois* (for *trois*), and *toujours*. He had a fifth word, which he pronounced when one asked his name. He then replied, 'Lelo' for Lelong, his real name. . . . . The first three words of his vocabulary corresponded each to a definite idea. In affirming or approving he said 'oui.' In expressing the opposite idea he said 'non.' The word 'tois' expressed all numbers, all numerical ideas. And lastly, whenever these three words were not applicable Lelong helped himself with the word 'toujours,' which, in consequence, had no definite meaning. I asked him if he knew how to write 'oui'; if he could do it. 'Non.' 'Try.' He tried, but he could not succeed in directing the pen." Broca gives details of his conversations with Lelong in illustration of these points.

His sight and hearing were good. The tongue was protruded straight and was moveable in all directions. There was no paralysis of the limbs.

This patient died twelve days after his admission, and at the necropsy the remains of an old hæmorrhage was found in the form of a small cavity filled with serum, and having its walls infiltrated with altered blood pigment, occupying the posterior third of the second and third frontal convolutions. This was evidently the lesion which occurred when he suddenly lost his speech about eighteen months previously, as other parts of the brain showed no focal lesion of any kind.

Tamburini and Marchi[1] have also recorded a case of destruction of the second and third left frontal convolutions which was associated during life with aphasia and agraphia.

Another case has been recorded by Simon[2] in which there was a traumatism affecting these same convolutions.

CASE XVII.—A perfectly healthy man fell from his horse. He got up at once, took hold of the bridle, and was about to vault into the saddle,

---

[1] "Rivista Sperimentale di Freniatria et di Medicina Legale," 1883, anno ix., p. 282.
[2] Quoted by Kussmaul, *Ziemssen's Cyclopædia*, vol. xiv., p. 735.

when a physician, who by chance accompanied him, approached and examined him. He was unable to speak, but made himself understood by signs. No paralysis whatever was present. On the head there was a small wound with depression of the bone.

Later, when death ensued from meningitis purulenta and inflammatory softening of the brain, a piece of bone that had been broken off was found in the left third frontal convolution which, together with the second and the island, was softened. In the skull there was only a perfectly round hole, no crack or fracture in any part.

Nothing is said here as to the presence of agraphia, but it would probably have manifested itself as soon as the inflammatory softening, caused by the original traumatism, had extended so as to involve the posterior part of the second frontal convolution. Probably the condition of the patient, however, was too serious to permit of this being tested.

A case has been recorded by Rosenstein[1] in which it is said that a hæmorrhagic softening involving the third frontal convolution only, gave rise to both aphasia and agraphia.

CASE XVIII.—A woman, aged 22, was admitted into hospital suffering from a fever of intermittent type, and from nephritis. She was ill for several months, and whilst recovering from this affection, she was suddenly seized with severe cerebral symptoms. The countenance was pale, the expression staring, the pupils dilated. She understood what was said to her, and put out the tongue when asked. She took a drinking-glass when it was offered to her. She tried to answer, but in reply to questions could only say "Ja, Ja." She nodded approvingly when one guessed what her wishes were. She had formerly been able to write well, but now when requested to do so she only made all sorts of scrawls, but no letters. She lived sixteen days longer without the aphasia disappearing, and then died from a thoracic complication.

At the necropsy a clot the size of a hazel-nut was found in the third frontal convolution, surrounded by a limited area of secondary softening. The precise site of the clot in the third frontal convolution is not stated.

In regard to this case it may be fairly said that the sudden irruption of a clot of blood of the size of a hazel-nut into the third frontal convolution may well have inhibited for a time the functions of the immediately adjacent portion of the second frontal convolution, and it is also by no means clear that the secondary softening may not have slightly extended into this part.[2] It will be of interest here, also, to quote a case of combined aphasia and agraphia in association with right hemiplegia, and therefore presumably the result of a much larger lesion. It is one of the cases recorded by Trousseau[3] but in which there was no necropsy. The patient

[1] *Berliner Klinische Wochenschrift*, 1868, p. 182.

[2] A case of aphasia in a boy five years old, resulting from a traumatic lesion limited to the third frontal convolution, has been recorded by Duval, and is cited by Bateman ("On Aphasia," second editon, 1890, p. 196).

[3] *Loc. cit.*, p. 237.

was, as is very often the case in true agraphia, able to write his own name, though nothing else. Trousseau says:—

CASE XIX.—" In August, 1863, a lady came to consult me with her son, aged 25. Four years previously this young man had for several days complained of headache, when he suddenly called out to his mother one morning, 'Oh! I feel something extraordinary inside me.' These were the last words he spoke. His right arm and leg became numb, and after a few hours the hemiplegia was complete. After a short period he regained some power of moving first his leg and then his arm; but when he came to me he still walked with difficulty, and could only use his hand for very rough purposes. The aphasia, however, which had from the first day been complete, had not diminished. He could articulate two words only—'No' and 'Mamma.' 'What's your name?' 'Mamma.' 'What's your age?' 'Mamma, No.' He yet knew that he did not answer as he ought. He had taught himself to write with the left hand, but had not got beyond signing his own name, Henri Guénier. He wrote it very legibly on a piece of paper which I gave him. 'Since you write your name,' I then told him, 'say Guénier.' He made an effort and said 'Mamma.' 'Say Henri.' He replied, 'No, Mamma.' 'Well, write Mamma.' He wrote 'Guénier.' 'Write No.' He wrote again 'Guénier.' However much I pressed him I could obtain nothing more. His mother informed me that he could play a pretty good game of cards or dominoes."

Here we do not know the extent of the lesion. In such a case, however, it would have been immaterial whether the second frontal was involved in the lesion or not, because, as I shall subsequently show, destruction of this centre together with Broca's centre would not prevent a patient learning to write with his left hand, provided the left auditory and visual word centres were not damaged at the same time. No sufficient details are given by Trousseau to enable us to decide whether the visual word centre was or was not affected. It certainly may have been, judging from the fact that this young man had been unable to do more than learn to write his own name during the four years that had elapsed since the commencement of his illness—this being generally the extent of the achievement in this direction of patients who are agraphic by reason of a defect in the visual word centre. Although this case, therefore, may have influenced the opinion of Trousseau, the modern development of knowledge enables us to say that it furnishes no evidence in elucidation of the question whether or not agraphia as well as aphasia is a necessary result of a lesion limited to the left third frontal convolution. I say this with all the more confidence because W. Ogle has recorded a case in which there was right hemiplegia and aphasia, though the patient could write with the left hand, and at his death the second left frontal convolution was found to be intact.[1] The essential details in this case are these:—

Loc. cit., p. 105.

CASE XX.—A man, who had suffered from rheumatism and was the subject of aortic and mitral disease, had three days before admission to hospital fallen down, without loss of consciousness, and had found himself hemiplegic and aphasic. On examination, his speech was found to be limited to the two words "yes" and "no." At first he had difficulty in deglutition and in putting out his tongue, but these symptoms passed away in a few days. He understood all that was said to him, and expressed himself well by pantomime. In regard to his writing W. Ogle gives us the following information: "He could write with his left hand with sufficient distinctness words which he could not pronounce when asked to do so. In his writing there was often a tendency to reduplication of letters. For instance, he wrote 'Testatament' for Testament. But I cannot say whether this was more than the result of defective education."

The necropsy showed the brain to be healthy, except at two limited spots, of which the chief was the posterior part of the third frontal convolution on the left side. Here was a softened and almost diffluent patch, about three-quarters of an inch in breadth, reaching from the highest part of the third convolution backwards and downwards to the fissure of Sylvius. The softened part was not actually the most posterior part of the convolution, for there was a narrow unsoftened strip between it and the transverse (ascending) frontal convolution. On cutting into the brain a second small patch of softening was seen in the centre of the left hemisphere, external to and rather above the corpus striatum, and extending towards the posterior extremity of the fissure of Sylvius.

Here, then, where we have the third left frontal affected whilst the second frontal convolution was proved to be healthy, there was aphasia, but no agraphia; and we may, perhaps, assume that a similar exemption from disease of the second frontal convolution existed in a case of my own in which there was a similar combination of symptoms.

CASE XXI.—An active and intelligent man, a builder by trade, had an apoplectic attack associated with convulsions, which left him paralysed on the right side and quite speechless at first. His mind was also much confused for a time, but after two or three weeks he became able to understand what was said to him. The only articulate sounds he was able to utter for the first nine or ten months were "bi-bi-bi," "poy-boy-ba," and "no." After fifteen months he became able to say both "yes" and "no" appropriately; though whilst he continued under my care I never heard him utter any other articulate sounds than those above cited, and according to the statements of his wife this was the extent of his vocabulary. He had regained almost the complete use of his right leg, though the hand and arm still continued very powerless. He had practised writing with his left hand, and wrote fairly well with it. He was asked to write a letter to me, and the next time I saw him he gave me a note, of which the following is a literal copy:—

"Jany. 5, 1869.

"Sir,—I am extremely obliged to you for the trouble you have taken about my affairs. You did not say thether (*sic*) you hav recovered from the severe cold and cough that you told me you were suffering from. I sincerely hope that you have quite recovered your health and strength.

"I am,

" ———."

This note was written, his wife informed me, in less than half an hour, and without any assistance. Yet since his attack this same patient had only been able to utter the two or three words and sounds above mentioned. He seemed to understand everything that was said to him, and appeared also to understand the newspaper equally well when he read it, which he did habitually. When told to count twenty by tapping on the table with the forefinger of the left hand, he did it rapidly and always correctly. . . . . He could add quite correctly columns containing five figures when the numbers were low, but often got puzzled and made mistakes when the numbers were higher. According to his wife's statement he had been much more irritable than he was before his illness. He had several epileptiform attacks after the one with which his illness commenced, and on each occasion the convulsions were almost limited to the right side of the body.

In each of these cases, even where there has been no agraphia, there has been a more or less distinct lowering of mental power, as compared with that occurring in the class of cases due to subcortical lesions which I include under the term "complete aphemia." For this there are two causes. First, there is the fact that in these cases of aphasia the brain may have been more considerably and widely damaged by the various structural lesions that have occurred than has been the case in the aphemic group of cases. Secondly, there is the fact that in the cases of aphasia one kind of word-memory is blotted out, whilst in aphemia none of the forms of word-memory are interfered with. Although the kinæsthetic memory of words is, as I have previously said, much less important for the carrying on of thought processes than the revivals which occur in the auditory and in the visual word centres, still the destruction of the glosso-kinæsthetic centre may interfere with the functional activity of the other two centres. And this interference with the customary completeness of the processes of association concerned with thought and speech might be expected to lead to more or less of mental deterioration.

The degree to which this interference with thought occurs may, however, in an uncomplicated case of disease limited to Broca's region be extremely slight—almost inappreciable in fact—and associated moreover with no impairment in the power of writing. A very remarkable case of this type has been recorded by Guido Banti[1] which will well repay the most careful attention, because it speaks most authoritatively against the views of Trousseau and the other writers already referred to who have taught that agraphia, as much as aphasia, is a result to be expected from an isolated destruction of Broca's region.

[1] "Afasia e sue Forme, Lo Sperimental," 1886, lvii., obs. ii., p. 270, and quoted by Prévost in the *Revue Médicale de la Suisse Romande*, June 30, 1895.

CASE XXII.—A right-handed man, aged 36, who was able to read and write correctly, had a sudden apoplectic attack in 1877. Recovering consciousness in a few minutes he was found to be suffering from right hemiplegia and loss of speech. The paralysis of the limbs disappeared almost completely during the following night, though the inability to speak persisted.

The next day he was admitted into hospital, and on most careful examination his condition was found by Guido Banti to be as follows :—
"The motility of the limbs on the right side had returned to their normal condition. There was no trace of paralysis of the face or of the tongue. The patient made ineffectual attempts to speak ; *he could not articulate a single word*, not even isolated syllables. He was much affected by this mutism, and sought to make himself understood by gestures. I asked him if he knew how to write, and after he had made a gesture in the affirmative I gave him what was necessary and told him to write his name, which he did immediately. I put various other questions to him, to which he replied similarly by writing. I told him to give me a description of his illness, and *he wrote without hesitation* the details above reported. I showed him various objects, pieces of money, etc., telling him to write their names, and he did so without making any mistakes. Then instead of giving him these directions by word of mouth, I wrote them for him in order to thoroughly convince myself that he was able to understand writing. He replied to these questions with perfect correctness. He always wrote very rapidly and did not seem to hesitate to choose his words. He made no mistakes in syntax or orthography. He could understand equally well ordinary writing and print, and when one spoke to him he grasped at once the meaning of the questions, and never wished to have them repeated. I next wrote some most simple words such as 'pain,' 'vin,' etc., and urged him ineffectually to read them aloud. I then pronounced myself some of the words, directing him to repeat them. He appeared to watch with great attention the movements of my lips whilst I spoke ; he made some ineffectual efforts to obey, but he never succeeded in pronouncing a single word."

This patient died in February, 1882, from an aneurysm of the aorta ; and a patch of yellow softening was found situated in the posterior third of the third left frontal convolution, and extending for some millimetres only into the white substance.

Here, then, we have a patient with a lesion destroying, but limited to, Broca's region, in whom there was loss of speech but no trace of agraphia, the patient being able to write well with his right hand. It is also of importance to note that there was here no more mental impairment than is found to exist where speechlessness is due to a sub-cortical or a bulbar lesion—that is, in a case of aphemia or of anarthria.[1] But for the result of the necropsy I should have regarded this case as one of complete aphemia. It shows, therefore, that there may be no clinical differences present to enable us to make this differential diagnosis, even though in the one case we have destruction of Broca's region, and consequently no kinæsthetic revivals of words possible, while in the other this kind of word-memory is in no way interfered with, and there

[1] See also a case recorded by Byrom Bramwell, *Lancet*, 1897, vol i., p. 1007.

is also less chance that the lesion would lead to any interference with the functions of the auditory and the visual word centres. For another very interesting case of aphasia in some respects similar to the last I am indebted to the kindness of Dr. Dickinson.

CASE XXIII.—A marine engineer, aged 49 years, was admitted to St. George's Hospital under Dr. Dickinson's care, having had a fit in a neighbouring public-house. He had landed a few days before, and had been robbed in some place of resort for sailors, and had been much disturbed in mind in consequence. One and a half hours after admission he had recovered consciousness and intelligence, but not his speech. He could not say anything intelligibly except the word "no," but the remarkable fact about his case was that he could express himself perfectly in writing, which he did with his left hand. He was hemiplegic on the right side, the right arm being completely powerless, the right leg partially so; there was also slight loss of sensibility over the whole of the right side of the body.

For three weeks his speech was limited to the word "no," a sound like "yong" which stood for "wrong," and the frequent repetition of the syllables "yah yah," which stood for everything. Told to repeat a sentence or phrase, "Great Britain and Ireland" or any other, he uttered the same "yah yah yah" for either. But he could express himself in writing as fluently and correctly as if his faculty of spoken language were uninjured. He wrote daily reports and an autobiography during the time that he was unable to speak. He understood perfectly all that was said to him. He retained his sense of music, and could whistle "God save the Queen."

On the twenty-fourth day after the seizure it was evident that his speech was improved; he could say "doctor" and "better." In the evening his full power of speech was rather suddenly restored. The limbs recovered more gradually, the leg sooner than the arm. He left the hospital six weeks after admission, able to walk, but not to lift his right arm.

Dr. Dickinson adds:—"The perfect ease and fluency with which the man wrote with his left hand showed that he had been in the habit of doing so. He was in common phrase left-handed, though he said he was not to be called so, since he used both hands equally."

Though in many respects different from the case previously recorded, this resembles it in the fact that we have to do with what certainly seems to be a simple case of aphasia without appreciable mental degradation and with the power of writing unimpaired. The case is also interesting since it constitutes an exception to the very general rule that left-handedness suffices to transfer the leading speech activity to the right side of the brain, and because of the further exceptional circumstance that being left-handed the patient had not been taught to write with his right hand. Another case that was recorded many years ago by Wadham[1] is in some respects even more remarkable, and all the more convincing because it was followed by a necropsy. It will be observed also that before

[1] *St. George's Hospital Reports*, vol. iv., 1868, p. 245.

death the patient had regained his power of speaking. The essential details are as follows:—

CASE XXIV.—A youth, aged 18, left-handed and ambi-dextrous, became partly hemiplegic on the left side and completely speechless after long exposure to cold, and was admitted to St. George's Hospital under the care of Dr. Wadham. Twelve days later, on being given a slate and pencil, he wrote readily the word "orange," and when asked his name also wrote that correctly with his right hand, although his mother asserted that she had never previously seen him do so. He and four of his brothers were left-handed.

About a week after this, being still absolutely speechless, when asked whether he tried to speak and was unable, he wrote "Yes." Asked if when well he wrote with his right or with his left hand, he wrote "Both," and then added, "Fight with left." In six weeks' time the left hemiplegia had much diminished, but he had still never spoken a word and continued to write all his wishes on a slate. His manner gave the impression of his "being very intelligent and rather facetious."

At the expiration of three months he left the hospital, and when seen at his own home two weeks later Dr. Wadham says: "I found that at his mother's dictation he repeated after her various words and sentences with the intonation employed by one who endeavoured to speak without moving his tongue." This power was gradually increased until at last he was able to talk with sufficient distinctness to be perfectly understood by those accustomed to him. He subsequently suffered from necrosis of the jaw, and three months after his discharge was re-admitted to the hospital. After an operation on the jaw he became able to move his tongue naturally. His articulation subsequently was indistinct, but he "had no difficulty in producing words, and always used the right ones."

He died twelve months after the onset of his illness. At the necropsy a large area of softening was found in the right hemisphere, involving part of the white substance beneath the Rolandic area, and completely destroying the island of Reil. The left hemisphere was found to be perfectly healthy.

It may be thought by some that these two cases have little bearing upon the question we have been considering. But this would be an erroneous impression. Certain authorities maintain, as I have said, that destruction of the third frontal convolution alone in the leading hemisphere for speech gives rise to agraphia as well as aphasia. They believe, for reasons that will be more fully specified presently, that destruction of Broca's centre hinders the revival of words also in the other word centres, and thus prevents writing by stopping the process at its source. But in the cases recorded by Barlow, Ogle, Guido Banti, Wadham (Cases xv., xx., xxii., xxiv.) —and possibly in those by Dickinson and myself (Cases xxi., xxiii.)—we see that the ability to write was not interfered with by such a lesion in the leading hemisphere for speech, and therefore that the activity of the corresponding auditory and visual word centres was neither suppressed nor gravely interfered with.[1]

[1] Dally (*Ann. Med. Psychol.*, 1882, p. 252) mentions the case of another left-handed man who had an apoplectic attack resulting in left hemiplegia

The case reported by Wadham is further of special importance because it has been contended by Déjerine and also by Miralli ́e,[1] upon the basis of four cases (without necropsies) of aphasia with left hemiplegia occurring in left-handed persons, in whom the power of writing was lost simultaneously, that this loss is the rule and only what might be expected. The cases I have cited show, however, that it is not the rule, and leave us free to suppose that those of Bernheim, Parisot, Magnan, and Déjerine (where the power of writing was lost) were not cases of simple aphasia due to lesions in Broca's region only. The lesions in them may also have involved the visual word centre in one or other hemisphere, or the commissure between them, and in either of these eventualities we might find an explanation of the co-existing agraphia.

In all the cases of left-handed persons in which the right hemisphere takes the lead in speech, whilst the left hemisphere executes the writing movements, it must be supposed that the right auditory word centre is the initial centre for the revival of words either for silent thought or for speech, and therefore also for spontaneous writing. Thus in such a person we must assume that the left visual word centre becomes educated concurrently during the process of learning to write (through the commissural fibres in the corpus callosum), and that the writing process is executed by this left side of the brain in the ordinary way by the passage of stimuli from the visual word centre on to the corresponding cheiro-kinæsthetic centre.[2]

There is, therefore, no reason whatever why some, and perhaps the majority of, left-handed persons when they become the subjects of left hemiplegia and aphasia should

---

and aphasia, who was not agraphic, though he wrote in a paragraphic fashion. Two years afterwards he had almost recovered from his hemiplegia, and he could write, speak, and read, though he was unable to spell properly. M. Mesnet mentioned also that he had formerly presented to the society a patient suffering from left hemiplegia and aphasia with preservation of ability to write. Again, Féré refers (*L'Encéphale*, 1885, p. 481), to a case of left hemiplegia and aphasia occurring in a man who was left-handed for all purposes other than writing, in the following terms :—" Le malade écrivait fort bien de la main droite ce qu'il ne pouvait pas dire."

[1] *Loc. cit.*, p. 85.

[2] The process of learning to write here would, in fact, be the converse of what occurs when an ordinary right-handed person suffering from right hemiplegia and aphasia learns to write with the left hand. The only other possibility would seem to be that the incitation for writing should pass direct from the visual word centre in the leading speech hemisphere (though in a diagonal direction) to the centre for writing in the opposite hemisphere, which seems to me much less probable.

not, after the shock attendant upon the brain lesion has passed, either be able to write at once, as was the case with Dickinson's patient and with the ambi-dextrous patient of Wadham, or else speedily learn to do so. Where there is not this ability after a time, as in the case of Trousseau and one subsequently to be referred to (p. 102) which has been recorded by Déjerine, there would, as I have said, be fair ground for suspecting that the lesion was not confined to Broca's convolution, but was also associated with some damage to the visual word centre, either on the same or on the opposite side of the brain.

Returning from this partial digression concerning the mode in which writing is performed in some aphasic persons, it will be seen that the case of Guido Banti, as well as that of Dickinson and of Wadham, is absolutely opposed to the views of Stricker and Hughlings Jackson. They show that for mere thought, apart from its oral expression, the memorial recall of kinæsthetic impressions may be even less necessary than might have been supposed by those who do not share their views. These cases are equally opposed to the views of Déjerine, Wyllie, Mirallié, and others, who, without looking upon this form of word-memory as the most important of all, nevertheless think that glosso-kinæsthetic revivals are necessary for silent thought, and that without such a process as an initiative there is an inability to recollect words (amnesia verbalis), together with an inability to read and write.[1]

Déjerine seems to have been led to these views in the first place by adopting, as does Wernicke, the opinion expressed by Trousseau, Jackson, and Gairdner, that a lesion limited to Broca's region carries with it agraphia as well as aphasia, and that the two defects are almost always proportional to one another. The frequent co-existence and parity of the two defects cannot be denied; but this, as we have seen, may be due to the common simultaneous damage to the glosso- and the cheiro- kinæsthetic centres, or to the commissures connecting these centres respectively with the auditory and the visual word centres.

A careful examination of the cases capable of throwing light upon these two interpretations, to some of the best of which I have above referred, tells distinctly against the view

---

[1] Thus Wyllie, who follows Déjerine, goes so far as to say (*loc. cit.*, p. 314): "There is reason to believe that in every case of severe motor aphasia that is due to destruction of the motor images amnesia verbalis is extremely well marked—even more so, perhaps, than it is in severe cases of auditory aphasia."

which has been adopted by Déjerine and others, that agraphia is to be regarded as one of the consequences of an isolated destruction of Broca's centre. The fact, however, of Déjerine adopting this view, has led, I venture to think, to his attaching an altogether undue importance to destruction of Broca's region, since he not only believes that it interferes with silent thought by hindering the revival of auditory word images, but that it leads to some amount of alexia, and is likewise a cause of agraphia.[1] It has also induced him to bring forth a variety of arguments tending, as he thinks, to disprove the existence of any cortical centre having the same relation to writing movements that Broca's centre has to speech-movements. All these positions have, moreover, been recently maintained by his pupil Mirallié in a valuable and interesting work entitled "De l'Aphasie Sensorielle." It will be necessary, therefore, to look at these various opinions in succession, in order to see how far they are warranted by existing knowledge.

(1) **Does destruction of Broca's region alone, entail Amnesia Verbalis?**—To this question I am disposed to give a negative answer. I rely upon the reasons previously adduced to show that words are primarily revived in the auditory word centre rather than in the glosso-kinæsthetic centre. I can also refer to the case of Guido Banti, as well as to those of Dickinson, Wadham, and others, seeing that the ability of these individuals to write clearly showed that they could recall words and were not suffering from amnesia verbalis.

It is true that Lichtheim also agrees with Déjerine and Wyllie in regard to this question. Thus he says :[2] "I think that in most motor aphasics—in those, among others, of the true Broca's type—the patient has lost the auditory word-representations—that is to say, cannot awake them voluntarily by an action of their higher centres." The only safe criterion, he thinks, of the preservation of the auditory symbols in these cases would be obtained by the existence of an ability to write, or, in default of that, the ability to realise the correct number of syllables in any word that cannot be uttered, and to signify this number by giving so many hand pressures. Both these powers are, he thinks, usually lost in patients suffering from motor aphasia.

[1] In these views he has been followed by B. Onuf, *Journal of Nervous and Mental Diseases*, 1897 (pp. 139 and 142), who seems indeed to think that words are primarily revived in Broca's centre, and that the auditory word centre is thence aroused in a secondary manner; also that a lesion in Broca's centre necessarily involves alexia (pp. 141 and 145).

[2] *Loc. cit.*, p. 471.

It has been shown, however, that this is not so in uncomplicated cases of destruction of Broca's region, since, as we have seen, the power of writing persists, and so doubtless would the other test of minor significance to which he refers. I am quite prepared to admit, however, that exceptions to this rule may be occasionally met with. The functional relations of the auditory word centre with Broca's region on the one side and the visual word centre on the other are so close that the destruction of either of these centres may be easily conceived to derange the functional activity of the others for a longer or a shorter time. The degree of this derangement may be considered to vary in different persons, perhaps in accordance with the relative potency, original or acquired, of the several centres. Thus in some persons an isolated destruction of Broca's region may disturb to a very slight extent the functional activity of the auditory and the visual word centres, whilst in others it is quite possible that the severance of this centre from the other two may greatly alter their functional readiness, so that the spontaneous revival of the auditory and visual images of words may, in such persons, be more or less hindered, and as a consequence the power of thinking and of writing proportionately lowered.

(2) **Does destruction of Broca's region entail Alexia ?**—I believe that in reading, as a rule, a proper comprehension of the meaning of the text requires a conjoint revival of the words in the visual and the auditory word centres, but that for this mere comprehension it is not necessary for the stimulus to pass on also to the glosso-kinæsthetic centre, as it must do in reading aloud. It may, however, be freely admitted that if the way is open, and this latter centre is in a healthy condition, it does commonly receive in reading to one's self a slight stimulus from the auditory word centre, a fact which is often enough shown by the occurrence of involuntary half-whispered mutterings when reading. It may also be admitted that the rousing of all three centres does give assistance in the comprehension of anything difficult, as is shown by the common practice of reading aloud any passage the meaning of which may be at all obscure. By this proceeding aid is obtained, however, not alone by the full rousing of the glosso-kinæsthetic centre, as there is also the full activity of the auditory word centre following upon the spoken words. This kind of aid is, indeed, commonly needed, as Wernicke pointed out, by an uneducated person for the full comprehension of almost all that he reads. Therefore it may easily happen when such a person becomes aphasic

from destruction of Broca's region that alexia may go with the aphasia.[1]

Similar aid may be necessary in cases in which there is functional weakness of the visual word centre. Thus there has been recently under my care in the National Hospital for the Paralysed and Epileptic a woman, apparently suffering from a tumour in the left hemisphere, who was hemiplegic on the right side and understood readily all that was said to her, but who did not seem to gather the meaning of the simplest written or printed request unless she read it aloud, which she was able to do without hesitation. She was quite unable to write spontaneously or to copy with the left hand, but then she had made very few attempts in this direction since her right hand became paralysed.[2]

It would seem, however, that in some persons the comprehension of writing or printing does not even require the associated activity of the auditory word centre, so long as the visual word centre is in a healthy condition and its other associational fibres are intact. This is proved by the fact that in a case recorded by Wernicke and Friedländer,[3] in which there was aphasia associated with word-deafness (due to a lesion in Broca's region and also in the upper temporal convolution), the patient was able to read well and even to write; there is also a valuable case of word-deafness (with lesions in both upper temporal convolutions) recorded by Mills, and to be quoted further on (Case liii.), in which the patient was able to understand what she read; and a still more remarkable case recorded by Pick, with very similar lesions (Case lvii.), in which, although the patient was word-deaf, he could write and comprehend perfectly what he read.

Long before I was aware of the crucial evidence afforded by these cases and that of Guido Banti I did not believe that the comprehension of writing or of print was appreciably interfered with by an isolated destruction of Broca's region—that such a lesion, in fact, did not lead to alexia. For independently of the general views discountenancing such a notion there were Case xxi., and two recorded by Broadbent, as well as others, pointing in the opposite direction. Some particulars concerning one of these cases recorded by Broadbent may well be reproduced here.[4]

[1] Kussmaul, *loc. cit.*, p. 775.

[2] This patient died within a fortnight, when a tumour about two inches in diameter was found in the midst of the left centrum ovale, not extending into the cortex, but which may have caused some slight but undue pressure upon the left visual word centre.

[3] *Fortschritte der Médicin*, No. 6, 1883.

[4] The other is recorded further on (Case xxviii.) under the head of Agraphia.

CASE XXV.—A man, aged 29, was admitted to St. Mary's Hospital on April 3, 1870. He had had fits when a child, the last at the age of seven years; and rheumatic fever twelve months previously to his admission. He had not been well, and had left off work for two days. During these two days he talked incessantly all sorts of nonsense. In the night his brother awoke and found him in violent convulsions, mostly of the legs, not more on one side than on the other. The attack lasted two hours; when it was over right hemiplegia with loss of speech was established.

On admission the paralysis was complete, the loss of speech absolute, and he could not protrude his tongue. He gradually gained some power in the leg so as to be able to walk about, but the arm remained motionless and flaccid. He could now also say " yes " and " no," but was not always right in his use of these words. On one occasion he twice said " Here " to his sister, when impatient that she had not complied with his wish for her to come to him. He could write his name, but nothing else ; and wrote it when asked to write an answer to a simple question.

Broadbent adds : " The case presented no unusual features, and is here recorded partly as an example of a rather unusual type of attack in embolism ; partly for the purpose of noting that the patient while in the condition described read the newspapers much with apparent interest, and at my request he pointed out a Crystal Palace advertisement of fireworks and an account of pigeon shooting at Hurlingham, the latter very promptly and with a pleased expression, as if he had already read and enjoyed it."

Many of the cases in which there has been alexia co-existing with aphasia are, I think, clearly cases in which there have been rather wide lesions extending far beyond the third frontal convolution and involving also the visual word centre. This is the interpretation that I attach to a celebrated case recorded by Trousseau, which evidently influenced his opinion very much. I allude to the wealthy patient whom he saw in consultation in the department of Landes.[1] On this point Trousseau says : "As he always answered ' Yes,' I asked him whether he knew how that word was spelt, and on his nodding assent I took up a large quarto volume with the following title on its back, ' History of the Two Americans,' and requested him to point out the letters in those words which formed the word ' Yes.' Although the letters were more than one-third of an inch in size, he could not succeed in doing as I wished. By telling him to seek for each letter in turn, and by calling out its name, he managed, after some hesitation, to point out the first two, and was very long in finding the third. I then asked him to point out the same letters again, without my calling them out first; but after looking at the book attentively for some time, he threw it away, with a look of annoyance which showed that he felt his inability to do as I wished him." Few, I think, would now suppose that we had to do here with a case in which there was no lesion about the posterior extremity of the Sylvian fissure as well as of the third frontal convolution.

[1] Lectures, Translation by Bazire, pp. 229-232.

Lichtheim is inclined to give much the same interpretation to this class of cases. He points out that the examination of an old case is apt to be confusing in regard to its correct pathogenesis, owing to the fact that some of the defects originally existing have cleared up and disappeared. In this relation he calls attention[1] to the "usually rapid disappearance of word-deafness, and the greater persistence of the troubles connected with written language." He adds: "This fact appears to me to throw light upon a not unfrequent type which does not readily fall in with our classification; I mean that of motor aphasia in which alexia co-exists."

It is, of course, possible that this explanation by Lichtheim may hold good for certain cases, but that it is not necessary to have recourse to it in order to explain those cases of apparently simple aphasia with which alexia co-exists I have become convinced by two cases that have recently come under my observation. What we have to ask ourselves when in the presence of such cases is whether there is any evidence to show that the case is not one of simple aphasia, and whether a lesion exists also in the region of the posterior extremity of the Sylvian fissure. We may inquire, therefore, for the previous existence of word-deafness as Lichtheim suggests, and also examine the patient carefully for the present existence of some amount of hemianæsthesia. The finding of this latter would show two things: first, that the lesion was not limited to the third frontal convolution; and, secondly, that either it or a second lesion also existed not far away from the visual word centre, which consequently may have been also involved. Since I resolved to apply this test two cases of aphasia combined with alexia have come under my care, and in both of them there was the co-existence of a more or less complete hemianæsthesia, as may be seen from the abstracts of these cases which I now give.

CASE XXVI.—A woman, aged 32, was admitted into the National Hospital for the Paralysed and Epileptic under my care on May 19, 1896. She had been married fifteen years, and had eight children. She had a loud, mitral regurgitant bruit; the pulse was small, and irregular in force and frequency. She had been in good health till October 25, 1895, when she had a fit with loss of consciousness lasting three hours. For two or three days afterwards she seemed quite dazed and did not know anyone. The fit was followed by loss of power in the right arm and leg and loss of sight in the left eye. She has regained some power in the right arm and leg, but is still blind in the left eye. Ophthalmoscopic examination shows simple white atrophy of the left optic disc with evidences of past embolism of some branches of the central artery of the retina. She is naturally right-handed. There is distinct, though slight, right hemiplegia, including the face, arm, and leg. She can walk with dragging of the right foot and

[1] *Loc. cit.*, p. 467.

perform all movements with the right leg and arm, though much less forcibly than with the left limbs. The tongue is protruded straightly. There is incomplete right hemianæsthesia; sensation is dulled to touch and pain in the mid-line of the face, head, and trunk (back and front), also over the arm, hand, and upper two-thirds of the thigh.

She is almost completely speechless. She can answer "Yes," "No," and say "Yes, I know," in answer to questions. Once she said "Good-night." Her articulation of these few words is good. She cannot speak voluntarily, and cannot repeat the simplest word when bidden. The patient understands more or less complex matters mentioned to her, and performs some spoken commands, but is slow at it, and often makes mistakes. When told to show her tongue she generally shows her teeth. There is almost complete alexia. She cannot recognise words or letters as a rule, except short names of external objects, such as "pen," "lamp," "cat," "dog." She will, when shown the name, point to such objects in the ward or in pictures. She does not recognise the word "man," and she cannot understand simple written commands. When shown four or five capital letters she can very rarely point correctly to one that may be named. She does, however, recognise and point out numerals; she can add two numbers together, but cannot subtract or multiply. She can also pick out objects named; she cannot name them, but shows by her expression that she recognises their names when she hears them. She cannot write letters with either hand spontaneously or from dictation, but she can copy single letters printed or written—the former better than the latter. She cannot do transfer copying.

This case, though interesting, is perfectly simple and needs no comment. The next, though equally cogent, is a little more complex.

CASE XXVII.—A young woman, aged 17 years, was admitted into University College Hospital under my care on June 20, 1896. She had had rheumatic fever when a child, since which she had been always ailing, principally by reason of a heart affection. She had a loud presystolic bruit and some hypertrophy of the left ventricle. On June 8, 1896, on awaking she found herself paralysed on the right side of the body. During the following day, without the occurrence of any fit, she lost her speech. During the first week of her illness she was very dull mentally, and there was incontinence of urine and of fæces. She also suffered from pains in the left side of the head.

When admitted, on June 20, the paralysis of the right arm and leg was complete, together with slight paralysis of the lower half of the face. There was well-marked right hemianæsthesia, but no hemianopia. There was ankle clonus, with increased knee, wrist, and elbow jerks on the right side. The limbs were flaccid.

She was almost speechless. She understood all that was said to her, looked intelligent, and often attempted to speak, but then and since her illness the only words she had uttered had been "yes," "no," and "oh dear!" She often said "yes" inappropriately when asked a question, and then immediately shook her head in correction. She could not repeat a single word when bidden. She recognised objects such as pencils, paper and pens, but could not name them. There was complete alexia; she did not understand written or printed words or short sentences; nor could she point correctly to individual letters.

On June 22 she could repeat short words uttered before her in a slow, hesitating manner, and with indistinct utterance. On June 27 she was much better; she could count up to ten, and say the days of the week when started; she could also name most common objects which she was shown. She could also read aloud short words in print or writing, and understand their meaning. On July 2 she talked a little in short sentences, though her pronunciation of words was still indistinct, and she had difficulty in recalling the words she wanted. On July 8 she was still improving. She could then move her right leg a little. She talked slowly and indistinctly, but still had difficulty in recalling words, though she did not utter wrong words. She read aloud a few lines of short words from a book slowly, passing over some and miscalling other words. She seemed to understand what she read. Some right hemianæsthesia still existed.

When first admitted this patient seemed to present a typical case of aphasia, but for the existence of the alexia. The fact of the presence of hemianæsthesia showed, however, that the lesion extended back to the posterior third of the hinder segment of the internal capsule, or else that another lesion existed there—that is, in a region very close to the visual word centre. This, of course, carried with it the possibility that the visual word centre itself might have been temporarily disabled. The subsequent progress of the case— the speedy clearing up of the aphasia, owing probably to a restoration of the circulation through the first cortical branch of the left-middle cerebral, which may have been temporarily blocked, and the revelation then of some amount of amnesia verbalis—made it all the more likely that the alexia was to be explained by some temporary disablement of the visual centre. That the auditory centre was only slightly (though more durably) affected was shown by the fact that there was never word-deafness and by the patient's subsequent ability to read even better than she could speak spontaneously.

The probability is, therefore, that in the cases of associated aphasia and alexia recorded by Trousseau and others the latter defect is not to be regarded as a consequence of the lesion in Broca's region, but rather as a result of some co-existing damage to the visual word centre.

(3) Does destruction of Broca's region entail Agraphia ?—The writers whom I have previously named say yes, that agraphia as well as aphasia is entailed by an uncomplicated lesion in the posterior part of the third frontal convolution. But with this doctrine, as already stated, I cannot agree. In writing spontaneously or from dictation there is, I believe, needed for most people a primary revival in the auditory word centre, and secondarily in the visual word centre, whence combined stimuli pass to the cheiro-kinæsthetic centre, and thence well over in coördinated form to the motor centres in the cord by which the

actual writing movements are effected. The evidence that I have already recorded[1] is quite in harmony with this view, whilst it is entirely opposed to the doctrines on this subject formerly enunciated by Trousseau, Hughlings Jackson, and Gairdner, and recently advocated by Déjerine, Wyllie, Mirallié, and others.

(4) **Is it right to deny the existence of a Special Centre for the registration and regulation of Writing Movements?**—We are now brought to the consideration of Déjerine's objection to the very existence of a special centre for writing movements—a "cheiro-kinæsthetic centre" as I name it. His reasoning here seems to me rather surprising, and is of such a nature as to make it necessary to premise that the case of writing is almost exactly comparable with that of speaking—and, indeed, of all other habitual voluntary movements, whether complicated or simple. Each set of movements must be associated with a set of ingoing impressions (kinæsthetic), which are registered in different portions of the Rolandic area of the cortex. This must be as inevitably true for writing movements as for speech-movements; and, just as re-excitation of Broca's region under stimulation from the auditory word centre is necessary for speech, so a re-excitation of the centre in which the impressions generated by writing movements are registered is, under stimulation from the visual word centre, needful for the production of writing movements.

Yet both Wernicke and Déjerine say that the intervention of a special centre for writing movements is not necessary, on the ground that writing consists in a simple copying of the visual images stored up in the visual word centres. This, however, is extremely vague. It must be asked, How is the copying achieved? That is, By what cerebral mechanism is it brought about? I maintain that the visual word centre cannot act directly upon the motor centres in the cord, and that the conjoint activity (for coördinating purposes) of a kinæsthetic centre is quite as essential in the case of writing movements as is the activity of Broca's region for the production of speech-movements. By parity of reasoning Wernicke and Déjerine might just as well say that Broca's centre is not needed, on the ground that speech consists in a simple copying of the auditory images stored up in the auditory word centre, and hold that the auditory word centre can act directly upon

[1] To this I am now able to add two cases recorded by Byrom Bramwell (*Lancet*, 1897, vol. i., Cases i. and xii.). The latter of these two cases again shows that complete destruction of Broca's centre does not necessarily entail agraphia.

the motor centres in the bulb for the production of speech. But I contend that in each case another regulating centre must intervene, and that such a centre is just as essential for the translation of visual images into writing movements as it is for the translation of auditory images into speech-movements.

Again, How can it possibly be alleged as a valid argument against the existence of a cheiro-kinæsthetic centre in a man accustomed to write with his hand that the same man is able to write on sand rudely with his foot, or that he can after a fashion write with his elbow, or with a pencil placed between his teeth? Clearly in these other crude modes of writing he would simply be dependent upon the activity of other parts of the general kinæsthetic centre. Such objections—for they are scarcely worthy of the name of arguments—are therefore altogether futile. The question that has to be considered is whether a man who has learned to write with his right hand, and who habitually writes in this fashion, calls into play a cortical centre the activity of which in the production of those movements is comparable with the activity of Broca's centre in the production of speech-movements. To this question there can be but one answer. And the fact that in those cases of word-blindness in which the patient is able to write spontaneously he can subsequently succeed in reading what he has written, by tracing the outlines of the letters with his finger must also be taken as direct evidence of the existence of a cheiro-kinæsthetic centre, as I pointed out many years ago.[1] The impressions produced by the writing movements thus occasioned pass from their centre of registration back to the visual and auditory word centres, and thus enable the writing to be read.

Of course, a patient would not be able to read in this way if his visual and auditory word centres were destroyed; therefore the objection of Déjerine and Mirallié to this evidence falls to the ground.[2]

Another argument adduced by Déjerine against the existence of a special centre for writing has been based upon an interesting case which he cites very briefly[3] of a patient, left-handed for all acts save those of writing, who became hemiplegic on the left side and aphasic, and at the same time lost the power of writing with his right hand. The patient continued in this condition for four years, and, judging from the symptoms (there being no necropsy) Déjerine regarded it as

---

[1] "The Brain as an Organ of Mind," 1880, p. 646, note.
[2] Mirallié, *loc. cit.*, p. 75.
[3] *Mémoires de la Société de Biologie*, 1891, p. 97.

a simple case of "motor aphasia." He thinks that the foot of the right third frontal was destroyed, and that this patient was as incapable of writing as of speaking because he had no longer a "notion of the word," owing to his supposed inability to revive verbal images. The fallacy of this inference has already been shown in part by the cases of Dickinson, Wadham, and others previously recorded; and in part by the evidence I have brought forward as to the non-occurrence of alexia as a consequence of destruction of Broca's region. This latter evidence goes to show that in many cases of what at first sight appear to be cases of simple "motor aphasia" with alexia, there is reason for believing that a lesion exists in the hinder part of the hemisphere which may have damaged the visual word centre. A lesion in this situation may therefore have existed in the case under the care of Déjerine, and thus have accounted for his patient's inability to write as he had been accustomed to do with his right hand.

For if his previous power of thinking in words had depended upon the activity of the word centres in the right hemisphere, then spontaneous writing would have required the activity of the auditory and visual word centres of this side as a first stage in the process, and secondly that they should be in relation (through the corpus callosum) with the similar centres in the left hemisphere, so that the proper stimuli might be passed on to the left cheiro-kinæsthetic centre for the bringing about of writing movements by the right hand. A damage to the visual word centre of the right side would therefore have caused this patient to lose his previous power of writing with the right hand. The case under the care of Déjerine therefore admits of an explanation quite different from what he has given to it, and which is also much more in accordance with existing knowledge. Clearly it cannot be deemed to support his notion that the kinæsthetic impressions of writing movements are not registered in a definite part of the cerebral cortex, and that this cortical region is not called into activity during the act of writing. That such impressions must be definitely registered in the cerebral cortex seems to me a self-evident truth.

It is, however, altogether another question, and one which does seem open to discussion, whether the cortical centre in which the sensory impressions produced by writing movements are registered exists altogether apart, or whether its structural elements are inextricably mixed up with others pertaining to less special movements of the hand and arm.

At the commencement of the section concerning "The Localisation of the different Word Centres" I expressly said (p. 13) that I was not a believer in their "complete topographical distinctness." All I claimed was "that there must be certain sets of structurally related cell and fibre mechanisms in the cortex, whose activity is associated with one or with another of the several kinds of sensory endowments." And I added:—"Such diffuse but functionally unified nervous networks may differ altogether from the common conception of a neatly defined 'centre,' and yet for the sake of brevity it is convenient to retain this word and refer to such networks as so many 'centres.'" Even if the nervous arrangements concerned with writing, constituting part of the general kinæsthetic centre, are intimately intermixed with those concerned with less special hand-movements, this does not make it seem to me undesirable to speak of such nervous arrangements as constituting a special centre for writing movements (which I have termed a "cheiro-kinæsthetic centre"). And certainly because such nervous arrangements are not topographically separate, that is no reason for denying the existence of a special writing centre—whether separate or not, the kinæsthetic nervous arrangements concerned with writing must be there.[1]

If there is a completely separate seat of registration, destruction of such a centre should produce loss of the power of writing, independently of paralysis of other less special movements of the limb. If otherwise, the loss of power of writing from damage to the cortical region in which its kinæsthetic impressions are registered would never be able to occur alone, but would always be merged in a more general paralysis of the limb. In other words, no agraphia of this type would be recognisable as such; it would be useless to look for it, because the co-existing paralysis of the hand and arm would naturally be supposed to be, and would in fact be, a sufficient cause of the inability to write.

[1] As will be seen further on (p. 137) we are by no means free from similar doubts as to the existence or not of a topographically separate glosso-kinæsthetic centre—that is how far, in man, the kinæsthetic nervous mechanisms concerned with speech are topographically separate from the kinæsthetic nervous mechanisms concerned with ordinary less special movements of the lips, tongue, palate, and larynx. The fact that, as a rule, these less special movements of bilaterally acting muscles may be equally well called into play by either hemisphere while speech-movements are evoked only by one, very considerably complicates this question. For instance, the special and less special nervous mechanisms just referred to might be intimately intermixed in the left third frontal convolution, and yet destruction of this part, whilst it would paralyse the special movements, would not paralyse the more common movements, seeing that they could be called into play by the opposite hemisphere.

It will be remembered that we came to a similar conclusion (p. 79) as to the cause of the absence of agraphia as a separate symptom resulting from damage to some of the pyramidal fibres.

On the whole it must, I think, be said that no proof has yet been given that the centre for writing movements is topographically distinct—in other words, that Exner's or any other separate localisation of such a centre has yet to be definitely proved. The evidence that I have previously cited, though it has been sufficient to discountenance the idea that destruction of Broca's centre of itself gives rise to agraphia, has no pretence to be adequate for the localisation of a separate centre, and the cases that have hitherto been cited in support of Exner's localisation are all of them open to very legitimate objections.

There is, for instance, a case recorded by Bar[1] in which an isolated lesion existed in the foot of the second frontal convolution, but, instead of giving rise to agraphia only, there were aphasia and agraphia, and the agraphia was not total (as it should have been if a topographically separate centre for writing had been destroyed, since the patient's penmanship was good); and, though he could not express his ideas either in writing or in speech, he wrote over and over again a certain phrase or word. This case, therefore, though cited by Exner in favour of his localisation, seems to me decidedly opposed to it. Another case, recorded by Henschen,[2] has been cited as favouring Exner's localisation, which has, however, no real value. Here the patient suffered from word-blindness and agraphia, and after death there was found not only a destruction of the foot of the second frontal, but also a softening which destroyed the angular gyrus.

A third case has been recorded by J. B. Charcot and A. Dutil,[3] and is thought by Pitres to favour Exner's localisation. Here the patient was a woman aged 64 years, who, twenty-nine years before her death, had a slight cerebral attack which left her completely agraphic. Subsequently she had three other attacks, which led successively to embarrassment of speech, left hemiplegia, and finally to pseudo-bulbar paralysis. The agraphia persisted throughout her life without appreciable diminution. She could copy letters and figures after a fashion, but was absolutely unable to write spontaneously. There was no word-blindness or word-deafness.

[1] *La France Médicale*, 1878, p. 609.
[2] "Klinische und Anatomische Beiträge zur Pathologie des Gehirns," Upsala, 1890, p. 273.
[3] *Mémoires de la Société de Biologie*, 1893, p. 129.

At the necropsy five small foci of softening were found in the right hemisphere and two in the left hemisphere, one of the latter being situated exactly in the foot of the second frontal convolution.[1]

Then, again, one of the causes which led Déjerine to deny the very existence of a centre for writing movements seems to have been based upon the characteristics of two cases that were brought forward by Pitres, and which, upon the basis of his authority, have been commonly cited by others as affording evidence of the correctness of Exner's localisation. They were cases in which the patients were unable to write spontaneously or from dictation, and yet were able to copy writing. If the centre is destroyed, says Déjerine, the possibility of executing any kind of writing should be abolished. That at first sight seems perfectly true; and it may be a valid criticism against Pitres for quoting any such cases as instances of "motor agraphia."[2] But Déjerine and his followers have overlooked the possibility that Pitres may not have been quite correct in his interpretation of these cases—in which there was no necropsy (see Cases xc., xci.).

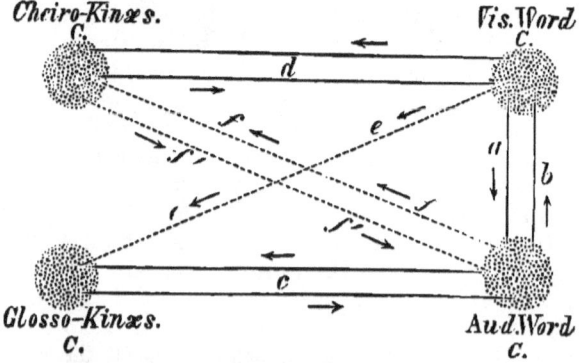

Fig. 5.—A Diagram illustrating the relative positions of the different Word Centres and the mode in which they are connected by Commissures.

The connections represented by dotted lines indicate possible but less habitual routes for the passage of stimuli.

They have not recognised the fact that this particular combination of symptoms may be present without the ex-

---

[1] Two other cases which have been cited as favouring this localisation, namely—one by Balzer (*Gazette de Médecine de Paris*, 1884, p. 97), and another by Shaw (*Brit. Med. Jour.*, February, 1892)—have in my opinion no value in this connexion.

[2] Pitres attempts to get over the difficulty standing in the way of his interpretation, arising from the fact that the patients were able to copy

istence of what is called "sensory aphasia" in any of its forms. It can be shown, however, that such a combination of symptoms may be easily explained by a cutting across of the audito-visual commissure that transmits impressions from the auditory to the visual word centre (Fig. 5, *b*). This I believe, after careful study of his memoir, to be the explanation of the partial agraphic defects in the two well-known cases cited by Pitres as examples of the effects produced by damage to the cortical centre for writing movements.[1] It is obvious that damage to the commissure above mentioned would be capable of preventing the passage of stimuli from the auditory to the visual word centre such as would be necessary in writing spontaneously or from dictation, whilst it would leave the power of copying writing intact so long as the visual word centre, the commissure connecting it with the writing centre, and this centre itself, remained undamaged.

At present, then, it would seem that there are no uncomplicated cases of agraphia upon record produced by a small lesion in the cheiro-kinæsthetic centre, or of agraphia as an isolated symptom certainly resulting from damage to some part of the visuo-kinæsthetic commissure. It seems clear, however, that if the centre for writing is topographically distinct, complete agraphia—that is, an inability to write spontaneously from dictation or to copy with anything like facility—should be producible in either of these ways, and that in each case the disability would be uncomplicated with word-blindness, whilst there would be nothing to prevent the patient learning to write with his left hand. Slow copying, stroke by stroke as a drawing is copied, may still be possible; and also, as there is no word-blindness, the patient may be able to form words quite correctly with separate or block letters (the so-called "typographic writing").

The reason for this extreme rarity of agraphia as an isolated

writing, in the following manner. He says (Congrès Français de Médecine Interne, 1894, " Rapport sur la Question des Aphasies," p. 10): "Copier, en effet, est toute autre chose qu'écrire couramment. C'est un acte de motilité générale qui n'a pas de centre d'exécution spécialisé, pas plus d'ailleurs que l'acte de tracer sur le sol avec le bout de sa canne, ou avec le pied, des traits ayant la forme de lettres. Ces actes là impliquent une attention soutenue, une surveillance constante de l'esprit, qui manque absolument dans l'écriture courante. C'est pourquoi l'agraphique dont les membres supérieurs ne sont pas paralysés, peut encore copier, bien qu'il ne soit plus capable d'écrire spontanément."

[1] *Revue de Médecine*, 1884, p. 864. These cases are also quoted more briefly by Wyllie (*loc. cit.*, p. 357), though in the first of the two cases he omits the very important detail that the patient was able to copy writing easily.

symptom doubtless is that often, in what might otherwise have been a simple case of inability to write, this very special disability becomes hidden and unrecognisable, owing to the lesion being large enough to cause paralysis of many other movements in the hand and arm. Thus the mere inability to write is hidden by the production of a partial brachial monoplegia. No similar causes intervene to obscure the recognition of simple, uncomplicated aphasia, and yet it must be borne in mind that even this is decidedly rare.

If, then, there are no simple, uncomplicated cases of agraphia that can be quoted, either from disease of the cheirokinæsthetic centre or from damage to the commissure between it and the visual word centre, all that can be done is to cite some of the best instances available in which agraphia has been associated with aphasia, but where the former defect has been more marked than the latter. This was, in fact, the class of cases originally described by William Ogle in his paper on " Aphasia and Agraphia." One of his examples (in which there was no necropsy), where right hemiplegia with aphasia was succeeded by a condition in which amnesia and agraphia were marked, still remains one of the most typical of this kind that can be cited. But two simpler and more typical cases have been recorded by Broadbent[1] to which I would first direct attention. One of them is notable also on account of the remarkable amount of intelligence and energy displayed by the patient, notwithstanding her very defective speech and inability to write. The following is an abstract of this important case.

CASE XXVIII.—A lady, aged about 70 years, had an apoplectic attack on July 26, 1867. She remained unconscious for nearly twenty-four hours, and was then found to be hemiplegic on the right side and quite speechless.

After two or three months she had " quite recovered the use of her limbs and regained a few words." Her relatives treated her as if she were utterly incapable of looking after her own affairs, and especially of taking charge of her own money. This was a great grievance, and when her husband's relatives visited her she tried to make it known by showing her purse and repeating excitedly, " Oh, shameful! shameful!" She had an intense dislike to her attendant, and when she left the room would often look in the direction in which she had gone, shaking her head and exclaiming with great energy, " Oh! nasty! nasty!" following it up by a long story in which there was not a single word intelligible.

At length she went to live near her husband's relatives. " A house was now taken and furnished for her, and she showed herself quite competent in making all arrangements about it, and when settled in it with a new housekeeper she seemed quite contented. Her principal topic of attempted

[1] *Transactions of the Royal Medical and Chirurgical Society*, 1872, pp. 146 and 160.

conversation for several months was her niece's and brother's behaviour to her. She laboured hard to make the friends among whom she now was understand her grievances. She would point out her relatives' names or refer to their letters, or in some other way show to whom she referred, and then in great excitement would exclaim, 'Oh, shameful! shameful!' Then she would turn over letters and newspapers, and when she had tried in vain for hours to convey what she wished, she would shed tears and say, 'Oh, pity! pity!' At length her banker's pass-book was obtained from her niece, and she scarcely knew how to express her joy. 'I tried, and I tried,' she said over and over again. But there was still something she wished to explain, and one day, while trying very hard as usual, she made the figures 40 quite distinctly, and it was understood at last that it was something about £40; but it was not till a friend came to see her and this happened to be mentioned, that it was discovered she had put £40 into a bank before her attack, and believed her niece had the acknowledgment of it, and some promissory notes with other papers. These she wished to be sent for, and eventually they were obtained. This was nearly a year after her removal, and all the time she had persevered in her attempts at an explanation.

"During the same time she had been making a new will, and week after week would have her sister-in-law to write out the different clauses. It often took a long time to do it to her satisfaction, but she never rested till every particular was exactly as she wished it; when the right guess was made, she showed her satisfaction quite unmistakably by her gestures, and by saying, 'Yes, yes, that's it.'"

Broadbent adds:—"That she knew what word she wanted was shown by her resorting to letters and sometimes to a dictionary; and in one of my visits, wishing to recommend a linseed-meal poultice and failing to convey her meaning, she went for the housekeeping book and found an old entry of linseed-meal. She also pointed out to me newspaper paragraphs which interested her. She frequently read, but preferred being read to. . . . . She was quite unable to write or even to sign her name. She once wrote down the figures 40, as related, and sometimes wishing to divide money would make the proper number of marks to indicate the amount of a share."

On August 7, 1871, she had another apoplectic attack, after which she was found to be paralysed on the left side. She died on August 12, and at the necropsy no cause for the last apoplectic seizure was found, but two old apoplectic cysts were discovered in the upper border of the fissure of Sylvius. One "which might hold a nut" was in the situation at the lower end of the ascending frontal gyrus, "and a very small part of the adjacent posterior end of the third or inferior frontal gyrus." "Again, in the supra-marginal lobule over the posterior end of the fissure of Sylvius was another hollow filled with connective tissue, and when the fissure itself was opened out it was found to contain at this part a superabundance of connective tissue, and the convolutions of the posterior end of the island of Reil were completely atrophied." There was some atrophy also of the tail of the intraventricular portion of the corpus striatum.

In addition to the evidences of intelligence and energy displayed by this patient, there is also the fact that there was no alexia associated with the aphasia and agraphia. The latter defect was much more marked than the former; and the fact that the patient could use more words and sentences than are customary may perhaps be explained by the small extent of the lesion in the left third frontal gyrus. Although the paralysis of the right upper extremity had been quite recovered from, the agraphia was absolute. The posterior extremity of the second frontal gyrus was uninjured, but the posterior of the two old focal lesions discovered would probably have destroyed the commissural fibres between the visual word centre and the cheiro-kinæsthetic centre not far from the former of these centres. In the next case there was a very similar inequality between the ability to speak and the ability to write.

CASE XXIX.—A German, aged 34 years, was admitted into St. Mary's Hospital in February, 1870. He had suffered from rheumatic fever in 1865. On the 4th he fell off his stool whilst at work. Two days after he was found to have lost the use of his right hand and arm, and to be unable to speak.

On the evening of the 8th he was brought to the hospital, and was then found to be partially hemiplegic on the right side and unable to answer questions intelligibly. He soon began to improve, and Broadbent's notes on the 14th are as follows : " All his answers are in a low, smooth tone, without modulation of voice; he could not be induced to utter loudly even a word he could say. Answered 'Yes' and 'No'; told me his age correctly and his name, but indistinctly. Asked how he is replies, 'Better' or 'Quite well,' rather indistinctly. If he had pain in the head since the attack? 'No.' Before? 'Yes, wonderful.' If he slept well? 'Nicht' .... (something or other which could not be made out). If appetite good? 'No'. . . . It was impossible to understand him when he attempted a phrase or when he tried to say anything to which I had no clue."

A strong German accent was distinguishable in all he said, whereas before the attack he spoke English particularly well. " A peculiarity of his attempts to read or speak was his persevering effort to master a word, syllable by syllable, by trying on and on again. For example, when asked to read the words 'diet-card' his attempt was something like this, 'Diget, dicht, dite, dite, dite car.' The next moment he pointed to his name, saying, though not very distinctly, 'Dat my name.' Up to this time and a week later, February 21, he could not sign his name or write the simplest word, such as 'yes' or 'no,' when asked to do so, or copy a word pointed out to him." He remained in the hospital some weeks longer, improving very slowly, and before he left he could " with great effort write his name."

Nothing can be said as to the situation of the lesion here or in the next case, which is the best of those that William Ogle was able to bring forward in his paper. We must be guided only by the clinical facts recorded.

CASE XXX.—James Simmonds, aged 54 years, after a heavy blow on the left side of the head, seven years ago was obliged to give up his work. He spoke without difficulty or hesitation, but miscalled things strangely. He then had a fit one morning whilst dressing, which left him speechless and hemiplegic on the right side. For a fortnight he could not speak at all, though he was quite sensible. He could not say so much as " Yes " and " No." From this he gradually recovered, but always, as before, miscalled things. . . . A month ago he had a second fit, which left him with less power than before in his right side, but made little or no change in his speech. There is now partial paralysis of the right side, which does not prevent his walking. The facial muscles on that side are slightly affected as well as the limbs. His speech is very hesitating and imperfect. He often stops suddenly at a loss for a word and then frequently uses a wrong one, as, for example, he substituted "barber" for "doctor," "two-shilling piece" for "spectacles," "winkles" for "watercress." He can, however, pronounce any word perfectly when prompted. He says that he generally knows when he has used a wrong word, but not always.

Before his illness he wrote a good hand and was above his lot as regards education. Now he cannot form a single letter. Even with a copy before him he makes only uncertain up-and-down strokes. I gave him some printed letters and asked him to pick out his name. After a long time he arranged JICMNOS. Clearly he had some slight notion of the letters which composed his name. According to his wife, before his illness he spelt well, and was very particular about the spelling of his own name, which is one admitting of many variations. When a copy was before him he quickly picked out and arranged his name correctly. He can read; but he says that reading makes him very giddy and causes great pain in the head. His general understanding seems good and up to the average of men in his class.

The conditions here recorded represent the remainders of an attack of aphasia, as it would seem that the amnesia was perhaps not much more than had previously existed. We may conclude that in this case also the most severe or durable lesion was in the track of the commissural fibres between the left visual word centre and the cheiro-kinæsthetic centre (Fig. 5, *d*), but that in all probability there was also some damage to the former word centre itself. This latter defect is indicated from the difficulty he had in spelling his name with separate printed letters placed before him; for inability to spell—that is, inability spontaneously to recall the letters composing a word—probably depends in the main upon some defect in the visual word centre, and although he could read, what is said seems to imply that his powers were perhaps not great in this direction. It is therefore by no means a clear case, but one in which the defects were somewhat mixed.[1]

Although we are unable as yet to cite any simple uncomplicated case of agraphia due to disease in the posterior part of the second left frontal convolution, yet, according to Binet and Féré,[2] it is easy to produce agraphia by suggestion in hypnotised persons. Speaking of their experiments Ballet[3]

---

[1] For reference to other possible cases in which the agraphia present was due to damage to the visuo-kinæsthetic commissure, see chap. xi. And for an abstract of a case published by Estridge in which a lesion is said to have existed in the foot of the second left frontal convolution, associated with inability to spell correctly during the act of writing, see Case lxxxiv. Here it seems to me possible that there may have been an incoördinate activity of the cheiro-kinæsthetic centres, and something of the same kind may have existed in Cases xlviii. and lxxxiii.

[2] "Les Paralysies par Suggestion," *Revue Scientifique*, 12 Juillet, 1884.

[3] "Le Langage Intérieur," p. 134.

says: "They show that the loss of the motor memory for writing is of the same order as the abolition of other motor [kinæsthetic] memories, such as those which preside over the ability to smoke, to sew, or to embroider. For each of these acts which we are accustomed to perform there exists, in fact, a motor memory, so much the more easily disturbed, in all probability, as the acts themselves are more complicated and require more delicate muscular combinations. As an example of motor amnesia connected with special movement associations I may recall a curious case related by M. Charcot in one of his lectures: it was the case of a player on the trombone who had lost the memory of the associated movements of the mouth and of the hand necessary for playing on the instrument. All the motor memories in this patient were intact save this one. This musician had forgotten how to handle the trombone, as others had forgotten how to handle the pen." If we substitute in the above-quoted passage for "motor memory" the term "kinæsthetic memory" we may believe that the loss of the different kinds of kinæsthetic memory referred to, which can be brought about in hypnotised persons, must be due to a temporary functional inertia produced under the influence of suggestion in the several parts of the general kinæsthetic centre in relation with this or that movement.

It seems best now, at the close of our discussion concerning aphasia and agraphia as combined and separate symptoms, to enumerate in tabular form the different modes in which complete or partial agraphia may be produced. Although I shall thus be anticipating results subsequently to be developed, it will be advantageous to show now how other different forms of agraphia are to be distinguished from those we have been considering and from one another.

AGRAPHIA.

Complete.
- Uncomplicated.
  1. Destruction of the cheiro-kinæsthetic centre.
  2. Destruction of the visuo-kinæsthetic commissure.
- Complicated with word-blindness.
  3. Destruction of the visual word centre.

Partial.
- No spontaneous writing or from dictation, but can copy; no word-deafness.
  4. Destruction of the audito-visual commissure.
- No spontaneous writing, or from dictation, but can copy; word-deafness present.
  5. Destruction of the auditory word centre.
- Can write spontaneously or from dictation, but cannot copy; word-blindness present.
  6. Destruction of the visual word centre (in some educated auditives).
  7. Isolation of the visual word centre from all afferent fibres.

The supposition is that in the large majority of cases the various lesions that have been above referred to as causes of complete or partial agraphia will be in the left hemisphere in right-handed and in the right hemisphere in left-handed persons. Of these various forms Nos. 1, 2, and 7 should occur in all individuals alike, in association with the lesions named; but Nos. 3 and 6 are two degrees of agraphia that may be met with in association with word-blindness according as the individuals in whom the lesions specified occur are "visuals," or strong "auditives" much accustomed to write. On the other hand, Nos. 4 and 5 are two cases in which similar degrees of agraphia may be met with, but differing by the absence of word-deafness in the one case and its presence in the other. I say *may* be, because such symptoms might not occur in a strong "visual." It will be observed also that each of the partial forms of agraphia except the first is associated either with word-deafness or with word-blindness.

CHAPTER VII.

MERE FUNCTIONAL DISABILITIES IN THE GLOSSO-KINÆSTHETIC CENTRE.

HITHERTO, in dealing with defects of speech referable to lesions in the glosso-kinæsthetic and cheiro-kinæsthetic word centres, one important section of this subject has been omitted—namely, the consideration of the defects due to so-called functional disabilities in the glosso-kinæsthetic word centre. This portion of the subject, therefore, will now be briefly dealt with.

Aphasia occurs not unfrequently in association with various sets of conditions as a temporary disability, and sometimes with a marked tendency to recurrence, where the brief duration and the complete recovery make it improbable that it can have been due to any gross lesion.

What the proximate causative conditions have been in these different sets of cases it is often impossible definitely to say. Instead of softenings of brain tissue or hæmorrhages, abscesses, new growths, or traumatisms of the brain as causes of loss of speech, we can only, in order to account for the defects of which we are now about to speak, appeal to ischæmia (from irritative congestion, temporary thromboses, minute embolisms, or spasms of vessels) affecting the posterior part of the left third or of the second and third frontal convolutions; to "inhibition" variously brought about; or to other little understood causes leading to degradation of function in the same parts. Whatever the precise cause of the disability happens to be, this may—as with functional paralysis of limbs—persist for variable periods (hours, days, weeks, months, or even two or more years), and yet permit ultimately of a perfect restoration of function more or less rapidly brought about.

In these cases we have to suppose that from one or other of the causes above referred to the molecular mobility of the glosso-kinæsthetic centre is so reduced that it is unable to send down to the bulbar speech-centres incitations strong enough to call them into activity.

The principal causative associated conditions may now be referred to in succession.

(a) **Irritative Congestion or Thrombosis following upon excessive literary work or worry, or as a result of exposure to cold.**—Trousseau recorded several interesting cases of this type,[1] and many others have since been published by subsequent writers. The case of his colleague, Professor Rostan, which Trousseau quotes, may be here reproduced as a typical instance of the kind of defect with which we are now concerned.

CASE XXXI.—"Rostan had been confined to his bed for a few days in the country in consequence of an injury to one of his legs, and, being alone there, he read all day and thus fatigued his brain. He was engaged one day reading Lamartine's 'Literary Conversations,' when he suddenly noticed that he did not clearly understand what he was reading. He paused for a short time, but on beginning again he made the same remark. Alarmed by this, he tried to call for help, but to his extreme surprise he was unable to utter a word. Believing himself to be struck with apoplexy, he then executed various complex movements with both his hands and with his uninjured leg, and thus found that he was not paralysed. He then rang the bell, and when a servant came into the room Rostan could not speak a single word, and yet he could move his tongue in every possible direction, and was perfectly aware of this singular discrepancy between the facility with which his vocal organs were moved and his inability to express his thoughts in words. He made signs that he wished to write, but on pen and ink being brought to him he was as incapable of expressing his thoughts in writing as in speaking.

"On a medical man arriving two or three hours afterwards Rostan pulled up his shirt-sleeve, and, pointing to the bend of his arm, clearly intimated that he wished to be bled. No sooner was the bleeding over than he was able to say a few words, although incoherently and imperfectly. Still some of these words clearly expressed an idea, whilst others seemed to have no relation to the principal idea. By degrees the veil was removed, he had a greater command of words to express his ideas, which also became more numerous, and at the end of twelve hours recovery was complete."

Long before anything definite was known about aphasia a very similar case was recorded by Scoresby Jackson[2] of Pennsylvania. The case is quoted by Bateman,[3] from whose work I take it.

CASE XXXII.—"The Rev. Mr. ——, aged 48 years, endowed with intellectual powers of a high order, of a sanguine temperament, with latterly a strong tendency to obesity, having exposed himself to the influence of the night air, received a check to the cutaneous perspiration. The next morning he awoke with a headache, and when a friend went into his room to inquire after his health he was surprised to find Mr. —— could not answer his questions.

"Dr. Jackson having been summoned, found the patient in full posses-

---

[1] Lectures (translation by Bazire), 1867, p. 219.
[2] *American Journal of Medical Sciences*, 1829, p. 272.
[3] "On Aphasia," 2nd ed., p. 83.

sion of his senses, but incapable of uttering a word; the tongue was not paralysed, but could be moved in every direction; all questions were perfectly comprehended and answered by signs, and it could be plainly seen by the smile on his countenance, after many ineffectual attempts to express his ideas, that he was himself surprised and somewhat amused at his peculiar situation. The face was flushed, the pulse full and somewhat slow, and to the inquiries if he felt pain in the head he pointed to his forehead as its seat. When furnished with pen and paper he attempted to convey his meaning, but he could not recall words, and only wrote an unintelligible phrase, ' Didoes doe the doe.'

" Forty ounces of blood were drawn from the arm, and before the operation was completed speech was restored, though a difficulty continued as to the names of things, which could not be recalled. The loss of speech appearing to recur in fifteen minutes, ten ounces more blood were abstracted, and sinapisms applied to the arms and thighs alternately. These means were speedily effectual, and no further return of the affection took place."

In each of these cases we see that the aphasia was associated with agraphia; and that the functional disability was not limited to the third frontal gyrus seems further shown by the fact that during recovery these patients also displayed some degree of verbal amnesia or paraphasia.

(*b*) **Minute Embolisms.**—Nothnagel[1] has called attention to cases of this kind in which patients suffering from valvular heart disease were suddenly attacked with aphasia, unattended by other symptoms, the loss of speech disappearing in the course of one or two days. Hammond[2] also mentions the case of a woman who, after repeated attacks of acute rheumatism, with marked aortic insufficiency, was seized with headache and vertigo whilst conversing with a friend, and whose speech was abruptly cut short. She was rendered completely aphasic, but recovered within forty-eight hours.

(*c*) **Spasm of Vessels.**—Sometimes the temporary ischæmia which leads to aphasia is due neither to irritative congestion, temporary thrombosis, nor minute embolism, but seems to be occasioned rather by spasms of vessels. A good instance of this supposed mechanism leading to " recurring attacks of transient aphasia and right hemiplegia" has been recorded by E. O. Daly,[3] of which the following is an abstract :—

CASE XXXIII.—A man, aged 68 years, was suddenly seized, on March 11, at 6 a.m., with aphasia and right hemiplegia. The attack lasted about five minutes and passed off as rapidly as it came on. He had suffered occasionally for years from gout, most frequently affecting his bladder.

[1] *Ziemssen's Cyclopædia* (Translation), vol. xii., p. 673.
[2] " Diseases of the Nervous System," p. 129.
[3] *Brain*, July, 1887, p. 233.

The first sound of his heart was found to be feeble; his arteries were thickened, and his urine was pale, with a well-marked trace of albumin, and a specific gravity of 1010. His temperature was normal and his pulse 80 per minute.

On March 13 he had two attacks of a precisely similar kind to the first, each lasting about half an hour.

On March 14 there were ten attacks, the shortest lasting about ten minutes, the longest almost an hour.

On March 15 there was one attack of ten minutes' duration, and three in succession with about a ten minutes' interval between, each extending over three hours.

After March 15 he remained quite free from an attack of any kind. The treatment consisted of perfect rest, a few grains of calomel, and a mixture containing small doses of iodide of potassium and citrate of lithia.

Dr. Daly says: "I saw this gentleman during several of these attacks. They came on perfectly suddenly without any premonitory symptoms, and passed off as rapidly. He would suddenly say, 'I am all right again.' There was never any kind of convulsion. . . . . When asked to put out his tongue he did so at once, and also closed his eyes and performed intelligently all actions asked of him. The aphasia was not complete, but the words he had at his command were limited and frequently misapplied; and in some of the attacks he was quite unintelligible. The hemiplegia was also partial, more marked in the arm than in the leg. He could move the right hand to a limited extent, and was able to grasp my hand feebly. There was also slight anæsthesia of the right side. . . . . The attacks in every instance ended abruptly, the speech becoming instantaneously perfect. The power in his right side returned almost as rapidly, but after the long attacks a few minutes elapsed after the return of speech before he could use his arm and leg perfectly."

This is certainly a remarkable case, and it seems more explicable on the supposition that the recurring attacks were due to recurring spasms of vessels, perhaps under the influence of gouty and uræmic poisons in the blood, than in any other way. Brissaud refers to another case of recurrent aphasia with right hemiplegia which showed itself in a woman suffering from grave heart failure.[1] At each return of the cardiac asystolie there was a corresponding return of the paralytic phenomena in the right limbs with transitory aphasia.

(*d*) **Narcotic and other Poisons introduced from without.**—Many cases are on record in which poisoning by stramonium, belladonna, cannabis indica, tobacco, and opium have, among other symptoms, given rise to temporary aphasia, and sometimes to amnesia verbalis. References to cases of this kind, and also as to the effects of snake poison, will be found in Bateman's work.[2] Concerning the latter he says: "M. Ruftz stated at a meeting of the Paris Anthropological Society that

[1] *Progrès Médical*, Dec. 30, 1893.
[2] "On Aphasia," 2nd ed., p. 273.

he had seen a certain number of persons who had completely lost their speech in consequence of a bite from a serpent (Fer de lance); sometimes aphasia was produced instantly, and at other times some hours only after the bite; but, what was most remarkable, those who survived the poisoning remained permanently aphasic. Van der Kolk quotes the case of a gunner in the Dutch Indies who was bitten by a serpent called by the natives 'oeloer'; in a few minutes he became giddy and lost the power of swallowing, there was total loss of speech, but consciousness was unimpaired; death occurred four hours and a half after receipt of the injury." Wm. Ogle has also written an interesting paper[1] on this subject, showing that loss of speech often precedes loss of consciousness, and that in cases of recovery it may persist for weeks after other symptoms have disappeared. These facts seem to show that snake poison exercises a specially deleterious action upon the speech centres of the brain.

In addition to vegetable and animal poisons, however, it has been found that one mineral poison at least—namely, lead—may also produce aphasia. Thus, cases have been recorded by Heymann[2] and others showing that among the various nervous symptoms producible in different persons by lead-poisoning aphasia is to be included.

(e) **Poisons engendered within the System.**—Under this head we have to include a very miscellaneous group of cases in which aphasia occurs as a temporary symptom in the course of certain specific fevers, in the puerperal state, in diabetes, Bright's disease, gout, etc.

The specific fevers in association with which aphasia is apt to manifest itself as a temporary symptom are typhoid fever more especially, and more rarely in typhus fever, small-pox, measles, and yellow fever. Numerous instances of this kind have been recorded.[3] It has also been met with in association with intermittent fever.[4]

Transitory aphasia may likewise occur in association with the puerperal state, and more frequently after than before delivery. It is then, in all probability, mostly due to minute thromboses. Instead of being transitory, however, it may be

---

[1] *St. George's Hospital Reports*, 1868, p. 167.

[2] *Berliner Klinische Wochenschrift*, Band ii., 1865, pp. 195, 208, and 223.

[3] See A. Clarus, *Jahrbuch für Kinderheilkunde*, 1874, p. 369; R. Kühn, *Deutsches Archiv für Klinische Medicin*, 1884, p. 56; and J. R. Longuet, *L'Union Médicale*, 1884, p. 717.

[4] M. E. Boisseau, *Gazette Hebdomadaire de Médecine*, 1871, p. 717.

permanent and associated with thrombosis of the middle cerebral artery, a good instance of which, as it appears to me, has been recorded by Bateman.[1]

In diabetes, in Bright's disease, and in gout we also have poisonous or other products accumulating within the body, and in each of these conditions nervous symptoms of various kinds are apt to show themselves in different individuals, among which a transitory aphasia is to be included.[2]

In these different cases of poisoning (whether of extrinsic or of intrinsic origin) the precise mode of production of the aphasia probably varies a good deal. In some cases (1) the altered quality of the blood may favour the occurrence of small or temporary thromboses in the vessels of the cortical centres at fault; in others (2) the poisons circulating in the blood may lead to contractions of the small arterioles, thus temporarily cutting off more or less completely the blood supply of these same cortical centres; while in other cases still it may be due (3) to the direct action of the various poisons upon the nerve elements of the cortical centres affected.

(1) That the first mode of production of the transitory aphasia in these cases is a frequent one there can be little doubt. But such minute thromboses would, strictly speaking, belong to the category of organic causes, and the cases in which they occur are only liable to be confounded or classed with cases of functional disease by reason of the brief duration of the trouble to which they give rise, this brief duration being dependent upon one or other of two conditions; either (*a*) that the thromboses occur only in small vessels, so that the functional effects thus caused can soon be minimised by the opening up of collateral anastomoses; or else (*b*) by the fact that the small thromboses are temporary, soon becoming resolved. These cases in which the thromboses are minute are connected, however, by insensible gradations with others where larger vessels are occluded and remain permanently blocked, so that the disabilities thus caused not only persist, but are of a much graver kind. And in all the previously enumerated causes of blood poisoning of intrinsic origin it is well known that thrombosis of larger cerebral vessels is apt to occur. In this way aphasia may be associated with hemiplegia, or hemiplegia may occur alone, and they may both

---

[1] *Loc. cit.*, p. 131. See also Poupon, *Gazette des Hôpitaux*, March 8, 1866, for an analysis of twelve cases of puerperal aphasia.

[2] See Bernard and Féré, *Archives de Neurologie*, 1882, p. 370, and Dunoyer, *Gazette Médicale de Paris*, 1884, tome i., p. 461.

persist long after convalescence from the original malady has been established.

(2) Cases of transitory aphasia due to spasm of the arterioles would have a more distinct right to come into the category of functional cases, but with what frequency this mode of production occurs we are quite unable to say. Uric acid in the blood is said by Haig[1] to have a tendency to contract the small arterioles, so that this mode of pathogenesis may obtain in some of the cases of transitory aphasia occurring in gouty subjects—the case of frequently recurring aphasia recorded by Daly, and previously quoted, being perhaps a notable example.

(3) Instances of transitory aphasia due to the direct action of poisons upon the nervous centres would have an equal right to be placed in the category of functional cases. We may suppose that the poison for the time being annuls or badly hampers the functional activity of one or more of the cortical speech centres, though what determines the noxious influence of the poisons upon these particular sets of nerve elements rather than upon others must remain as obscure in this as in other cases where general causes produce particular local effects.

In other classes of temporary aphasia, about to be referred to, the actual mode of pathogenesis would seem to be no less obscure, though in some of them it would be commonly said to be due to " exhaustion " of nerve elements following upon epileptiform discharges ; or else, in still other cases, to an even more mysterious process known as " inhibition," to which some are very prone to appeal where explanation is difficult.

(*f*) **Before or after Epileptiform Convulsions.**—It is now a familiar fact that in cases of epileptiform convulsions of the Jacksonian type, whenever the convulsion begins with twitchings about the right side of the mouth (in right-handed persons), or with sensations and twitchings in the tongue, aphasia may precede and follow, for a more or less brief period, the occurrence of each of such fits.[2]

---

[1] *The Lancet*, March 7, 1896.

[2] Occasionally after an epileptic fit an abiding aphasia, without hemiplegia, may be produced, an instance of which has been recorded by Gairdner (*Glasgow Medical Journal*, May, 1866) ; whilst in another case recorded by Llewellyn Thomas (*British Medical Journal*, September 18, 1875), a child 3½ years old became dumb, and also absolutely deaf, after an attack of convulsions.

Delasiauve[1] has recorded the case of an epileptic woman in which aphasia alternated with epilepsy. She would be aphasic for a week, when on the occurrence of a fit of epilepsy the power of speech would return, paralysis of the bladder, however, ensuing. After a time she would again lose her speech, and the same sequence of symptoms would recur.[2]

Sometimes slight temporary loss of speech has been known to recur two or three times as early events in a chain of symptoms which have subsequently developed into those of general paralysis of the insane. I have recently met with two such cases, and subjoin a few details concerning one of them.

CASE XXXIV.—A merchant, aged 31 years, had experienced a great shock and prolonged anxiety in connection with his business affairs four years previously. This was followed by insomnia, which had become much worse during the last eighteen months.

I saw this patient first on February 14, 1894, and was informed that one month previously there had been, without any apparent cause, loss of speech for a time. During this attack he could not get any words out, only making abortive attempts to do so. After about two hours this difficulty ceased rather suddenly, and afterwards he felt even better than he had been before.

Again, five days before he came to me he had another slight attack of the same kind, which lasted for about half an hour. No cause could be assigned for this attack, except that the previous night had been an extra bad one.

On February 24, after a little extra work in his office, he had another attack of loss of speech (though the loss was less absolute than before) which lasted about one hour. Up to this date there was no paresis of the limbs and no mental symptoms, though there was very slight thickness of speech during the intervals between the attacks.

This patient improved greatly for a time in all ways; but early in April he began to have difficulty in writing letters—omitting words, or repeating and spelling them wrongly. By the end of April he became more restless, over-exerting himself, and would not be controlled in this respect. By the middle of May he began to develop delusions of grandeur, which rapidly increased; and on May 26 he was sent to an asylum in a very excited state and full of delusions of all kinds.

It should be borne in mind, likewise, that a temporary aphasia may occur in the course of general paralysis of the insane, as an immediate sequence of one of the so-called "congestive attacks" to which these patients are liable.

In other cases, as will be seen presently, a different form of speechlessness—namely, "hysterical mutism"—is found to follow some attacks of convulsion of an hysterical type.[3]

---

[1] Quoted by Bateman, *loc. cit.*, p. 269.

[2] See also another case recorded by Bateman (*loc. cit.*, p. 198).

[3] Megrim is another functional periodical affection which is occasionally associated with temporary speech defects, though for the most part they are of an amnesic rather than of an aphasic type.

(g) **In association with Insanity, Catalepsy, and Ecstasy.**—
A condition of mutism of more or less duration is not at all
uncommon in patients suffering from chronic dementia or
from melancholia. It is often a lack of will to speak which
leads to taciturnity extending over long periods, though
various other forms of speech defect are also met with among
the insane.[1]

Not unfrequently this persistent mutism is the result of
some hallucination or delusion. Speaking of this class of
cases Séglas[2] says: "Often it is a special hallucination which
is the origin of the patient's mutism. He hears, for example,
an imperative voice which forbids him to speak; and in spite
of all entreaties he keeps silence. In other cases the mutism
is the consequence of a delusional idea, which, moreover,
may vary in character. Sometimes it is an idea of unworthi-
ness, of humility; the patient believes himself fallen from his
position as a man and unworthy to communicate by speech
with his fellow creatures. Sometimes it is an idea of expia-
tion; he keeps silence to expiate the imaginary sins that he
reproaches himself with. In other cases it is the fear of
hurting some one—of compromising, by speaking, some one
that he loves—that makes him keep silence. A patient under
the care of M. Falret, who had shut himself up in absolute
mutism, avowed at intervals that it was for fear of compromising his son by speaking. Sometimes this mutism has its
source in an idea of a hypochondriacal nature; if the patient
does not speak any more it is because he has the idea that he
has no longer a tongue or that his larynx is destroyed."

Many years since an instance came under my own observation in which a semi-demented man had not uttered a single
word for several years; he then had a slight attack of
pleurisy, and, while suffering from the pain of this affection,
he talked pretty freely for several days, answering all questions addressed to him. On another occasion, while suffering
from the pain of toothache, he spoke and answered questions
for a time, but after the extraction of the tooth at once
relapsed into his usual silent and stolid condition. Bateman
says that Brierre de Boismont has recorded "the clinical
history of a lunatic who had not spoken for thirty years, and
who, when perseveringly interrogated, gave a kind of grunt
and ran away. About fifteen days before his death he regained
the use of speech, and answered perfectly well all the questions
put to him."

[1] See Séglas. "Les Troubles du Langage chez les Aliénées," Paris, 1892.
[2] *Loc. cit.*, p. 29.

No mention is made as to the ability to write possessed by the patients above referred to, but in a case cited by Clouston this additional detail is given. This example is instructive also because it illustrates the difficulty of discovering the reason for the silence that exists in many of these cases of what we may call *pseudo-mutism*. Clouston says :[1] " I have a man in the asylum, D. T. K., who for ten years has never spoken a word, but who, I may say, in all other respects behaves sanely, showing no symptom of morbid pride or suspicion. He is about the best joiner we have. We know he has a delusion which prevents him speaking, but what it is we can't find out. If he wants instruction about his work he writes, but nothing will induce him to write why he wont speak."

In reference to catalepsy and ecstasy it is only necessary to say that during the continuance of either of these states the patient remains completely mute, probably from the mere suspension of the will to speak. These cases, in fact, as well as those previously referred to, where the speechlessness is attributable to a lack of will to speak owing to the existence of some determining hallucination or delusion, may well constitute a category apart under the name of "pseudo-mutism."

(*h*) **From Fright and other powerful Emotions.**—Many cases both of temporary aphasia and of mutism have been recorded as resulting from these causes. One cited by Kussmaul[2] is as follows :—

CASE XXXV.—A girl, aged 13 years, was run over by a wagon. Although she escaped with some slight scratches she became speechless, and remained so for thirteen months. After various methods of treatment had been tried without success Dr. Wertner at last gave her bromide of potassium. One day, immediately after taking the medicine, she threw herself into her mother's arms and whispered, " Mother, I will speak again." After a few weeks she had entirely regained the use of language.

It should always be borne in mind that in the case of an accident the importance of the mental shock and its consequences often far transcends, as in this instance, the mere physical injury.

Another case of this type has been recorded by Popham[3] which is worth reproducing here.

CASE XXXVI.—A boy, aged 15 years, received a kick from a cow between the nose and the forehead which stunned him, but left apparently

---

[1] " Clinical Lectures on Mental Diseases," 1883, p. 260.
[2] *Loc. cit.*, p. 799.
[3] Quoted by Bateman, *loc. cit.*, p. 258.

at the time no other injury than a few scratches and slight epistaxis, so that he walked after it some miles to a fair. On the fourth day he was seized whilst at work with vertigo and loss of speech, his hearing, taste, and sight, as well as deglutition, remaining unaffected. A variety of remedies, amongst others mesmerism, were tried, but without any benefit. He continued for twelve months as servant to a medical man, although totally mute, when he got extensive inflammation of the anterior part of the scalp, followed by suppuration, and regained his speech as suddenly as he had lost it eighteen or nineteen months before.

But mere emotion—apart altogether from traumatism—may produce very similar results. Thus Todd[1] refers to the case of a man, aged about 55 years, of an irritable temperament, who in a very animated conversation became excited to such a degree that his power of speech was completely lost. There was no paralysis, and the mental faculties were unaffected. He continued in this condition for about one week, but soon afterwards the power of speech had completely returned.

(*i*) **Reflex irritation from Neuralgia, Intestinal Worms, etc.**—The pathology of these cases is very obscure, but that such associations are at times met with cannot be denied.

Bateman[2] refers to three cases in which temporary loss of speech occurred in association with facial neuralgia, and also to three other cases where the presence of worms in the intestine was associated with loss of speech which disappeared more or less quickly after their expulsion.

Where a temporary aphasia has been associated with a fæcal accumulation in the large intestine, which has disappeared with its expulsion, instances of which have been recorded by R. Jones[3] and by M. Mattei,[4] the cause cannot in all probability be ascribed to mere intestinal irritation. Such cases are rather to be explained as belonging to group (*e*)—"Poisons engendered within the system." They may be supposed to result from the effects of auto-intoxication upon the brain, due to the absorption of alkaloidal poisons from the overloaded intestinal canal.

(*j*) **By Hypnotic Suggestion.**—Charcot and others have been able to produce typical "hysterical mutism" in some hysterical patients who are capable of being hypnotised. Referring to this subject Charcot[5] says: "I ought to tell you how the

[1] "Clinical Lectures on Diseases of the Brain," p. 278.
[2] *Loc. cit.*, p. 270.
[3] *Edinburgh Medical Journal*, 1809, p. 281.
[4] *Gazette des Hôpitaux*, June 15, 1865.
[5] "Diseases of the Nervous System" (Sydenham Society), vol. iii., p. 372.

phenomena of mutism may be artificially produced. The patient being plunged into the somnambulic stage of hypnotism, you commence by conversing with her for a few minutes, then gradually you approach closer and closer to her, and finally pretend neither to hear nor to understand her. She makes further efforts to speak louder, but you continue to practise the same ruse, and appear not to understand any better than before. Then it happens that the voice of the subject becomes progressively lower, and in the last stage aphonia becomes complete, and there is an impossibility of articulation. Artificial mutism obtained during the somnambulic period persists as you see in the waking state. I dare not allow this experiment to be prolonged too much, for I have remarked on many occasions that hysterical symptoms artificially produced during hypnotism are more difficult to be made to disappear in a waking state in proportion as they are allowed to persist for a longer time."

(*k*) **Hysteria.**—The cases of temporary speechlessness that occur in hysteria belong almost solely to the type which I have previously referred to as "*hysterical mutism.*" These cases are now commonly known by that name, though they were formerly included, with others, under the more general head of "functional aphasia."

The name "hysterical mutism" was first given to such cases by Revilliod, of Geneva, in an important memoir[1] dealing with this subject, though both the condition and the name were subsequently made more widely known by Charcot[2] in his lectures, as well as by a collection of twenty cases published by his assistant, M. Cartez.[3] Subsequently another series of cases was recorded by Bock,[4] of Berlin, and a still more important collection of eighty cases by Natier,[5] of Bordeaux.

The subject has also been recently considered by Wyllie,[6] who has thrown further light thereupon, and has lent his support to a view first advanced, I believe, by Liouville and Debove,[7] to the effect that hysterical mutism is only a rarer

[1] *Revue de la Suisse Romande*, 1883.
[2] *Progrès Médical*, 1883.
[3] This paper, as well as Charcot's lecture, is most accessible to English readers in the latter's "Clinical Lectures on Diseases of the Nervous System," vol. iii. (Sydenham Society), 1889. See also Charcot's "Leçons du Mardi," 1888-89, p. 247.
[4] *Deutsche Medicinische Zeitung*, December, 1886.
[5] *Revue de Laryngologie, d'Otologie, et de Rhinologie*, Nos. 4, 5, 8, and 9, 1888.
[6] "The Disorders of Speech," 1895, p. 41.
[7] *Progrès Médical*, Feb. 26, 1876.

and more aggravated form of a condition which is familiar enough as hysterical aphonia. This doctrine was also adopted by Charcot. As Wyllie[1] says, Charcot shows "that there are many features shared in common by hysterical aphonia and hysterical mutism. They occur in the same type of patients, are produced by similar exciting causes, and are not unfrequently found to alternate with each other in the same patient, the aphonia in some cases passing into mutism, and the mutism in others passing into aphonia. Obviously the mutism is the more severe condition of the two. It implies absence even of the power of whispering, as well as that of speaking aloud."

Some of the leading peculiarities of hysterical mutism are these. Its onset is generally sudden, and often after a fright or some strong emotional disturbance. Sometimes it follows an hysterical seizure, either with or without paralysis of limbs. At other times it occurs without assignable cause, or it may be induced, as already stated, in some hypnotised persons by suggestion. The subjects of this disability are completely mute—presenting in this latter respect a notable contrast to ordinary aphasics, who so frequently make use of recurring utterances or articulate sounds of some kind. The intellect seems unimpaired, and they are able freely to express their thoughts by writing.[2] Though the common movements of the lips, tongue, and palate are preserved, these parts (constituting the oral mechanism) are unable to act in the particular combinations needful for speech movements, in association with the other combinations of muscular action pertaining to the vocal mechanism. Some of the muscles of the larynx, however, are often found to be more or less paralysed, though the particular muscles involved have varied in different cases. Sometimes it has been the crico-thyroids alone, or these with the thyro-arytenoids; in others the arytenoids, or even one only of the adductors has been at fault; and Cartez adds[3] that in certain cases "one may see variations of the laryngeal troubles supervene in the course of the malady without any modification in the mutism."[4] There is also often more or less complete anæsthesia of the larynx and of the pharynx.

Sometimes hysterical mutism occurs in association with many hysterical stigmata; at other times it may be almost the sole manifestation of this neurosis.

---

[1] *Loc. cit.*, p. 45.

[2] Taken alone these latter two characters are by no means such important diagnostic features as was supposed by Charcot and Cartez (*loc. cit.*, p. 431), since they are met with also in complete aphemia due to a subcortical lesion.

[3] *Loc. cit.*, p. 431.

[4] This was so in Case xxxvii. quoted below.

Recovery may occur suddenly, perhaps after a fit or a recurrence of some strong emotion; or it may be more gradual, under treatment; and in some of the latter cases there may be a sort of stammering articulation for days, or even weeks, before speech is completely restored.

In illustration of the condition generally, and of the very important fact of the alternation of aphonia and mutism that exists in some patients—which of itself helps to show that the two conditions, if not different degrees of the same affection, are at least very intimately related—the two following cases may be quoted from Cartez,[1] the first of them having been recorded by Thermes,[2] and the second by Liouville and Debove.[3]

CASE XXXVII.—A woman, aged 21 years, on Feb. 15, 1876, after exposure to damp cold—at least, according to the patient's account—was taken with a fit of coughing, and soon after the voice became modified both in quality and intensity. Laryngoscopic examination (by Isambert) did not reveal any organic lesion or inflammatory condition, and the diagnosis was "paralysis of the vocal cords from defective innervation of the muscles of the larynx, and particularly of the crico-thyroids." Consequently, the induced current was advised and applied by Isambert himself. But instead of the usual amelioration, as expected, the aphonia rapidly developed into a mutism. Many varied medicaments were employed, but without effect.

During treatment at the thermo-resinous baths we had the opportunity of examining the patient, and certain objective and subjective symptoms caused us to suspect the case to be one of mutism grafted on to hysteria of a non-convulsive form; or rather, a case of hysteria the manifestation of which had invaded the laryngeal region and particularly the tensor muscles of the vocal cords.

The laryngoscopic examination then made (February, 1877) by Krishaber revealed that the left vocal cord was immobile, and that the free border occupied the median line and divided the glottic space like the perpendicular in an isosceles triangle. The corresponding arytenoid did not perform movements of rotation on its axis. The left vocal cord seemed shorter than the other, because of its laxity and because it was hidden by the arytenoid. The pharynx was slightly hyperæmic.

Hydrotherapy was prescribed. At the first application there was a cry of surprise; the mutism was changed into incomplete aphonia. After a dozen local and general douches, the aphonia gradually disappeared, and a fortnight after the first douche the voice resumed its normal character and intensity.

Under the influence of a fall a convulsive attack occurred, on coming out of which the voice was again lost and the mutism again became complete. This happened in 1877.

In February, 1878, the mutism was still complete. Hydrotherapy, as on the first occasion, produced the same result. The patient cried out and instantly regained her voice, but only for a moment. However, incomplete aphonia succeeded the mutism. She was able to whisper, the words being weak and low, but she could be well understood. Amelioration progressed

---

[1] *Loc. cit.*, pp. 422 and 423.
[2] *La France Médicale*, 1879, p. 290.
[3] *Le Progrès Médical*, Feb. 26, 1876.

to cure, and the patient at the time the case was published had not relapsed for ten months.

This is a particularly interesting case, not only because of the transitions that were shown between aphonia and mutism, but because of the varying condition of the larynx and the relapse after a convulsive attack. The next case is also an interesting and important one.

CASE XXXVIII.—The patient was a girl, aged 18 years, hysterical, but generally of good health. Her mother had attacks of hysteria major. A sister, 13 years old, had frequent attacks, and for two months she had been affected with trembling, which several medical men had regarded as hysterical chorea. The father was very nervous.

Until the last few years the hysteria had only been manifested by incomplete attacks. For eighteen months the patient had been painfully impressed by the quarrels and violent scenes between her father and mother, and it was to this cause that she—probably with reason—attributed her symptoms. About this time, in fact, she became aphonic, not being able to speak above a whisper, and in the course of two months the aphonia grew into mutism. In the house where she lived they named her "the mute." She communicated with those around her by means of a slate, which she habitually carried. The patient came several times into hospital. All those who examined her were agreed in the diagnosis of hysterical paralysis of the vocal cords. Different methods of treatment were adopted without success.

On November 10, 1875, she was taken to the Hôtel Dieu. She had no globus hystericus, no hemianæsthesia, and no affection of the organs of special sense. The ovaries, especially the left, were tender on pressure; but, briefly, apart from the laryngeal troubles and the ovarian pain, the patient presented nothing abnormal.

The laryngeal paralysis was not simply a paralysis of movement, it was also a paralysis of sensation. There was not the least pharyngeal reflex. Laryngoscopic examination by Dr. Moura revealed a paralysis of the vocal cords; these made an almost imperceptible movement when the patient tried to emit a sound.

Pressure over the ovary brought on attacks of dry cough and a few stifled cries. The patient was able to articulate these words in an almost imperceptible voice, "You hurt me." The following day the compression was continued (from five to ten minutes each time), and intonation became more and more distinct; she first ceased to be aphonic, then to be mute. She became able to speak, though in a low voice, hissing out her words.

Two cases may also with advantage be quoted here which were recorded by the late Dr. Ernest Jacob.[1] They will serve as instances of the affection occurring in adult males, as well as of its ready cure under the influence of ether.

CASE XXXIX.—"A man, aged about 50 years, applied for relief at the out-patient department of the Leeds Infirmary, bringing a paper on which was written: 'I have pain in my shoulder; I can hear, but not speak.' The history, obtained subsequently, showed that he had had fits in infancy, but

---

[1] *British Medical Journal*, September 13, 1890.

enjoyed good health till he was 34 years of age, when after monetary losses he began to show signs of mental instability, violent temper, etc., and when agitated was unable to speak. He gradually became perfectly dumb, though he retained for some years the power of saying 'yes' and 'no.' For the last five years he had not spoken, communicating with his wife by means of signs and writing, but understanding perfectly what she said. He seemed quite intelligent, could cough and blow out a candle. He resisted laryngoscopic examination, but I was able to ascertain that the cords could be approximated, and occasionally a hoarse sound uttered. There was no attempt at articulation, though he willingly made attempts when urged to do so.

"The pain in the shoulder was found to be due to dislocation. No account could be given of this, except that six days before he had wakened in the night and found his arm in pain. (In all probability he had had an epileptic fit and fallen out of bed.) Thinking to investigate the aphasic symptoms later, I asked my surgical colleague, Mr. W. H. Brown, to reduce the luxation. To facilitate this ether was administered, and on recovery from the narcotism he began to speak clearly and volubly, evidently much pleased at recovering his power of speech, and determined to make up for lost time.

"Some weeks after, when I saw him again, he was speaking well. I wish to point out here that there was no wilful dumbness, such as is not uncommon in cases of insanity. The man was greatly pleased at recovering his power of speech; nor was there the entire loss of phonation which obtains in functional aphonia, though there was great want of tension of the vocal cords. The patient was not morose or melancholic, though occasionally giving way to passion."

This case is remarkable on account of the rapid and complete recovery of speech, although the dumbness had existed for five years; that is, there was none of the stammering or hissing articulation that sometimes exists for a time when hysterical mutes first begin to regain their speech. In the next case the mutism had been of very brief duration.

CASE XL.—"A healthy-looking miner, aged 34 years, applied on January 13, 1890, suffering from slight hoarseness, and his cords were injected. He gave a history of having had some epileptic fits eight years ago (six in all), and at one time he had been intemperate. Some zinc chloride was applied to his larynx and a pine-oil inhalation prescribed. A week later he appeared with a paper on which was written 'Five days ago I lost my voice. I feel as well as ever I did, but I have not been able to talk, and I cough.'

"Inquiry showed that he had been quietly talking with some friends on no very exciting topic, feeling rather hoarse, when he was suddenly seized with dumbness. His larynx was now rather difficult to examine, but the cords could be approximated without tension and an occasional hoarse grunt elicited. After a few trials he managed to utter one or two vowel sounds. Remembering the effect of the anæsthetic on the former case I had ether administered. After he had taken one or two breaths the inhaler was removed and he articulated 'Yes,' in answer to a question, and after recovery from narcotism could talk freely. He remained an out-patient for a few weeks, but there was no return of the dumbness."

Here the mutism was not absolute, and being only of brief duration the principal interest of the case lies in the prompt restoration of speech after the inhalation of ether.

In the instances cited examples have been given of the recurrence of mutism on two or three occasions. At times, however, the attacks recur again and again, and the duration of each attack is then often very brief. Thus Delasiauve[1] has related the case of a woman who for three years was at each menstrual period affected with mutism and partial paraplegia, being at these times only able to make herself understood by signs. Bateman[2] records an interesting case of his own occurring in a young man who in the course of six months had nine attacks of mutism, most of which lasted only a few days, though one of them remained for over six weeks. Each attack began with sudden pain in the occiput and nape of the neck of an extremely severe character, and the pain was associated with contraction of the muscles at the back of the neck. A still more remarkable case has been recorded by Hun,[3] in which frequently recurring attacks of mutism were generally associated with deafness or blindness, one or both; these disabilities often ceasing after an attack of hysterical convulsions. The following are the most important details of this extraordinary case.

CASE XLI.—A livery-stable keeper, married, aged 43. In December, 1863, while at work in his stable about 4 p.m., he felt some nausea; he walked home and during the evening lost his speech; an hour later he lost his sight and hearing, but retained his consciousness; at the end of another hour he had several severe convulsions. After the convulsions he was perfectly conscious, but did not regain his sight, hearing, or speech for twenty-four hours.

Six months after he had another attack, which was not preceded by nausea; he lost his speech, sight, and hearing at 6 p.m., at nine he commenced to have convulsions, which lasted until 6 a.m. There was no unconsciousness either during or after the fits. His sight returned the next day, but he remained speechless and deaf for forty-eight hours. Three weeks later he had a similar attack.

He now continued well till April 20, 1865, when he was startled and excited by the discharge of a cannon, causing the horse he was driving to become difficult to manage. Soon after this he had a headache, and in a few hours became speechless, blind, and deaf, but retained his consciousness. He wrote upon a slate what he wished to communicate; and when his hand, holding the pencil, was guided by another person, he understood what was written;[4] he could open his mouth and move his tongue freely, but swallowed with some difficulty; he made signs as if he felt some obstruction in his throat; the pulse was slow and regular. In the course of the night he had an hysterical convulsion with opisthotonos, which lasted about five minutes, when the rigidity suddenly relaxed. At the end of twelve hours his speech,

[1] *Journal de Médecine Mentale*, 1865.
[2] *Loc. cit.*, p. 186.
[3] *The Quarterly Journal of Psychological Medicine*, January, 1868, p. 119.
[4] That is, the other person communicated the writing movements to his hand, and he, by means of kinæsthetic impressions, was able to read what he had been made to write.

sight, and hearing suddenly returned, and he regained his usual good health.
   In October, 1865, he had another attack; this time he lost his speech and hearing, but retained his sight. He went about town, communicating with those he met by means of his slate and pencil. This continued for several days, when, after some excitement, he lost his sight, and had several severe convulsions of the kind above mentioned. After the convulsions he regained his speech, sight, and hearing, and felt perfectly well.
   On March 14, 1867, at 5 p.m., he complained of headache and nausea, and an hour later he was seized with convulsions, during which he lost his speech, sight, and hearing. One of these convulsions was seen by Dr. Hun; it was of the type before mentioned, and when it was over he was found to be hemianæsthetic on the left side. The next morning he was able to see, but was unable to speak or hear. A week afterwards Dr. Hun met him walking about the streets, communicating with others by means of his slate and pencil, but he could still neither speak nor hear. On March 26, he had four severe convulsions after which he still remained speechless and deaf until June 6, when, after severe headache, he recovered his voice for fifteen or twenty minutes, but could not hear.
   On June 8, while in New York upon business, early in the day he was attacked with headache and with twitching of the muscles, but no regular convulsions; in the evening he began to talk, and half an hour afterwards his hearing was restored, but at the same time he became blind. The next morning when he awoke he felt perfectly well, and was able to speak, see, and hear.
   He now remained well till November 16, when he had a dull headache during the morning, which increased in severity till 6 p.m.; he then had convulsions with loss of sight, speech, and hearing; at 11 o'clock the convulsions ceased, and his sight returned, but he remained deaf and speechless till November 29, when he had headache, and a slight twitching of the muscles during the morning; at 6 p.m. he had convulsions lasting two hours, after which his speech and hearing again returned. After that time he continued perfectly well and was able to attend to his business as usual.

This case is remarkable from the fact of the recurring attacks of mutism being associated with blindness and deafness, and also from the fact that the restoration of the lost powers usually followed more or less immediately upon the occurrence of an attack of convulsions. In Cases vi. and xvi. recorded by Cartez there was also some association of deafness with mutism; while in Case xi. there were alternating attacks of mutism and blindness.

During the present year I have also had under my care a patient who had suffered from an extraordinary number of attacks of mutism extending over a period of more than eighteen months. They occurred in a sailor who had not hitherto suffered from nervous symptoms and had previously led an active life in all parts of the world. He was sent to me by Dr. Allan Mahood of Appledore, and the following is an abstract of notes received from him, and of others taken whilst the patient was under my care at the National Hospital for the Paralysed and Epileptic.

CASE XLII.—The patient, a sailor aged 31 years, had for a time been out of work, worried, and having insufficient food, when on July 13, 1895,.

he had in the evening a sudden attack of "shaking," characterised by small, rapidly repeated contractions of the muscles of the face, body, and limbs, lasting some hours, but without loss of consciousness. Under treatment and with rest in bed he improved much during the next week, but on July 20 a political procession went past his house, cheering wildly, after an election. This awoke him and gave him a great fright, resulting in complete loss of speech for one hour, together with much twitching of the muscles on the right side of the face.

During the next month he had very many brief attacks of complete mutism. He was rarely many days without one, and sometimes had two or three on the same day.

On August 13 irregular spasmodic contractions of the limbs occurred without loss of consciousness, lasting off and on for nine hours.

On September 4 there were violent convulsive movements of the shoulders, arms, and upper half of the trunk for one and a half hours.

During the next six months he had similar attacks lasting from ten to fifteen minutes, several times daily. The "shaking" was mostly on the right side of the body. He generally lost his speech during the attacks—though on several occasions the mutism lasted for a week at a time, when he communicated with others only by writing and signs. Sometimes he has lost his speech without having an attack of "shaking"—especially under the influence of excitement from any cause. He had a good deal of worry during this time.

On May 8, 1896, he seemed quite well. There had been no return of the old symptoms during the last month. He went, therefore, for a month's drill with a corps to which he was attached, and after that was about to resume his sea-life, when suddenly his old symptoms returned, and he continued to have "shaking" fits and frequently recurring attacks of mutism up to the date of his entry, under my care, into the National Hospital for the Paralysed and Epileptic on January 13, 1897. In the early days of this latter month he had the most violent attack of all. He had to be restrained for four hours. He did not hurt himself, but bit his brother severely in the face. Three days before admission he lost his speech entirely on being told he was coming to the hospital, and did not regain it till twenty-four hours after admission.

On admission he was found to be completely mute, though he made a few inarticulate noises on attempting to speak. He was perfectly sensible in all respects—could understand everything said to him quite readily and write his answers on a slate, as well as understand quite well what was written by others. There was no motor or sensory paralysis. His gait was normal. There was some facial overaction on attempting to speak. The eye movements, pupils, and optic discs were normal. There was no contraction of the visual fields.

During the first six weeks of his stay in the hospital he was treated with tonics and tepid needle baths; then with sulphur baths together with bromides and other drugs internally. In this period he had only one slight "shaking" attack two days after admission, and nothing notably wrong, apart from some slight tremors about the hands and face at times, was noticed, except that he continued to have frequent attacks of mutism lasting for some hours or even a day or two.

I then left instructions that during the next attack he should be put under the influence of ether, and on March 2 I received a report from the resident medical officer (Dr. W. J. Harris) announcing that on the previous afternoon the patient, after some slight difference with another patient, had again suddenly lost his speech. When seen about half an hour afterwards he could not make any sound, not even "Ah." His larynx was examined again and found to be, as before, very hesitating in adduction, with never complete closure of the rima glottidis. The administration of ether was then commenced. "He took it quietly for about three minutes, until just beginning to feel its effects, when he suddenly bounded off the bed, upset everything around, fought wildly, and was with difficulty subdued—uttering no sound, however. He then lay quiet till let go, when he suddenly attacked us again, and when again subdued he began to talk quite well, declaring he could not help it, etc."

Though he remained in the hospital subsequently for over five weeks he had no further attack of mutism and only one threatening, after some slight annoyance. On April 11 he discharged himself, against advice, on the plea that another patient was noisy at night.

The Pathogenesis of Hysterical Mutism.

We must now look a little more closely into the pathogenesis of mutism.

From the point of view of the physiology of the two conditions aphonia and mutism, Charcot attempted to draw too sharp a distinction. He adopted the doctrine that has been taught by Marey and other physiologists, that whispered speech is the product of the oral division of that mechanism alone, and that the larynx takes no part in its production. According to Charcot, therefore, aphonia (in which the power of whispering is preserved) is a result of a partial paralysis of the adductor muscles of the larynx; while as to hysterical mutism, he writes, "If the individual suffering from the affection is unable to whisper, it is not because he is aphonic, or rather because his larynx does not vibrate; it is not because he has lost the common movements of tongue and lips—you have seen that this patient was perfectly able to blow and to whistle; it is because he lacks the ability to execute the proper specialised movements necessary for the articulation of words. In other terms, he is deprived of the motor representations necessary for the calling into play of articulate speech."

Wyllie, however, contends (and he is supported by the opinion of Michael Foster) that the larynx does take part in whispering, and that there can be no speech without the coöperation of both the oral and the laryngeal mechanisms. He consequently holds that "the disablement of either the one mechanism or the other may produce mutism." In opposition to Charcot and Natier, therefore, who believe the oral division of the speech mechanism only to be at fault in hysterical mutism, Wyllie maintains that whilst this may be so in some cases, in a second group it is the laryngeal division of the speech mechanism which is at fault; and in a third set of cases both oral and laryngeal mechanisms are simultaneously disabled.[1] He refers[2] to various cases in Natier's list in support of his opinion.

I formerly thought that these cases of mutism were instances of complete aphemia, dependent upon a functional defect in the outgoing fibres just below the left third frontal gyrus,[3] because cases of sub-cortical defect, where there was no damage to either of the word centres, at that time seemed to me alone capable of producing the particular combination of symptoms met with in hysterical mutism. And, as in hysteria the afferent fibres in the "internal capsule" appear to be so frequently affected in the production of hemianæsthesia, there seemed no valid reason why the same kind of disability (whatever its nature) might not at times affect some of the bundles of fibres proceeding from, rather than going to, the cortex. Though this was the hypothesis formerly advanced by me, I always recognised that it would have been more satisfactory if one could have supposed that in hysterical mutism portions of the cortex itself were at fault. The bar to this explanation, however, was the undiminished ability to write which these patients displayed and the supposition that an affection of the glosso-kinæsthetic centre would cause some mental impairment of which there was no trace in hysterical mutism.

An examination of the records that have been published of actual cases of sub-cortical lesions, and the finding that in some of them where there had been destruction of the first part of the geniculate fasciculus (Fig. 14) a mutism was produced almost exactly similar to that met with in the hysterical cases, seemed to be a further justification of my previous view. Our Cases xi. and xii. are examples of this type.

On the other hand, a remarkable case recorded by Guido Banti has since convinced me that the symptoms may not be different, even when an organic lesion exists in the posterior part of the left third frontal convolution itself, so long as it is absolutely limited thereto, and leaves the cheiro-kinæsthetic centre uninjured. This very important case, which I have previously quoted,[4] shows that in some individuals (besides the preservation of ability to write) the destruction of Broca's region alone may not interfere with the general intelligence any more than does a lesion in the bulbar speech centres or

---

[1] His views as to the cortical localisation of these mechanisms will be subsequently considered.
[2] *Loc. cit.*, p. 487.
[3] See *Brit. Med. Jour.*, Nov. 5, 1887.
[4] See p. 89, Case xxii.

in the internuncial fibres between them and the third frontal gyrus. But if such may be the case with regard to organic lesions in Broca's region we may all the more readily admit that a similar absence of agraphia, and a similar preservation of intelligence may result from a functional disability affecting Broca's region. We have seen, in fact, that the supposed existence of such a disability was Charcot's explanation of the condition. He considered hysterical mutism to be an instance of pure "motor aphasia," resulting from a functional trouble in Broca's region.

This, however, is by no means an adequate explanation of the condition. As we have seen, Charcot too sharply distinguishes hysterical mutism from aphonia, and believes mutism to depend only upon a disorder involving the cortical centre for the oral division of the speech mechanism. The probabilities, however, are against this view, as Wyllie has shown. Aphonia and mutism are most intimately related, differing in degree only; and the oral and vocal speech mechanisms are probably concerned in all speech, whether it be sonorous or whispered. Still we must suppose that the cortical speech centre in Broca's region, which I name the glosso-kinæsthetic, is, in reality, composed of two parts most intimately related, both functionally and structurally—one of them being the centre for the oral mechanism, and the other the centre for the vocal speech mechanism.

At first it was thought, from the experiments of Semon and Horsley[1] upon monkeys, that the adductors of the larynx were alone represented in the cortex, and that in the anterior part of the foot of the ascending frontal; but Risien Russell[2] has since ascertained that the abductors are also represented in a region slightly removed therefrom. In answer to a query of mine, Russell writes:—"In the dog and cat the abductor centre is situated in front of, and a little above, the level of the adductor focus, so that probably the same holds good for man; but I do not wish to commit myself on this point till I have had more opportunities of ascertaining the relative position of these two centres in monkeys." He, however, confirms the statement of Semon and Horsley that there is a complete bilateral representation of the laryngeal movements in the cortex, so that destruction of the centre on one side will not produce paralysis of the muscles of the opposite side of the larynx, seeing that they can still be called into activity by the uninjured centre in the opposite hemisphere.

[1] *Philosophical Transactions*, vol. clxxxi., 1890, p. 187.
[2] *Proceedings of the Royal Society*, vol. lviii., 1895, p. 237.

If a similar bilateral representation of the laryngeal movements concerned with speech exists in man[1] it would follow that to account for hysterical aphonia, characterised by paresis of laryngeal adductors, we must suppose that the corresponding centres of both hemispheres have been simultaneously affected, and somewhat degraded in functional activity.

As I have already pointed out, however, it has been shown that hysterical mutism is intimately related to aphonia; and therefore mutism might be supposed, even for this reason alone, to be also due to a bilateral affection of the cortex. But we shall now see that there is another reason pointing in the same direction. Charcot's explanation of hysterical mutism does not account in any way for the fundamental difference that exists between it and ordinary simple aphasia. The subject of the former affection is, as we have seen, absolutely mute, and has no "recurring utterances" of any kind. It is commonly supposed that in ordinary aphasia, due to a lesion in Broca's convolution, the recurring utterances are brought about through the instrumentality of the right hemisphere. The same thing would, therefore, doubtless occur in the functional affection if the left third frontal convolution were alone involved, but as instead there is absolute dumbness, we are, I think, on this account bound to assume that in this disorder, as in aphonia, there must be a simultaneous affection of the posterior part of the right as well as of the left third frontal convolution.

Charcot does not attempt to explain the above-mentioned important difference between the two affections. As I have already indicated, he regards hysterical mutism as due to a functional defect in a part of Broca's centre in the left hemisphere—namely, the part having to do with the oral mechanism for speech. The only additional remark that he makes on this subject occurs towards the close of his lecture, where he says: "It is in the grey cortex of the cerebral hemisphere that we must seek for the dynamic lesion whence emanate the symptoms in question, and the mechanism that is to be invoked in such conditions is none other than that which acts in the production of psychical or, if you like it better, mental paralysis." This is rather vague, and introduces no further explanation of the pathogenesis of the condition. Will, except in regard to speech, does not seem

---

[1] This cannot, however, with certainty be considered to follow from the fact that experiments upon lower animals show that common movements of the larynx are bilaterally represented.

to be defective or hindered in its manifestation, and we have evidence also that in these patients the other word centres—namely, the auditory, the visual, and the cheiro-kinæsthetic, are in a condition of unimpaired activity.

Thus the clinical differences between simple aphasia and hysterical mutism force us to the same kind of conclusion as to the pathogenesis of the latter affection, as that to which we are also led by a consideration of the intimate relations existing between it and hysterical aphonia. It impels us to believe in the existence of a bilateral cortical disability in Broca's region in each of these affections, though varying in intensity in the two cases. This explanation, I believe, has not hitherto been adduced for either of these affections.

It will be observed that hitherto I have neither adopted nor definitely referred to the hypothesis of Wyllie as to the existence of certain subdivisions of what he calls Broca's convolution. I say "what he calls Broca's convolution," because instead of limiting this region, as is customary, to the foot of the third frontal convolution, Wyllie seeks to include under it also the foot of the ascending frontal and the foot of the ascending parietal convolutions.[1] He suggests that in these two latter sites the "executive cortical motor mechanisms" concerned with speech are to be found—comprising a "centre for phonation" and a "centre for the oral articulative mechanism;" while in the foot of the third frontal are stored up the "guiding psycho-motor images for spoken speech," or, in other words, what I have termed "glosso-kinæsthetic impressions." These views have been accepted and still further elaborated by Elder,[2] who also reports one case which he thinks "goes a long way to support the hypothesis of Wyllie." These doctrines are also being accepted by others, as I find from a paper by B. Onuf[3] entitled "A Study in Aphasia," in which, in my opinion, various erroneous views are supported on the authority of others, though the paper is otherwise able and interesting.

Now in regard to the impressions registered in the foot of the third frontal convolution, it would appear, as I have said, that Wyllie's "guiding psycho-motor images" correspond with what I term kinæsthetic impressions, and further that these impressions resulting from speech-movements must partly correspond with impressions from the oral and partly

---

[1] "The Disorders of Speech," Edinburgh, 1894, pp. 300-302, and 309.
[2] "Aphasia and the Cerebral Speech Mechanism," London, 1897, p. 31.
[3] *The Journal of Nervous and Mental Disease*, 1897, p. 94.

from the vocal subdivision of the speech mechanisms. These, according to my view, are the guiding sense impressions which (in conjunction with auditory word-images) act directly upon, and are capable of evoking the proper activity of, the bulbar motor centres. I therefore do not believe even in the existence of the "cortical executive motor mechanisms" for speech which Wyllie and Elder assume to exist at the bases of the ascending frontal and ascending parietal convolutions; and further, I consider that the reasoning which induced Wyllie to suppose that such centres might be thus situated is of itself open to very grave objections.

He seems to have been guided altogether too much by the results obtained by experimental physiologists (Ferrier, Semon, Horsley, and others) upon monkeys and dogs, and not to have kept distinctly in mind that the movements of the larynx resulting from stimulation of the cortex in these dumb animals could not have been such movements as were concerned with speech, though they might have been the reproductions of much less specialised movements of the vocal cords (adduction and abduction) associated with respiratory acts; and, again, that the movements of the lips and tongue similarly induced in these dumb animals could not afford any evidence as to the site of the "oral articulative mechanism," although they would serve to determine the seat of registration in such animals of the impressions resulting from the common movements of these parts concerned with biting, mastication, etc.

The very fact, however, that such common movements are registered in these sites (if it may be considered to hold good for the human subject) would of itself lead us to look elsewhere for the registration in man of the multitudinous impressions derived from the highly specialised movements concerned in speech—for instance, they might be found in the contiguous foot of the third frontal convolution. And, as Wyllie himself admits[1] (though in a rather hesitating way), the clinical evidence points to the importance of the foot of the third frontal convolution, and not to the bases of the ascending frontal and parietal convolutions, as the region most concerned with the production of speech. It ought to be almost superfluous to enforce the point that it is upon clinical evidence alone that we must rely for the localisation of speech centres, and not upon the "conclusions of the experimental physiologists" drawn from experiments upon dumb animals.

[1] *Loc. cit.*, p. 301.

An attentive study of the case above referred to that has been recorded by Elder, seems to me to show that the lesion was an essentially sub-cortical one, and that all the defects in articulation met with could have been produced by damage to some of the pyramidal fibres proceeding from the foot of the third frontal convolution, even though the lesion also destroyed portions of the bases of the ascending frontal and parietal convolutions. Then, again, the patient died on the sixth day from the onset of his illness, when the effects of mere shock or functional derangement caused by the lesion may not have quite disappeared. It is very unsafe to build new doctrines upon cases which live so short a time.

No real evidence is therefore forthcoming in favour of the hypothesis of Wyllie; and that the distinction which I have dwelt upon between the common movements of the lips, tongue, and palate, and the common movements of the vocal cords, as contrasted with the highly specialised combinations of movements of these parts concerned in speech, is a perfectly valid and important one is shown by the fact that in hysterical mutism, though the latter movements are impossible, the common movements of the lips, tongue, and palate are quite unaffected—just as in aphonia the vocal cords can be perfectly approximated during the act of coughing, though they cannot be brought together as a component of the movements necessary for speech.

CHAPTER VIII.

STRUCTURAL DISEASE IN THE AUDITORY AND IN THE VISUAL WORD CENTRES.

SINCE the publication in the year 1874 of an important memoir by Wernicke entitled "Der Aphasische Symptomencomplex," it has been the fashion to speak of the defects of speech due to lesions in the auditory and the visual word centres as "sensory aphasia," in contra-distinction to that produced by damage to Broca's region, which, in accordance with then prevalent notions, was and has since been very commonly spoken of as "motor aphasia." This mode of distinguishing these defects, though it has a certain convenience and has been widely adopted, is not in accordance with my views, as I hold the latter to be as much a sensory region of the brain as the former.

It would, I believe, be much better if the term "aphasia" were restricted to the defects of speech produced by lesions in Broca's region, and the term "aphemia" to those dependent upon sub-cortical lesions in the course of the pyramidal fibres, leaving the speech defects produced by lesions of the convolutions around the posterior extremity of the Sylvian fissure to be grouped as so many varieties of "amnesia." Under this latter generic name would be included many forms of speech defect due to defective recall of the auditory and the visual images of words, and produced either by lesions of the auditory and the visual word centres themselves or of the commissures by which they are united to one another and to corresponding centres in the opposite hemisphere.

In all these cases there would be more or less interference with the recall of auditory and visual images of words. And whether we call the case one of sensory aphasia or of amnesia, in each instance alike the precise degree and nature of the defect or defects would have to be settled by a systematic examination, so as to determine whether we had to do with mere diminished recollection of words, with complete loss of their auditory or visual images, or with other combinations of symptoms pointing, either to partial isolation of these centres from one another, or to their isolation from the general auditory or visual word centres of which they form part.

It is true that such a nomenclature involves a slight inconsistency, seeing that aphasia and agraphia are also, in accordance with my views, forms of amnesia, due to the non-revival of glosso-kinæsthetic and cheiro-kinæsthetic images respectively. But these kinæsthetic images, as I maintain, play only a small part in thinking processes, and neither of them is subject to independent conscious recall like the auditory and visual images of words which constitute our habitual thought-counters. The inconsistency is, moreover, much less than that which is entailed by speaking of " motor aphasia " and " sensory aphasia," as though the former belonged to a radically different category, and really depended upon the lesion of a motor centre.

The advantage would be great of confining the term "aphasia " to its original signification, and not including under it various types of speech defects which are radically different in nature and produced by lesions in totally different cerebral regions; this is especially desirable when the objectionable generic term " sensory aphasia " can be replaced by another having a very similar general connotation.

It has been commonly supposed, and repeated over and over again by workers in Germany and France, as well as by those writing in our own language both here and in America, that Wernicke was the first to call attention to and explain that form of speech defect that is now known as " word-deafness "—this particular name having been given to the condition shortly afterwards by Kussmaul. It is admitted that the companion defect of the visual centre, now known as " word-blindness," had been previously recognised by several writers, though it had not been clearly explained;[1] but "word-deafness" was supposed to have escaped observation by writers on speech defects, the individuals suffering from it having been previously thought to be insane, or at least demented.[2]

The merit of having determined the region of the brain at fault in cases of word-deafness (viz., the hinder half of the upper temporal convolution, with perhaps a portion of the hinder extremity of the middle temporal convolution) clearly

[1] On this subject see Bernard, "De l'Aphasie," pp. 74 and 145.
[2] This statement has been repeated by Wyllie in his recently published work on the "Disorders of Speech" (p. 295), although de Watteville had previously pointed out in *Brain* (1885, p. 267) that I had, five years before the appearance of Wernicke's memoir, fully appreciated the nature of the defect as well as that of "word-blindness."

belongs to Wernicke; though, as I shall subsequently endeavour to show, he was far from correct in saying that the complex of symptoms resulting from such a lesion was word-deafness, paraphasia, alexia, and agraphia. This view as to the symptomatology of the lesion was founded upon too narrow a basis of observed cases with necropsies.[1] A wider experience and knowledge now necessitates its complete revision.

That it is not correct, however, to say that word-deafness and word-blindness were not well understood till Wernicke defined their nature, may be shown by the following quotation[2] from my paper "On the Various Forms of Loss of Speech in Cerebral Disease," published in 1869 :—

"Most aphasic patients can *understand perfectly what is said* to them and can follow and feel interested when they hear others read aloud. In these cases we may presume that the afferent fibres connecting the auditory centres of the medulla with the auditory perceptive centres of the cerebral hemispheres, and also these latter centres themselves, are intact, so that the spoken sounds revive their accustomed impressions in the hemispheres, these being perceived as words symbolic of things or ideas, which, being duly appreciated by the individual as they are conjured up, suggest to him the thoughts which they are intended to convey. In certain of the severe cases of aphasia, however, as in that noted by Dr. Bazire, recorded at p. 16, and in Dr. Gairdner's case (*Glasgow Medical Journal*, May, 1866, p. 13, and *Transactions of the Philosophical Society of Glasgow*, 1866), it is distinctly stated that the patient either did not gather at all, or with difficulty and imperfectly, the import of words when he was spoken to, though he could be made to understand with the utmost readiness, by means of signs and gestures. Must we not suppose in such a condition either the communication of the afferent fibres with the auditory perceptive centres is cut off, or that this centre itself, in which the sounds of words are habitually discriminated and associated with the things to which they refer, is more or less injured? In either of these cases, though the sound is not appreciated as a *word* having its definite meaning, we must not expect that there would be deafness; the sound would be still heard as a *mere sound*, only it does not call up that superadded intellectual discrimination by the ingrafting of which upon it, it can alone be made to serve as a symbol of thought. Hence the individual does not adequately comprehend when spoken to, though he may be quite capable of receiving and appreciating fully the import of signs and gestures which make their impression upon his visual perceptive centres. . . . . And where the individual *cannot read* I am inclined to think this must be owing either to some lesion of the afferent fibres to the visual perceptive centre, of the visual perceptive centre itself, or of the communications

---

[1] It appears that Seppilli looking into this subject ten years later (*Revista Sperimentali*, 1884), collected seventeen cases of "Sensory Aphasia," though he discarded all but two of the ten cases collected by Wernicke as not being free from objection. In the following year Amidon reproduced (*New York Medical Journal*, 1885) these cases of Seppilli, and added four new cases. I shall subsequently refer to Amidon's list as being more accessible and convenient for reference by English readers than the work of Seppilli.

[2] *British and Foreign Med.-Chir. Review*, April, 1869, pp. 482 and 484.

between the cells of this centre and those of the auditory perceptive centre. If lesions existed in either of the first two situations the visual impression could not receive its intellectual elaboration, and consequently it could not call up its associated sound (word) in the auditory centres, and hence no meaning would be conveyed by the hieroglyphic marks of the printed or written pages. They would be to the person mere meaningless strokes, just as we have assumed that if similar defects existed in the auditory perceptive centres, or in the afferent fibres with which they were connected, the individual could not appreciate the meaning of spoken words—these would be to him mere sounds."

At the date when this paper was written I knew nothing about what I now term "kinæsthetic centres" and had not fully realised the importance of kinæsthetic functions. At this period, moreover, the notion that definite "motor centres" exist in the cerebral cortex had not been mooted. This explains the reason of the absence in the above quotation of any reference to the needed activity of kinæsthetic centres.

Turning now to the question of the symptomatology of the condition named "sensory aphasia" by Wernicke, and referred by him to a lesion of the posterior extremities of the upper and middle temporal convolutions, it should be observed that Kussmaul shortly afterwards broke up into two sets of symptoms what were originally described as a single group by Wernicke. According to Kussmaul, what he first termed "word-deafness" is the primary and essential result of destruction of the hinder extremities of the upper temporal convolutions; while what he termed "word-blindness" holds a similar essential relationship to destruction of the angular and parts of the supra-marginal gyri. This is the view now commonly held, and both varieties are spoken of as forms of "sensory aphasia," though Wernicke's original doctrine as to the very complex results of a lesion in the auditory word centre is held at the present day by Déjerine as well as by Mirallié. They both speak of word-deafness, paraphasia, alexia, and agraphia as the results of such a lesion.

It must be admitted that the functional relations of the auditory and the visual word centres are so intimate, and their sites so close to one another, that a lesion occupying the one is apt more or less to interfere with the functions of the other for a longer or shorter period, which may vary with the nature and abruptness of the lesion, as well as with the different endowments of individuals. Disturbance to a very marked extent of the functions of the visual word centre as a result of a lesion in the upper temporal convolutions is, however, by no means so universal as Wernicke's and Déjerine's statements as to the symptomatology of this lesion would imply. Thus, out of sixteen recorded cases of "sensory aphasia" in which the lesion was pretty closely limited to the hinder part of the first, and more or less of the second temporal convolutions, in only five of them is there any mention of the existence of some amount of word-blindness. Doubtless, had the patients been studied more minutely (after the manner practised by Thomas and Roux [1]), some minor

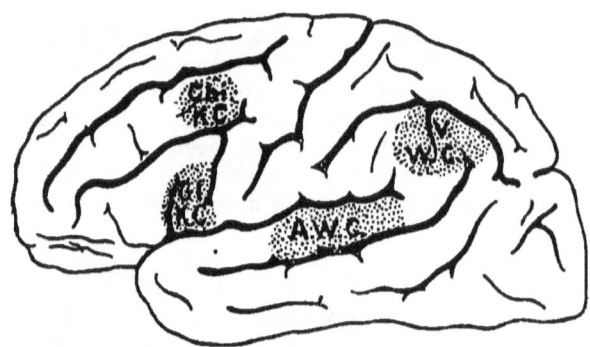

Fig. 6.—Diagram showing the approximate sites of the four Word Centres in the left Cerebral Hemisphere.

amounts of word-blindness might have been found in others; but that is not what Wernicke meant, and is not other than what might be expected if we look to the close functional and topographical relations of the two word centres.[2]

It will subsequently be shown, moreover, that paraphasia is met with in only a little more than one-third of these cases,

[1] *Bulletin de la Société de Biologie*, 22 Février, 1896.

[2] The cases to which I refer where the lesion has appeared to be pretty completely limited to the hinder part of the first or of the first and second temporal convolutions are those which in Mirallié's list ("De l'Aphasie Sensorielle," 1896, p. 135) are numbered, 1, 3, 4, 6, 7, 10, 12, 14, 19, 21, 25, and 32; together with those in Amidon's list (*New York Medical Journal*, 1885, pp. 113 and 181) numbered, 7, 14, 17 and 18. Of these, the five cases in which some amount of word-blindness co-existed are those which in Mirallié's list are numbered 12, 14, 19, 21 and 25. In Amidon's No. 11, moreover, there was the co-existence of aphasia and word-deafness, owing to simultaneous lesions in Broca's region and in the upper temporal convolutions; nevertheless, the patient showed neither alexia nor agraphia. I have purposely alluded to these two lists, which include, I believe, almost all the complete cases with necropsies of "sensory aphasia" hitherto published (many of them being included in both lists), because they are both fairly accessible, and a reference to them will thus enable any one interested in the matter to verify the correctness of my statements and criticisms with comparative facility.

so that Wernicke's "symptomen-complex" does not prove to be at all in accordance with the data at present available, although his views are still adhered to by Déjerine and Mirallié.

It must, indeed, be admitted that the defects in the auditory and the visual word centres giving rise to word-deafness and word-blindness respectively may occur separately or together, and in the latter case the defective action in the two centres may be unequally developed. Hence the very marked clinical variations that are met with as a result of lesions in the convolutions surrounding the posterior extremity of the Sylvian fissure.

Another cause of clinical variations, even with similar lesions, is perhaps to be found in the different degrees of education of the persons attacked, and their consequent greater or less facility in reading and writing. Still another cause of variability is to be found in the varying individual endowments of patients in regard to the relative activity of their different word centres—in other words, according as the patients are marked "auditives" or "visuals," respectively. Connate individual variations of this kind may give rise to notable clinical differences, even as results of similar lesions.

All that can be done here, therefore, is to point out the most common combination of symptoms, dealing with the various parts of the subject in the following order: (1) defects resulting from abnormal conditions of the left auditory word centre; (2) defects resulting from destruction of the auditory word centres in each hemisphere; (3) defects resulting from destruction of the auditory and the visual word centres of each hemisphere; (4) defects resulting from isolation of the left auditory word centre; (5) defects resulting from abnormal conditions of the left visual word centre; (6) defects resulting from isolation of the left visual word centre; and (7) defects resulting from combined lesions of the left auditory and visual word centres, together with some remarks on the condition which has been named "psychical blindness" or "object blindness."

1. Defects resulting from Abnormal Conditions of the Left Auditory Word Centre.

Of all the word centres the integrity of the auditory is of the most importance, and the defects due to different degrees of functional disability therein are the most varied, because this is the centre in which, in the great majority of individuals, words are first revived during thought, whether this be silent or whether it constitute the first stage in the processes of speaking or of writing. The effects of functional degradation or of partial damage to this centre may be considered first, and afterwards those resulting from its complete destruction.

(*a*) **Effects produced by functional degradation or partial damage of the left Auditory Word Centre.**—In the slighter degrees of damage and functional degradation we have to do with the most typical forms of verbal amnesia, in which various words fail to be recalled as they are needed in ordinary speech. The term "amnesia verbalis" applies especially to this defect, and the objections that have been raised to it on the ground that loss of auditory images constitutes only one kind of verbal amnesia are of no practical value because the two kinds of kinæsthetic images are, as I maintain, not spontaneously revivable as primary thought-counters, and because it is only in rare cases that visual images of words are primarily revived. The term "amnesia verbalis" is therefore especially applicable to this particular functional disability of the auditory word centre.

This failure to recall words is always most marked with the names of persons, places, and things, these being the most specialised units of speech. The most familiar type of this defect is that which occurs as a result of defective nutrition, either with advancing years or during convalescence from prostrating diseases. Such persons are often noticed to halt in their speech, owing to their inability to recall some such words. Occasionally, however, a similar or more marked defect of the same order occurs as a result of some more or less distinct lesion of the brain. A good example of amnesia of this type, where the "volitional" recall of names was impossible though their "sensory" recall was preserved, is thus referred to by Trousseau in his Lectures.[1]

CASE XLIII.—"The patient does not speak because he does not remember the words which express ideas. You recollect the experiment that I often repeated at Marcou's bedside. I placed his nightcap on his bed and asked him what it was. After looking at it attentively he could not say what it was called, and exclaimed, 'And yet I know well what it is, but I cannot recollect.' When told that it was a nightcap he replied, 'Oh! yes, it is a nightcap.' The same scene was repeated when various other objects were shown to him. Some things, however, he named well, such as his pipe. He was, as you know, a navvy, and therefore worked chiefly with the shovel and the pickaxe, so that these are objects the names of which a navvy should not forget. But Marcou could never tell us what tools he worked with, and after he had been vainly trying to remember, when I told him it was with the shovel and the pickaxe, 'Oh! yes, it is,' he would reply, and two minutes afterwards he was as incapable of naming them as before."

[1] Translation by Bazire, 1866, part i., p. 267.

As Lichtheim points out,[1] evidences of amnesia are " more easy to demonstrate when the patient is made to name objects than when he is engaged in ordinary talk ; names which occur without effort in fluent speaking arrest him when he has to find them for objects or persons shown to him." There is general agreement as to the fact that in amnesia words are lost in a tolerably definite order. First there are failures in the recall of proper names, then of other nouns ; and only much more rarely of verbs, adjectives, and pronouns. Of this fact different explanations have been given. I will quote two of the best. Kussmaul says :[2]—

" The more concrete the idea the more readily is the word to designate it forgotten when the memory fails. Probably the only reason for this is that the conceptions of persons and things are more loosely connected with their names than the abstraction of their circumstances, relations, and properties are. We easily represent persons and things to ourselves without names : the image of sense is here more important than the symbol—*i.e.*, the name. which conduces but little to our comprehension of personages or objects. More abstract conceptions, on the contrary, are attained only with the aid of words, which alone give them their exact shape. Hence verbs, adjectives, pronouns, and, still more, adverbs, prepositions, and conjunctions, possess a much more intimate relation to thought than nouns. We can conceive that the processes of excitation and the combinations in the cellular network of the cerebral cortex must be much more numerous for the creation of an abstract than of a concrete conception, and that the organic tracts which connect the former with its name must be correspondingly much more numerous than those of the concrete."

The other explanation, which may be considered as supplementary to, and not incompatible with, Kussmaul's, is that which was given by Ross. He said :[3]—

" The science of language teaches unmistakeably that the language of aboriginal man consisted almost entirely of verbs, demonstrative pronouns and a few adverbs of time and place, and that the names of even common objects are always derivative, and consequently of much later growth than the roots themselves, and it is only what might have been expected that, in the dissolution of language caused by disease, nouns should disappear from the vocabulary of the patient before the parts of speech which have been first developed, and therefore most deeply organised."

In the slighter forms of amnesia the efforts at recollection of a person who is " at a loss for a word " tend also to call the visual word centres into an incipient activity. Graves[4] placed on record an excellent illustration of this fact, though he quotes the case merely as " a remarkably exaggerated

---

[1] *Brain*, January, 1885, p. 478.
[2] *Ziemssen's Cyclopædia*, vol. xiv., p. 759.
[3] *Loc. cit.*, p. 112.
[4] *Dublin Quarterly Journal*. 1851.

degree of the common defect of memory observed in the diseases of old age, in which the names of persons and things are frequently forgotten, although their initials are recollected."

CASE XLIV.—" The man was a farmer, aged 50 years, who had suffered from a paralytic attack from which he had not recovered at the time of observation. The attack was succeeded by a painful hesitation of speech. His memory was good for all parts of speech except noun-substantives and proper names; the latter he could not at all retain. This defect was accompanied by the following singular peculiarity: he perfectly recollected the initial letter of every substantive or proper name for which he had occasion in conversation, though he could not recall to his memory the word itself.

"Experience had taught him the utility of having written on manuscript a list of the things he was in the habit of calling for or speaking about, including the proper names of his children, servants, and acquaintances; all these he arranged alphabetically in a little pocket dictionary, which he used as follows: if he wished to ask anything about a cow, before he commenced the sentence he turned to the letter C, and looked out the word 'cow,' and kept his finger and eyes fixed upon the word until he had finished the sentence. He could pronounce the word 'cow' in its proper place so long as he had his eye fixed upon the written letters; but the moment he shut his book it passed out of his memory and could not be recalled, although he recollected its initial and could refer to it when necessary. He could not even recollect his own name unless he looked out for it, nor the name of any person of his acquaintance; but he never was at a loss for the initial of the word he wished to employ."

The fact, moreover, that in those cases where we cannot "get out" a particular word we often seem to know something of its length, and can say that it consists of about so many letters, also seems to testify to an abortive revival of the word in the visual centre. But the fact that this partial revival is not associated with full consciousness of the word and does not enable it to be written is one of considerable significance, because it seems to show how all-important in the majority of cases is the primary revival in the auditory centre, not only for the accomplishment of speech, but also for that of writing—the visual word centre being probably called into play in writing spontaneously as well as in writing from dictation through the intermediation of the auditory word centre.

It seems reasonably certain that in the great majority of cases in reading aloud there is first the excitation of the visual word centre, and then the passage of stimuli through commissural fibres to related portions of the auditory word centre, before they pass on to rouse the glosso-kinæsthetic centre.[1] This affords the explanation of another peculiarity

[1] Several of the writers on Aphasia, however, in their diagrams assume that the stimulus passes direct from the visual word centre to Broca's region (see Ballet, *loc. cit.*, p. 164).

in the class of cases of which we have been speaking, as well as of others in which the amnesia has been even more profound. Many cases have, in fact, been recorded in which the patients' speech has been so disordered that they could scarcely say more than three or four consecutive words, and could perhaps recall no nouns; yet when a book is placed before them they are capable of reading aloud correctly and with ordinary facility. I have seen three such cases. One of them was a typical instance, and the following are some details concerning it.

CASE XLV.—A lady, aged 81 years, had a slight stomach derangement, followed by diarrhœa, in the spring of 1885. She had been rather deaf for many years, but after this slight illness an amnesic difficulty in speech very gradually became more and more pronounced, and after about eighteen months became associated with marked paragraphia.

Two years after the commencement of this speech trouble I found this patient capable of understanding all that was said to her in a sufficiently loud and distinct voice. She was thoroughly capable also of superintending her household affairs, correct in her accounts, and also knew exactly what change she ought to receive after having given money for the purchase of food or other necessaries. She made use of brief sentences, generally stopping short when she came to a noun, or she would then substitute the word "things" for any such part of speech. Thus, she said, "I will go up and see what we can," meaning that she would go upstairs and see what she could find; or "Will you have the things?" meaning some refreshments for a journey.

At this time she was accustomed to read novels with large print for several hours daily. I found that she could read aloud as well as ever, and she did actually read to me, with only two or three slight mistakes in pronunciation, nearly half a column of a newspaper. I was able to satisfy myself, also, that she understood the meaning of what she was reading; her laughter when I gravely gave her a short amusing paragraph to read left no doubt as to this point.

During the next six months her condition became steadily but markedly worse. Here is a letter written to her son about the end of June, in which for the first time she omitted to put the name of the month and his Christian name at the commencement of the letter. "29, 87.—MY DEAR PAPA.—I hope you are well, and think you are well for something for the Queen, and two things for the Queen [evidently referring to two papers which had been sent to her with details of Jubilee proceedings]. Pills for the father [an intimation that she wished some pills to be sent to her]. I hope you are well.—Yours affectitory." [The surname was spelt wrongly; she had previously always signed with the Christian name alone.]

Here are a few details concerning another of these cases, in which, however, the amnesia was much less severe.

CASE XLVI.—A lady, aged 65 years, was sent to me on February 18, 1893. She had had no fit of any kind, and no pains in the head—only a feeling of "confusion in the occiput" for several years. She had suffered much from throat affections in early life, and there had been some deafness for nearly thirty years, which had slowly increased. She did not hear the ticking of a watch on either side even with contact. She heard a tuning fork on both sides with contact, and also for about two inches away from the ear on each side. Her husband died four years previously; she then became weak and suffered from dyspeptic symptoms for a short period.

There was no illness after this, but eighteen months ago she began to suffer from difficulty in expressing herself. There was no defect of articulation, but a difficulty in finding words, and this trouble had become much more marked during the last five months. There had been no unilateral weakness and there is no trace of any now. She speaks in broken sentences, omitting smaller words, and is often at a loss for a noun. Articulation is distinct. She reads from a book fluently, correctly, and with no appreciable hesitation.

A good instance of this degree of lowered excitability of the auditory word centre was also reported by Ross.[1]

Lichtheim has likewise recorded an extremely interesting case of this type. It occurred in the person of a busy medical practitioner who met with a carriage accident and was carried home unconscious. There was paresis of the right arm and leg, and he was confined to his bed for about a week. The following are some of the particulars given by Lichtheim as to the patient's condition.

CASE XLVII.—Speech was much affected; the first day the patient only said "Yes" or "No," but quite appositely. Gradually more and more words returned, at first imperfectly. Whilst his vocabulary was still very meagre, it was observed that he could repeat everything perfectly. Soon after the accident he began to read with perfect understanding. It was established beyond doubt that he could read aloud perfectly at a time when he could scarcely speak at all. The statements of his wife are most positive and trustworthy upon this point, though he himself does not recollect what took place just after the accident. She states that after much difficulty in making himself understood by gestures he obtained a newspaper, and to the great astonishment of all present he began to read fluently. She herself thought it most strange and inexplicable. . . . . . He could not write voluntarily at all; but this faculty returned slowly and imperfectly, as did speech. On the other hand, he could, soon after he left his bed, copy and write from dictation.

The meaning of this ability to read aloud in such a case is that though the auditory word centre is so much damaged as to be unable to act spontaneously (that is, under volitional stimuli), it is still capable of responding to the associational stimuli coming to it as a result of strong excitation of the visual centre.[2] Persons so affected are also quite capable of

[1] "On Aphasia," 1887, p. 40, Case 8. I am a little puzzled, however, in view of the simple explanation that can be given of this patient's defect, to find that in different parts of his work Ross describes cases essentially similar under different names—see the case of a boy referred to on p. 68 as an example of what he terms the "second degree of aphemia," and also his reference to another example (on p. 72), which he terms, after Kussmaul, "aphasia of recollection."

[2] Lichtheim's interpretation of this case (*Brain*, 1885, p. 447) is wholly different from mine. He accounts for it by supposing a damage of com-

responding to sensory stimuli passing direct to the auditory centre itself—that is, they can at once repeat words uttered before them.

Another very interesting but more complicated case illustrating the same degree of amnesia has been recorded by Dr. P. J. Cremen,[1] the essential details of which are as follows.

CASE XLVIII.—The patient, a man of intemperate habits, was admitted to the Cork North Infirmary on April 10, 1885. He was the subject of aortic and mitral disease, fully compensated, which had followed an attack of rheumatic fever in childhood. About October, 1883, he suddenly became giddy, with loss of speech, but no paralysis of limbs. He continued to work during the same day, and during five months his speech slowly improved very much. "He now commenced to do some light work, being able to make himself understood fairly well, occasionally using one word for another."

About Christmas, 1884, on awaking one morning, he was surprised to find that he had lost the sight of his left eye. (This was subsequently found to be due to embolism of the central artery of the retina.) "He continued otherwise the same as regards speech, until about four weeks before admission, when, after a hard day's work, whilst in the act of holding a candle, he suddenly allowed it to drop and commenced to cry. His speech again became very imperfect and his face was slightly drawn, but there was no appreciable paralysis of limbs." Since then no change had taken place in his symptoms.

On admission volitional speech was much affected. When asked of what he complained he pointed to his left temple and said he had a pain there, and on being asked if it was constant he said not. He also pointed to the left eye, and indicated that he could not see with it. His memory of names, places, and things was very defective. He did not recollect the names of his father, mother, or other near relatives. When asked to try, he made an effort to do so, sometimes repeating his own name instead, evidently knowing that he was wrong. When corrected, he repeated the name distinctly, having no difficulty whatever in articulation, the latter being very distinct. When asked to name the organs of sense he called the ear a "hair-pin," and named correctly the nose and tongue; he did not recollect the name of the eye, and called this also the "tongue;" but when corrected said, "Yes, eye; that's right." When further questioned he became confused, calling nearly everything that was shown to him "tongue." His vocabulary was, however, liable to variation from day to day. When asked to repeat the Lord's Prayer he failed to do so, but repeated another prayer instead; however, when given the first sentence he repeated it through correctly. His understanding of spoken and written language was perfect. He could repeat accurately and quickly words spoken before him.

missural fibres to exist, which pass between his postulated. "centre for concepts" and Broca's convolution, which for him also is a motor region rather than one of sensory type. These are the cases which Lichtheim includes in his Type IV. A defect closely allied to this which I have just been considering has been described by Broadbent in a paper "On a Particular Form of Amnesia; Loss of Nouns" (*Transactions of the Medico-Chirurgical Society*, vol. lxvii., 1884, p. 249). His explanation is also more in accordance with that of Lichtheim than with mine.

[1] *Brit. Med. Jour.*, January 2, 1886, p. 14.

Notwithstanding this great defect of memory of words, he could read aloud clearly and distinctly, without hesitation, a page of a book from beginning to end.

The following composition, in which it is impossible to discover any meaning, was written by him at my request as the history of his case: "Cork Molens.—I noscent nountg ani aminbys gosisbyoyey initwats yab I bet yas you me sent sml me good me much cocleped." He wrote his own name and residence accurately, and could write numerals to any extent without dictation. The following will exemplify defects in writing from dictation. When asked to write the word "just" he wrote "fugl," for "subject" "supfect," for "speak" "sery," for "found" "spunt." Strange to say, when told to spell those words he did so accurately in every instance, and when asked why he did not write them he explained that he had forgotten how to make the letters.

He remained in hospital for about three months, and left much improved in his speech, being able to carry on a conversation fairly well, though occasionally using one word for another. He could name objects better. Volitional writing was not improved; writing from dictation was slightly better.

Apart from the marked amnesia with preservation of ability to read aloud, this case is remarkable for the gibberish character of the patient's writing, combined with an ability to spell correctly—two characters that do not often go together.

In this relation it may be mentioned that it sometimes happens that the speech of patients is entirely limited to a mere imitative repetition of words spoken in their hearing, while they are without the power of volunteering any statement—that is, their auditory word centres respond only to direct sensory incitations, and not at all to those of an associational or volitional order. In these cases (usually included under the term "echolalia") a marked general mental impairment almost invariably co-exists (see Case lix.).

A defect of this kind (occurring in a woman who was hemiplegic from cerebral hæmorrhage) has been recorded by Professor Béhier.[1] She was born in Italy, and had resided both in Spain and France; of the three languages she had thus acquired she had completely forgotten the Italian and Spanish, and had only retained a most limited use of French. In this latter language *she only repeated like an echo* the words pronounced in her presence, without, however, attaching any meaning to them. But in the case of a woman seen at the Salpêtrière by Bateman the mimetic tendency was much stronger. She even reproduced foreign words with which she had never been familiar. It is clear that in such a case as this there must have been a mental degradation of a much wider kind than that which occurs when the auditory word centre alone is reduced to its lowest grade of functional activity.

[1] *Gazette des Hôpitaux*, May 16, 1867.

Having considered the different degrees of amnesia verbalis that may result from functional defects or partial lesions of the left auditory word centre, we must now consider the effects produced by more serious lesions of this centre.

(b) **Effects produced by destruction of the left Auditory Word Centre.**—Where, instead of partial or mere functional defects, we have to do with more or less complete destruction of the left auditory word centre by some organic disease a correspondingly complete word-deafness is produced, so that the patient is no longer able to comprehend spoken language. Spoken words become to such a patient mere meaningless sounds. As I have already pointed out, word-blindness (or alexia) is also an occasional consequence of such a lesion, though it is not to be regarded as a necessary accompaniment, whatever Wernicke and Déjerine may say to the contrary. These authorities, moreover, as well as Mirallié, hold that paraphasia is the kind of speech defect that is entailed by destruction of the auditory word centre. Wyllie also, in his recent work on "The Disorders of Speech," says (p. 289) that the extreme defect in an "auditive" in consequence of this lesion would be what is termed "gibberish aphasia," and adds: "In general, the destruction of the auditory centre leaves the patient in possession of considerable powers of expression." This question as to the nature of the alteration of speech that goes with destruction of the auditory word centre is a very interesting and important one, and will require careful consideration.

The views that I entertained on this subject some years ago were thus expressed :[1] "Suppose a person to be suffering from defective activity of the auditory word centre, so that names cannot be recalled 'voluntarily' or by 'association.' There would already be great hesitation and difficulty in the expression of thoughts both in speech and in writing. But suppose this mere defective activity to be replaced by actual destruction of the left auditory word centre so that its functional activity became actually lost, words could then, of course, neither be recalled 'voluntarily' nor by 'association,' and, still further, they could not be perceived and consequently could not be imitated. An individual thus affected would neither be able to speak nor to write—that is, he would be completely aphasic, with the super-added peculiarity that he would not readily comprehend spoken and perhaps written language."

[1] "The Brain as an Organ of Mind," 1880, p. 685.

It should be observed that these statements were made partly as *à priori* deductions and partly as a result of the observation of actual cases, though at this period there had only been too few of them. Still, it seemed to me then—and from several points of view it would still seem a legitimate conclusion—that loss of articulate speech should result from destruction of the auditory word centre, just as agraphia results from destruction of the visual word centre. Ballet[1] and Ross,[2] following me, expressed very similar views, though the majority of writers have rather followed Wernicke, and regarded paraphasia as the natural accompaniment of word-deafness.

Now that more cases have been observed we are in a better position for throwing light upon this point. An examination of the recorded observations in which the lesion has been limited to the hinder part of the first, and perhaps also of the second, left temporal convolutions (without perceptibly encroaching upon the visual word centre) shows surprisingly different results in regard to the nature and degree of the speech alteration met with. In some of the cases there has been aphasia; in others a more or less marked paraphasia; whilst in two or three of the cases the defect in voluntary speech has been less marked. Thus, looking to the sixteen cases of destruction of the auditory word centre to which (p. 144) I previously referred[3] when speaking of the frequency with which word-blindness and agraphia occur as additional symptoms, I find that what is described as "motor aphasia" existed in six of them (viz., in Mirallié's list Nos. 7, 11, 12 and 25; and in Amidon's list Nos. 14 and 18). In six of the cases, also, some amount of paraphasia existed (viz., in Mirallié's list Nos. 3, 4, 19, 21, and 23; and in Amidon's list No. 7). In one case (Mirallié's No. 14) both aphasia and paraphasia are said to have existed; while in the three remaining cases (Amidon's Nos. 6, 15, and 17) voluntary speech seems to have been rather less affected.[4]

[1] "Le Langage Intérieur," 1886, pp. 91 and 166.
[2] "On Aphasia," 1887, p. 119.
[3] Excepting that I omit Mirallié's Case 10, and include instead his Case 2, where the only lesion beyond the destruction of the first temporal convolution was a very minute "plaque jaune," no more than a millimetre square, situated in the upper border of the foot of the third frontal convolution.
[4] Numbers of the cases of "sensory aphasia" hitherto recorded are comparatively valueless for the purpose of any critical study of their symptoms, either on account of the nature, extent, and complication of the lesions, or from the paucity of the clinical details, or perhaps from two or three of these causes in combination. Among such more or less vague cases I

Thus it will be seen that we have all been more or less wrong; and I must confess that the revelation of these apparently contradictory results in the way of speech disturbance in association with word-deafness surprised me not a little. The meaning of such remarkable variations demands a careful consideration.

The three cases last referred to in which voluntary speech has been least interfered with are, I believe, altogether exceptional. One of them is a case recorded by Claus,[1] concerning which, unfortunately, there are only a few meagre details.

CASE XLIX.—A man, aged 68 years, understood nothing said to him. He answered always wrongly, while he spoke correctly. His intelligence was weakened. His hearing was good.

A focus of softening was found involving the left first temporal convolution, except its anterior third. This disorganised portion of the cortex, and also the corresponding border of the second temporal convolution, had a brown colour.

I have been unable to obtain any further particulars concerning this case, as the periodical in which it is reported is not contained in either of the libraries to which I have had access. It seems to be Case No. 20 in the "Table of Cases of Sensory Aphasia with Lesions and Symptoms" cited by Allen Starr,[2] and there also no further details are given. No weight, therefore, can be attached to this example, the details of which are so incomplete. The second case is one that was originally recorded by Wernicke,[3] and the following are all the essential details that are given.

CASE L.—A woman, aged 75 years, was admitted to hospital on October 7, 1873, showing well-marked indications of old age. It was ascertained from her relatives that her speech had suddenly become troubled and almost lost on November 2, 1872.

On admission her psychical condition seemed to be that of a crazed person complicated with aphasia. She mostly lay huddled up in bed, and suffered from incontinence of urine and of fæces. She answered all questions put to her quite wrongly, and did not attempt to do anything she was told. She paid little attention to her surroundings, and showed no desire to make any communications concerning her distressing condition. Her spontaneous speech was comparatively insignificant, though it was sufficient to show she was not suffering from motor aphasia. Her speech trouble was further shown by her alteration and displacement of words. Thus she often said

include thirteen in Mirallié's list (viz., Nos. 2, 15, 18, 20, 22, 27, 28, 32, 33, 34, 35, 36, and 37), and three in Amidon's list (viz., Nos. 8, 9, and 22). In four other cases—namely, Nos. 10, 11, and 17 of Mirallié's list, and No. 20 of Amidon's—there was a lesion in Broca's centre as well as in the auditory word centre.

[1] *Der Irrenfreund*, No. 6, 1883.
[2] *Brain*, 1890, p. 100.
[3] "Der Aphasische Symptomen-Complex," 1874, p. 43.

correctly, " I thank you heartily." and another time, "I thank you most gively," etc.; "I am quite ill"; "Ah! how cold I am"; "You are a kind gentleman." These were usual expressions; yet the medical man whom she had just called a kind gentleman she called soon after " mein Tochtel " or " mein Sohnel."

An ophthalmoscopic examination on November 5, 1873, showed grey atrophy of the right optic disc. Sensibility seemed intact, and the handgrip was weak, but equal on the two sides. More exact observations on these points were not made.

There was no improvement in her psychical or bodily symptoms, and death occurred as the result of intestinal troubles on December 1, 1874. At the necropsy there was found an increase in the sub-arachnoid fluid, and the convolutions generally of both hemispheres were somewhat wasted. There was also an excess of fluid in the lateral ventricles. All the brain arteries were highly atheromatous, and there was thrombosis of some of the terminal branches of the left middle cerebral artery. The whole of the upper temporal convolution and part of the supra-marginal lobule were in a state of yellow softening. The island of Reil was intact.

It will be seen from the above particulars that this patient's spontaneous speech was extremely limited, and that she made frequent mistakes in words and in their pronunciation. The statement contained in Amidon's abstract, therefore, that her "spontaneous speech was very good," certainly goes far beyond what there is any warrant for. It is worthy of note also that the auditory word centre may not have been entirely destroyed, since the second temporal convolution seems not to have been at all involved. In the previous case, reported by Claus, it should also be observed that the second temporal convolution was only very slightly involved. It is merely said that its upper border was affected ("had a brown colour").

The third case, curiously enough, has often been quoted as though it were a typical example of word-deafness rather than one of an unusual character. It was recorded by Giraudeau[1] and some of the principal details are these.

CASE LI.—The patient, a woman, aged 40 years, had previously enjoyed good health, though she had never menstruated. For three months before admission to l'Hôpital St. Antoine she had suffered from a constant headache, with nocturnal exacerbations of such a character as to render sleep impossible. There had been no vomiting or epileptiform attacks. For upwards of a month before admission she was obliged to give up work in consequence of the severity of the pain. At about the same time it was observed that she no longer understood what was said to her, and that she did not answer when spoken to. This information was obtained from the persons who accompanied her to the hospital.

On admission the patient was found to be very stout; there was no fever, the right pupil was slightly dilated, and there was severe headache. When asked her name she raised her head, but did not answer. When asked a second time she answered, "What do you say?" and on the same

Revue Médicale, 1882, p. 448.

question being put a third time she said, "I don't understand." When asked a fourth time she answered correctly, "Marie Bouquinet." On being asked, "How long have you been ill?" she evinced the same difficulty of understanding, but replied, after a time, "Three months." When asked to give her address she replied, "Perhaps for three months and a half." Being interrogated as to her occupation, she presented the prescriptions of the physician who treated her in the town, and added, "a white powder" (sulphate of quinine). On several occasions we changed the mode of our interrogations, but the replies of the patient were always analogous to those mentioned above. After having by great difficulty made her understand a question by frequent repetition, she answered; but whatever subsequent questions were put to her she followed her first idea, and her subsequent replies had no relation whatever to the questions put to her.

Sometimes, however, it was impossible to make her understand our meaning at all, and to every question addressed to her she invariably replied: "What do you say? I don't understand. Cure me." Her hearing, however, was unaffected; there was no discharge from the ear; she heard the ticking of a watch, and turned her head at any slight noise made near her. Vision was good in both eyes, and there was no word-blindness, for she could easily read the headings of the bed-ticket and written questions put to her, to which she replied verbally or in writing after a little reflection. Tactile sensibility was preserved, as well as the senses of taste and smell, and her motor power was unaffected.

Without entering into further details, suffice it to say that the psychical affection rapidly increased, and on the ninth day after her admission the word-deafness was complete, and whenever she was addressed she invariably said, "I do not understand," and then began to weep. The next day she fell into a state of coma and died.

At the necropsy a sarcomatous tumour of the size of a walnut was found occupying the posterior part of the first and second temporo-sphenoidal convolutions of the left hemisphere, the rest of the brain being healthy.

Here the details are fortunately more ample. Although this patient's speech was limited, there was clearly neither aphasia nor paraphasia, and this may, I think, be explained either by supposing that the lesion in the early days did not destroy the whole of the auditory word centre—which is more than possible seeing that the lesion was a tumour and also that the word-deafness was not absolute; or else that the patient was a strong "visual"—a person, therefore, in whom the excitation of Broca's centre might have been brought about directly from the visual word centre, both in spontaneous speech and in reply to written questions. And to the latter she is expressly said to have replied "verbally or in writing, after a little reflection. "This possibility might, of course, be applicable also in explanation of the other two cases.[1]

[1] Since writing this section I have noticed that in Allen Starr's table (*Brain*, vol. xii., p. 100), he cites two examples (Cases 1 and 7), one by Bateman ("On Aphasia," 1870, p. 73), and the other by Kussmaul (*Ziemssen's Cyclop.*, vol. xiv., p. 765), of word-deafness in which the speech was "good." A perusal of these cases, however, will show that they do not answer to the description given of them—either from the pathological or from the clinical point of view.

There is an interesting point worth noticing here in reference to cases of partial word-deafness, illustrating the increased power of recollection or comprehension when two or more bonds of association are simultaneously called into activity. Thus, in a case of incomplete word-deafness observed by Fränkel, when the patient was asked, " What is a fork ? " he replied, " I don't know what you say." But when a fork was shown to him at the same time that its name was pronounced he not only recognised the fork, but comprehended its name and repeated it, the visual image of the object here helping to revive the auditory image of the word. At other times when this same patient was questioned, at first he did not understand, but he would struggle to articulate the words used and would thus, after various attempts, arrive at the meaning of the question.[1]

Another instance in which visual impressions were habitually called upon to aid a damaged auditory centre was long ago recorded by Abercrombie in his " Inquiry into the Intellectual Powers," where he says (seventh edition, 1837, p. 158) : " The gentleman to whom it referred could not be made to understand the name of an object if it was spoken to him, but understood it perfectly when it was written. His mental faculties were so entire that he was engaged in most extensive agricultural concerns, and he managed them with perfect correctness by means of a remarkable contrivance. He kept before him in the room where he transacted business a list of the words which were most apt to occur in his intercourse with his workmen. When one of them wished to communicate with him on any subject he first heard what the workman had to say, but without understanding him further than to catch the words. He then turned to the words in his written list, and whenever they met his eye he understood them perfectly. These particulars I had from his son, a gentleman of high intelligence."

This slight digression may serve to illustrate one of the two suggestions I have made in explanation of the comparatively slight defects in speech presented by the three altogether exceptional cases of word-deafness to which I have above referred. But, before pursuing our attempt to explain the differences presented by the other cases, it will be well first of all to see what the results have been in two other sets of conditions involving destruction of the left auditory word centre as one component of the existing lesions—that is to say, in the two next sets of cases cited for consideration. These are

[1] See also Case lix.

(2) the cases in which there have been combined lesions of the auditory word centres in each hemisphere; and (3) the cases in which the auditory and the visual word centres have been destroyed or severely damaged in both hemispheres.

2. DEFECTS RESULTING FROM LESIONS OF THE AUDITORY WORD CENTRE IN EACH HEMISPHERE.

Cases of double lesion in the upper temporal convolutions are extremely rare. Only four instances are known to me, and, concerning these, clinical details in three are unfortunately extremely meagre, consisting of particulars obtained from friends or others, rather than from the examination by a skilled observer.

The first of these cases to which I will refer is one that has been recorded by Kähler and Pick,[1] though the details are very scanty and of the kind above mentioned.

CASE LII.—A woman, aged 42 years, was seen eighteen months after the attack. Her husband stated that she suffered from headache for three months, when loss of speech and hearing supervened, and symptoms of apparent dementia. She uttered inarticulate sounds almost all day long. She was after a time admitted to an asylum, where she remained for six months—to the date of her death. Whilst there she continued to mumble constantly, the only articulate sounds she uttered being "Tschen" and "Tscho." She occasionally moved her head when spoken to as if she heard, though she did not understand what was said. There was no paralysis or other motor disturbance.

At the necropsy there was congestion of the membranes, the frontal and parietal convolutions were somewhat wasted, the sulci being wide and filled with serum. Over the left temporal lobe the pia was jelly-like, whilst its convolutions were swollen, pressed together, and of soft, jelly-like consistence. The right temporal lobe was also slightly softened. Sections showed that the softening was confined to the cortex—on the right side of all the temporal convolutions except the basilar, and on the left side to the same region. The foot of the third frontal convolution was not involved.

The second case is one that has been recorded by C. K. Mills,[2] and although the details of the necropsy are full and all that could be desired, the patient only came under his observation a few days before her death, when she was very weak and ill, suffering from the effects of an old valvular disease of the heart. He says: "Some of the facts of her history were obtained from relatives who came to see her at the hospital, and others were procured after her death from a brother and sister-in-law, whom I sought out and carefully

---

[1] *Vierteljahresschrift für Praktische Heilkunde*, No. 1, 1879, p. 6.
[2] *Philadelphia University Medical Magazine*, November, 1891; and *Brain*, 1891, p. 468.

interviewed." The following are all the particulars given that are of any importance from our point of view.

CASE LIII.—A woman, aged 46 years, was admitted to the Philadelphia Hospital in August, 1891. Fifteen years before her death she had an apoplectic attack which left her word-deaf but not paralysed. Prior to the first attack of apoplexy her hearing had been good, but after it she could not by hearing understand anything that was said to her. She could, however, hear music and sounds of various kinds; for instance, when an organ or band had performed in the street she had at times called attention to the fact, and she had also come down from the second or even the third storey to open the front door in answer to a knock. She could hear such sounds as a bell ringing and a clock ticking. These facts were elicited from her relatives through various statements made by them, chiefly spontaneously.

When any one wished to communicate with her it was done by means of writing or signs, as she had fully preserved her vision and was evidently not word-blind either for writing or printing. She often read the newspapers, and could do so with intelligence up to a few weeks before her death. [If correct this is remarkable, as it would have been long after the destruction of both auditory word centres.] Her sister-in-law said that several times she had heard her try to read the newspaper aloud, and in so doing she had seemed to understand what she read, "but made a tangle of her words." From the time of the first attack she had never been able to speak well, her words becoming jumbled or tangled. From the description given of her manner of speech the defect was evidently a serious form of paraphasia and paralexia.

Her relatives spoke positively of her deafness as having been due to the "stroke;" but the apoplectic attack, although it had at once caused this word-deafness and paraphasia, had not in any way, as far as could be ascertained, affected either motion or sensation. She could write, but "sometimes mixed up her words in writing."

Nine years before her death she had another and more severe apoplexy, after which her deafness increased for sounds as well as words until it was almost total. This seizure left her also with partial left hemiplegia, chiefly affecting the arm. Dr. Mills adds: "She was first examined by me on August 24, 1891. Her condition then was one of almost complete helplessness. It was impossible to make her understand what was said to her, and, so far as could be determined by repeated tests, she was totally deaf; but notwithstanding her weakness, helplessness, and deafness, her face had a somewhat intelligent expression. She looked about her as if she knew what was going on. She was very emaciated; her heart's action was excited, and examination showed the presence of marked murmurs, both mitral and aortic. She became feebler day by day, and died August 28."

At the necropsy the posterior two-thirds of the left first temporal convolution was found to be shrunken to a thin strip, and the posterior fourth of the second temporal convolution was also atrophied. The other left temporal convolutions were healthy. On the right side there was an old and very extensive hæmorrhagic cyst, which had completely destroyed the first, and almost completely the second, temporal convolution, as well as the island of Reil and other parts in the immediate neighbourhood. (Very full details of the condition of the brain are given, and in the *Philadelphia University Medical Magazine* there are two excellent photogravures showing the extent of the lesions.)

This is undoubtedly a very important case, especially from the point of view of cerebral localisation, from which standpoint Mills principally deals with it. It is interesting to find that after the patient became word-deaf, owing to the first lesion on the left side, there was no word-blindness, as "she often read the newspapers and could do so with intelligence" —although some would have us believe that even the destruction of Broca's centre, which is so much further away from the visual centre, causes alexia. There was, however, paraphasia and paralexia.

Unfortunately, there is no definite statement as to what alteration took place in the patient's speech after the lesion occurred on the right side—that is, nine years before her death. I therefore wrote to Dr. Mills to ascertain whether he could supply any further information on this important point, and his reply is as follows: "She was admitted to the hospital some little time (probably two or three weeks) before I examined her, and my recollection is that it was reported to me that she had not spoken during that time, but more than this I cannot say . . . . and, unfortunately, I have no longer the means of verifying my recollection. She certainly was not able to speak, or at least did not speak, at the time of my examination." I will merely add that, from the fact of the word-deafness in this patient not being associated with word-blindness, it does not necessarily follow that the paraphasic speech should have been lost—that is to say, that there should have been speechlessness—immediately after the occurrence of the lesion in the right hemisphere. There is the possibility that she might have been one of those "visuals" capable of speaking, though in a disordered fashion, by the direct action of the visual word centre upon Broca's convolution. It seems pretty clear, however, from what Dr. Mills says, that at the time she went into the hospital she was incapable of speaking. Still, it is important to bear the above-mentioned possibility in mind, as it is one to which we shall have to recur.

The third case is one that was recorded by Wernicke and Friedländer,[1] and an abstract of it has been given by Ferrier[2] which I will now quote.

CASE LIV.—A woman, aged 43 years, who had never suffered from deafness or affection of vision, was attacked on June 22, 1880, with right hemiplegia and aphasia. She remained in the hospital until August 4, when she was discharged. At this time the patient could speak, but she spoke unintelligibly, and was sometimes believed to be intoxicated. She not only could not make herself understood, but she could not understand what was said to her.

[1] *Fortschritte der Medicin*, Band i., No. 6, 1883.
[2] "Lectures on Cerebral Localisation," 1890, p. 89.

She was received into the hospital again on September 10, with slight paresis of the left arm. The right hemiplegia had entirely disappeared. The patient was looked upon as insane. She was absolutely deaf, so that she could not be communicated with. She died from an attack of hæmatemesis on October 21.

An extensive lesion was found in each temporal lobe (of which figures are given). On the left side it involved the posterior half of the first and second temporal convolutions and a small part of the third; on the right side it involved the angular gyrus, the posterior extremity of the upper temporal convolution, and it encroached slightly on the occipital convolutions.

The fourth case has been described by Pick. It is one of great importance, but as it is also one of a very exceptional nature I reserve its consideration for the present.

3. Defects Resulting from Destruction of the Auditory and Visual Word Centres in each Hemisphere.

There is only one case known to me in which these combined lesions have existed in the two hemispheres, although there is another, with which it will be not uninteresting to compare it, where complete blindness and deafness had been produced in infancy by peripheral lesions.

The case definitely belonging to this category was reported many years ago by J. C. Shaw,[1] but unfortunately in this, as in other rare cases, many essential details are omitted. Still this case is so valuable by reason of its extreme rarity that we are glad to accept it notwithstanding the paucity of clinical details. All the essential recorded facts are these.

Case LV.—A woman, aged 34 years, was admitted into King's County Lunatic Asylum on September 25, 1879. A few facts concerning her were obtained from a sister. Two months before admission she complained of being unable to use the right arm. Soon afterwards she had a sudden attack characterised by loss of consciousness, inability to speak, and loss of hearing. They did not notice at this time that she had any loss of vision. She is said to have remained "unconscious" about two weeks, after which she was able to get up, and then began to talk. But she could talk sensibly only for a few minutes, then she would ramble and become incoherent. The friends were indefinite as to the extent of her sight trouble at this time, but there seems to have been marked mental disturbance. The certificates committing her relate that "she says some men are trying to kill her; she screams without cause; and she is at times violent."

The following meagre notes concerning her condition are then given:—
"On admission she is found to be very stupid; does not answer any questions, evidently does not hear them; is quite deaf. Pupils slightly dilated and equal; she is evidently quite blind. Physical condition poor; habits dirty. October 18 and November 21: Had a severe epileptiform fit on each of these days. December 11: Repeated tests of tactile sensibility and smell show them to be unimpaired, but she is absolutely deaf to all noises and also

[1] *Archives of Medicine*, New York, 1882, p. 80.

blind. December 16: Walks about the hall using disconnected and unintelligible words; at times speaking very loudly and screaming. January 31, 1880: Severe convulsions lasting seven hours. Muscular spasm most marked on left side (the right arm remained contracted afterwards during two days). February 14: Is up again; walks about; shows no sign of paresis anywhere. October: She has become quite noisy, talking and screaming unintelligibly night and day; is growing thinner and more stupid, if that were possible." The patient remained in about the same condition till September, 1881, when she developed a low form of pneumonia and died on September 19.

At the necropsy an almost exactly symmetrical patch of cortical atrophy was found in each hemisphere. It involved (as is shown by two figures) the greater part of the upper temporal and angular gyri, the marginal lobule, and the upper parietal lobule on each side. In the affected areas it was found "that the grey matter had entirely disappeared, leaving the very outer layer to which the pia was attached and the white matter below, so that there was really a cavity or space beween the two, formed at the expense of the grey matter. Microscopic examination showed this outer layer to be composed of dense connective tissue filled with nuclei." There was no atheroma of the great vessels, and, except that the patches of cortical atrophy mentioned involved on the left side a small portion of the upper part of the ascending parietal, and on the right side a small portion of the lower part of the corresponding convolution, other portions of the brain seem to have shown no appreciable lesion.

It is interesting to find also that in the celebrated case of Laura Bridgman, who was totally deaf and blind from about the end of her second year (owing however to peripheral rather than central disease), her power of uttering articulate sounds was not very different in kind, though less in degree, than that which was met with in the case above recorded. The following account of her condition in this respect is given by Kussmaul.[1]

CASE LVI.—The child lost her eyesight and hearing about the end of her second year, when she was beginning to talk, and in consequence soon lost the power of speech. Now, although it may be assumed that she retained no memory of sounds, which long experience teaches us is the rule with those who become deaf-mutes so early, yet she was able to produce a multitude of sounds. She took the greatest pleasure in giving utterance to these, and even went so far occasionally, in order to indulge herself in noises to her heart's content, as to lock herself up when her teachers endeavoured to restrain her from so doing. These noises were, in fact, inarticulate—*e.g.*, a kind of chuckling or grunting, an expression of contentment; while others —*e.g.*, "h-o-ph-ph"—were better formed, and served for an indication of astonishment. Other sounds, again, she elevated to the dignity of names, and bestowed them on certain persons. She uttered these latter when those so denominated approached her, or when she wanted to find them, or even at times when she only thought of them. She employed some fifty or sixty such sounds as names, many of which could be written down, such as fu, tu, pa, fifpig, ts, pr, lutt, etc.; many, however, it would be almost impossible to express in letters. She formed, however, only words of one syllable, but frequently re-duplicated them two or three times, as fu-fu-fu, tu-tu-tu.

Loc. cit., p. 634.

There was, therefore, slowly developed in this child something more of method in the use of the simple or complex sounds that were uttered than has been met with in the early stages, or for some time after the onset, of the cases of disease in which both auditory word centres have been destroyed. The sounds emitted were also less voluble and free, which is only to have been expected, seeing that in Laura Bridgman the glosso-kinæsthetic centres had never been developed as they have been in the case of patients who, by reason of the lesions above mentioned occurring in adult life, have been reduced to a condition in which their speech has become unintelligible. The powerlessness of Broca's centre when unaided for the production of anything like intelligible speech is, however, clearly shown.

After the narration of these additional cases in which there was a destruction of the auditory word centre in each hemisphere, we may now with more advantage return to the question (p. 155) why in a certain number of the cases of destruction of the left auditory word centre alone there is (1) an aphasic loss of speech, in others (2) paraphasia, and in still others (3) a comparatively slight impairment in voluntary speech.

My original contention was that an aphasic loss of speech is what is naturally to be expected as a result of such a lesion, if we are to suppose, as seems highly probable, that the auditory word centre in the great majority of persons holds a relation to Broca's centre similar to that which the visual word centre holds to the writing centre. It would appear, therefore, that what stands in need of explanation is the occurrence of the second and third class of cases—namely, those in which more or less marked paraphasia, or even fair speech, has existed.

The evidence in favour of the posterior half of the upper, with perhaps the posterior part of the middle, temporal convolutions being the chief, if not the exclusive, seat of the auditory word centre must be regarded as very conclusive.[1] The difference in voluntary speech power presented by the cases above referred to cannot therefore be explained by the supposition that the auditory word centre has not yet been properly localised.

The various possibilities that suggest themselves to me in explanation of the differences in the degree of impairment of speech in different cases in which lesions have been found in the left auditory word centre are as follows :—

---

[1] On this subject see Ferrier on "Cerebral Localisation," 1890, and C. K. Mills, in *Brain*, 1891.

(a) There is the possibility that in some of the cases in which the word-deafness has been complete whilst voluntary speech has been fairly good, the auditory word centre has been rather isolated from its afferent fibres than destroyed. These would be cases in which the amount of damage to the auditory word centre itself was small—cases approaching those named by Lichtheim "isolated speech-deafness," which will be subsequently considered (p. 174.)

(b) Then in some cases the auditory word centre may be damaged sufficiently to prevent perfect comprehension of speech, though it may still be sound enough, with the aid of the visual word centre, for the production of paraphasic speech. This might afford an explanation of certain cases in which there is only partial damage of the auditory word centre, together with incomplete word-deafness, as in No. 7 in Amidon's list, and Nos. 2 and 48 of Mirallié's list, as well as in a case by West,[1] which is not included in either of them.

(c) In other cases, where the individuals affected are strong "visuals," the paraphasic speech may be produced by the fact of the visual word centre taking the place of the auditory word centre—that is, by the passage of the stimulus directly from the visual word-centre to Broca's region across the commissure $e\ e$, as represented in Fig. 5. Such an explanation may be applicable to some of the cases where the destruction of the auditory word centre has not produced word-blindness as well as word-deafness. And this, as we have seen, is what has occurred in about two-thirds of the recorded cases.[2]

There is one extraordinary case that has been recorded by Pick[3] which seems, moreover, quite incapable of being understood except by the supposition of the adequacy of the direct action of the visual word centre upon Broca's region even for the production of correct speech—as it is brought about, in fact, in all deaf-mutes who are taught to speak. This is the case of lesion in the upper temporal convolutions of each hemisphere, the consideration of which was deferred because of its altogether exceptional character. Although this patient was perfectly word-deaf, he was able to speak correctly, to read correctly, and to comprehend writing perfectly, as may be gathered from the following details.

[1] *Brit. Med. Journ.*, vol. i., 1885, p. 1242.
[2] The next explanation is, however, equally applicable to this class of cases as well as to those in which word-blindness co-exists with the word-deafness.
[3] *Archiv für Psychologie*, 1892, p. 909.

CASE LVII.—A day labourer, aged 24, was completely word-deaf and behaved like a deaf person, taking no notice of ordinary sounds near him. It was found that he only noticed loud calls, clapping, or ringing of bells, and this not always readily. Yet if one shouted to him unexpectedly, he said, angrily, "Don't shout at me so;" and he often said spontaneously, "I hear quite well but I don't understand; I can hear a fly flying past me." His power of recognising airs previously known to him seemed to be also lost.

His speech was perfectly correct. He spoke fluently, and only occasionally hesitated about the right word. He named objects shown to him correctly. He could not repeat words or phrases. Writing was executed slowly but quite correctly, though he could not write from dictation. With regard to his power of copying nothing could be stated, as he could not be persuaded to make the attempt. He read aloud easily and quite correctly, and he understood both print and writing perfectly. Writing afforded the only means of communicating with him apart from gestures. The patient's condition in the above-mentioned respects remained essentially unchanged during the whole period of his stay in the hospital, from January 17 to May 12, 1891.

At the necropsy the upper parts of both temporal lobes were found to be shrunken, soft, and of a yellow colour. On the left side the posterior half of the upper temporal convolution and the supra-marginal gyrus were the parts that were softened. The island of Reil was intact. On the right side there was softening of the upper temporal convolution and a great part of the second temporal, as well as of the island of Reil, together with some small foci in the lower part of the ascending frontal, and in the third frontal, convolution.

It appears from the history of this patient that the first lesion occurred on the right side of the brain about ten years before his admission to the hospital. Six years later the lesion in the left hemisphere occurred, which resulted in the production of word-deafness. Speaking of this patient's condition during the four months that he was under observation, Pick says that his intelligence was distinctly weakened and his behaviour childish. It was difficult to fix his attention; he was very emotional and spent much time in prayer. When so engaged no sort of noise seemed to disturb him, for, notwithstanding what he said as to his hearing, it appears that, like Lichtheim's patient,[1] he was also really very deaf to ordinary sounds.

As above stated, there seems no way of explaining this case except by supposing that the left visual word centre was capable of acting directly upon Broca's centre, and without any aid from the left auditory word centre (the destruction of which had caused the word-deafness four years previously)—of acting so effectively that the patient was capable of reading aloud and of speaking, even without making mistakes in words. The word-deafness was clearly caused by the lesion in the left auditory word centre itself, and it seems equally

[1] *Loc. cit.*, p. 460.

certain, therefore, that this centre could not have been instrumental in enabling the patient to speak. It is superfluous, consequently, to look for a sub-cortical lesion.[1]

(*d*) Another possibility in explanation of the preservation of slight or disordered speech after destruction of the left auditory word centre is that *the auditory word centre of the right hemisphere acts, by means of commissural fibres passing obliquely through the corpus callosum, upon the left third frontal convolution.* This is a possibility which has not, I believe, been previously suggested by any writer on speech defects, though it seems to me almost the only means of accounting for the existence of speech of a paraphasic type in those cases where the destruction of the left auditory word centre leads to word-blindness as well as to word-deafness—that is, in about one-third of the total number of such cases. The addition of word-blindness as a symptom in these cases must, as I have said, be supposed to be due to the very close functional and topographical relations of the auditory and the visual word centres; but if the visual word centre has had its functional activity so deranged as to have led to word-blindness it cannot reasonably be supposed that it could, in that state, take on independent functions in the way of supplementing the activity of the destroyed auditory word centre.[2]

The argument in favour of the possible action of the right auditory word centre in the manner indicated is, however, even stronger than that just stated, because we shall find in a future section, when dealing with cases in which there has been more or less complete destruction of both the auditory and the visual word centres of the left hemisphere, that out of fifteen of such cases, in two speech is reported to have been fairly good, in seven there was jargon speech or paraphasia, whilst in six it was more or less markedly aphasic. But when the auditory and the visual word centres of the left hemisphere are destroyed speech, if it is to be produced by that hemisphere alone, would have to be effected solely by Broca's centre. We have, however, already seen that with double destruction of the auditory word centre, as well as with double destruction of the auditory and the visual word centres, speech

[1] Yet Mirallié actually quotes it as an example of a sub-cortical lesion.

[2] Since the publication of this paragraph it has come to my knowledge (see *Lancet*, 1897, i., p. 1408) that Byrom Bramwell had arrived independently at the view that the word centres (or "speech centres" as he terms them) in the right hemisphere have an actual existence, and an ability to take on increased functional activity far more readily than is commonly supposed.

has always been rendered unintelligible, except in one single case—that of Pick, above recorded. It thus seems to stand out very clearly that Broca's centre alone is not capable of acting as an instigator and adequate guide for the production of intelligible speech. We have good reason for believing, moreover, as I have previously pointed out, that it is not a centre whose images are much used as counters for thought, although the rousing of the sub-conscious memories of this centre is, under proper guidance, essential for the production of speech as a series of voluntary movements.

We are, therefore, absolutely driven to look for aid or guidance from the right hemisphere if we are to account for the production of even paraphasic speech in cases of damage to the left hemisphere of such a nature that word-deafness is associated with word-blindness, whether the latter be of structural or of functional origin. This same hypothesis enables us to understand the unintelligible character of the speech in three out of the four cases cited of double destruction of the auditory word centre, as well as in the case in which the auditory and the visual word centres were destroyed in each hemisphere. The old notion that intelligible speech can be initiated and produced by the influence of Broca's centre alone must be absolutely renounced.

By reference to other considerations it may also be shown to be by no means devoid of probability that both auditory word centres are actually in structural relation with, and therefore capable of acting upon, the glosso-kinæsthetic centre of the left hemisphere. It will not be difficult to show that the notions hitherto prevalent in regard to the understanding of spoken and written language and the utterance of speech have probably left altogether too much out of account that conjoint activity of the right and left hemispheres which almost certainly exists in connexion with perceptive processes, and which, therefore, we may suppose exists in connexion with derivative intellectual processes, as well as during the initiation of all sorts of motor reactions directly or indirectly following thereupon.

It is well known that each olfactory bulb and that each retina is connected with both cerebral hemispheres. Modern knowledge compels us to believe that in every act of visual perception the half-vision centres in each occipital lobe are called into concurrent activity. It would be contrary to existing knowledge for us to suppose that the cochlear fibres of the auditory nerve come into direct relation with each hemisphere, although Lichtheim does make an assumption to that effect.[1] On this subject of the intra-cerebral distribution

[1] *Loc. cit.*, p. 483.

of auditory fibres, Foster says[1] the evidence derived from "secondary degenerations" "suggests that the path of auditory impulses is along the cochlear nerve to the lateral fillet of the crossed side, and so by the posterior corpus quadrigeminum and median corpus geniculatum to the cortex of the temporal lobe of that crossed side." Still, this distribution cannot yet be considered quite settled, so that, as Foster adds, "the matter needs further investigation."

The probabilities are, however, that each auditory nerve is indirectly connected with each hemisphere through the intermediation of commissural fibres between the two upper temporal and other convolutions in each temporal lobe, constituting the cortical seat of this endowment—such commissural fibres forming part of the corpus callosum.[2] In this way, therefore, a provision exists for the conjoint activity of the two hemispheres in connexion with the sense of hearing. In healthy persons auditory impressions pass equally to the two sides of the brain, and it is commonly supposed that minute differences in intensity of the impressions, dependent upon the source of the sound in relation to us, enable us principally to judge of the direction whence the sound emanates.

Although we have no distinct proofs other than those of an analogical order that the general auditory centres are connected by commissures, we have proof that the specialised portions of them known as auditory word centres must be so connected. Had it not been for the existence of such a connexion, word-deafness would result from absolute deafness in the right ear (whether due to labyrinthine or to middle-ear disease), owing to the left auditory word centre being cut off from all afferent impressions, as I pointed out some years since.[3] We know, however, that this does not occur; therefore impressions must reach this centre through the left ear, the right auditory word centre, and thence across the commissure in the corpus callosum. And the fact that in such cases the comprehension of speech seems not to be diminished would point, I think, to the probability that the right auditory word centre is about as well developed as the left on its receptive side.[4] And if that be so, is it to be supposed that it

---

[1] "Physiology," 5th ed., 1890, p. 1089.

[2] The anterior commissure seems mainly to connect the convolutions of the tip and under surface of the temporal lobe, so that it is more probable that its upper and posterior regions should be brought into structural and functional relations through fibres of the corpus callosum.

[3] "Paralyses: Cerebral, Bulbar, and Spinal," 1886, p. 267.

[4] On this subject see what has already been said towards the end of Chapter i.

would remain functionally inert, save in the case of the accidental supervention of right-sided deafness?

Clinico-pathological investigations have long since taught us in an unmistakeable manner that the right auditory word centre usually exercises (in right-handed persons) little influence upon the glosso-kinæsthetic centre of its own side; and now I would urge that new clinico-pathological observations should strongly incline us to believe that it does, to some extent, coöperate with the left auditory word centre in the excitation of Broca's region for the production of articulate speech. We have seen that instances of paraphasic speech in cases of word-deafness associated with word-blindness are only capable of being explained in accordance with some such hypothesis, and there is the additional fact that speech has been found to be so much worse (with one exception) as to be quite unintelligible when both the auditory word centres have been destroyed.

A difficulty that presents itself in connexion with such views as I have just been expressing, is to account for the non-production of any amount of word-deafness from disease of the right auditory word centre alone. It might be thought that, even if the left auditory word centre exercises an altogether preponderant influence in connexion with the comprehension and the production of speech, still, the destruction of the auxiliary right auditory word centre ought to produce effects of some kind, of a temporary if not of a permanent nature. Extremely little evidence, however, is available upon this subject, and it is one to which most careful attention should be given in the future.[1] We know, of course, that in left-handed persons the predominant activity of the auditory and the visual word centres is in the right hemisphere, and that consequently in them we may get as a result of lesions in this hemisphere either word-deafness or word-blindness or both, just as we may get aphasia either alone or in association with left hemiplegia. Such cases of so-called "sensory aphasia" due to lesions in the right hemisphere have been recorded by Kussmaul[2] and also by Bernheim,[3] and there are doubtless many others. There was also a typical case of

---

[1] What has just been said concerning the right auditory word centre applies almost equally strongly to the right visual word centre. We want to know also to what extent and for how long the comprehension of written and printed words is impaired by destruction of the right angular gyrus and supra-marginal lobule.

[2] Ziemssen's "Pathologie und Therapie," Band xii., 1876, p. 168.

[3] *Revue de Médecine*, 1885, p. 625. Quoted in abstract as Case lxxvi.

amnesia verbalis due to a right-sided lesion recorded by Trousseau.[1]

In what I have said hitherto I have only been contending for the existence of another special illustration of a general principle which has previously been strongly advocated by others—namely, that of the double representation of movements in each hemisphere of the brain. Indeed, for many years Hughlings Jackson has gone further still, and has advanced the speculation[2] "that all parts of both sides of the body are represented in each half of the brain." Definite evidence on this subject was brought forward by Goltz, who, at the Physiological Congress in Basle in September, 1889, showed a dog from which almost the whole of the left hemisphere had been removed, and from whose behaviour, as demonstrated to the Congress, he felt able to announce the following important conclusions:[3] "Each cerebral hemisphere is in connexion with the voluntary muscles of both halves of the body. It appears, however, that the connexion between one cerebral hemisphere and the muscles of the opposite half of the body is more convenient than that on the same side. Hence a dog whose left hemisphere is destroyed prefers to use the left paw, because the will impulses from the right hemisphere flow by more convenient paths to the left limbs." He showed that sensibility was also present, and only somewhat impaired, in the right limbs of this dog. On the death of the animal the brain was removed, and was subsequently submitted to a most thorough examination by Langley and Grünbaum, who, at the conclusion of their report, say:[4] "We have seen that in this dog there was on the left side no part of the cortex remaining except a small portion of the hippocampal convolution, and a little grey matter round the optic chiasma; that the left corpus striatum had been removed, and that the nerve cells of the left optic thalamus had degenerated. Notwithstanding this, it appears from Professor Goltz's account that the animal retained voluntary power over all its muscles and could feel in every part of its body." So far as the production of movements is concerned, these results are harmonious with the well-known results obtained by Franck and Pitres,[5] who, after removing the so-called motor

[1] Lectures, translation by Bazire, pp. 253, 267. This is quoted in part as Case xliii.
[2] *Brit. Med. Jour.*, vol. i., 1884, p. 593.
[3] *Journal of Physiology*, 1890, p. 608.
[4] *Loc. cit.*, p. 634.
[5] *Archives de Physiologie*, August 15, 1883.

cortex on one side in a dog, found on faradisation of the opposite cortical region that a convulsion was produced which involved first the limbs of the opposite side and then those of the same side of the body. Although this evidence tends in the same direction, it is not nearly so conclusive as that of Goltz.[1]

There is likewise actual evidence tending to show that in man, also, all movements are to some extent represented in each hemisphere of the brain. The representation of bilateral movements, as Broadbent originally pointed out, is commonly pretty equal, whilst that of limb movements is very unequal, being strong for the independent movements of the opposite limb and weak for the corresponding movements on the same side of the body. In support of this double cerebral representation, there is the well known fact that in hemiplegia some amount of paresis exists also on the opposite side of the body (that is, on the side of the brain lesion). There is also the well-known fact that in cases of large lesions in one hemisphere occurring in early life, defects of sensibility are often even more thoroughly recovered from than the motor defects.[2]

What has already been said tends to show that bilateral arrangements of different kinds exist in the brain more or less analogous to that which I now postulate, namely, that Broca's centre is connected with both auditory word centres— very intimately with the centre of the same side, and less intimately with that of the opposite side. The fact, moreover, that there are half vision centres only in each hemisphere would seem to make it highly probable that for the voluntary performance of movements of one hand and arm, as in writing, there must, in undamaged brains, also be concurrent activity of these two centres brought to bear upon the cheiro-kinæsthetic centre during the conception and voluntary initiation

---

[1] It has been shown by Sherrington and others in their studies of "secondary degenerations" that some of these double motor effects may be accounted for by the distribution in the cord of the fibres of the pyramidal tracts. It is known also that Broadbent explains some of the bilateral movements by reference to commissures existing between functionally related bilateral centres in the cord. But the voluntary movements of the right fore-paw in Goltz's dog must have implied the development in the only existing right hemisphere of the animal of all the nervous arrangements needful for the "conception" of separate movements in this right limb—that is, the passage to it, and registration of kinæsthetic impressions from this limb, as well as the bringing of this centre into relation with the visual centre, in the manner that would be needful for the production of the voluntary movements in question.

[2] "Paralyses; Cerebral, Bulbar, and Spinal," 1886, p. 262.

of the movements. It has long been known that a functional activity of the visual centre is as essential for the conception and initiation of voluntary limb movements as a functional activity of the auditory centre is for the initiation of the voluntary movements concerned with speech. It would seem, therefore, as if, in each case, we have double auditory or double visual centres exercising their influence upon a kinæsthetic centre in the left hemisphere, in which speech movements have been registered in the one case and writing movements in the other.

The existence of the kind of nervous arrangements that I have been endeavouring to indicate in the two hemispheres would clearly necessitate the presence of numerous fibres in the corpus callosum having an oblique direction, in addition to those that are more directly transverse. It is well known that for long Meynert's view was generally accepted, to the effect that the fibres of the corpus callosum only serve to connect similar regions in the two hemispheres. By the careful study of the course of " secondary degenerations " in the corpus callosum resulting from lesions of different parts of the cortex, Sherrington has, however, shown that this was an erroneous view. He has found[1] that not only similar but dissimilar parts of the two hemispheres are connected through this commissure. In regard to some inquiries of mine concerning this subject, Sherrington writes : " My evidence was in part, as you suggest, the obliquity of the fibres, and also in part the spreading of them. . . . . I found the degenerations from even circumscribed lesions of the fronto-parietal cortex spread considerably, especially after crossing the corpus callosum. I saw nothing at all incompatible with what you urge, but anatomically everything in favour of it."

There is another point of great interest in this connection, and that is the greatly increased development of the corpus callosum in man as compared with that in the brain of the great man-like apes. We have it on the excellent authority of the late Professor Marshall[2] that the bulk of the corpus callosum, in proportion to the size of the brain, is twice as great in man as it is in the chimpanzee. It is only fair to assume that this relatively enormous increase in the comparative bulk of the corpus callosum is due in large part, either directly or indirectly, to man's acquirement of language and all the mental and moral development that has grown out of it.

[1] *Journal of Physiology*, 1889, p. 429.
[2] *Natural History Review*, vol. i., p. 296.

4. DEFECTS RESULTING FROM ISOLATION OF THE LEFT AUDITORY WORD CENTRE.

A clinical condition is produced by the isolation of the left auditory word centre—that is, by cutting it off from all its afferent fibres—which was first described by Lichtheim in 1884,[1] upon the basis of a single case of word-deafness that seemed to differ from all others that had been previously recorded. To this condition Lichtheim gave the name of "isolated speech-deafness." It has been spoken of also by Wernicke as "subcortical word-deafness," whilst lately a similar condition has been described by Déjerine (who seems to have been oblivious or unaware of Lichtheim's case) under the name of "pure word-deafness." The state itself is simple enough, though an extremely interesting one. It is the condition of a partially deaf man, the deafness being not general, but limited to word-deafness. The patient can hear ordinary sounds, though more or less badly; he can talk correctly, can write correctly, can read aloud, and understand what he reads. His three disabilities are that he is unable to comprehend spoken words, and that he is consequently unable either to repeat words or to write from dictation.

I cannot agree with the pathology of the affection as given by Lichtheim or with that advanced by Déjerine. Lichtheim says:[2] "Both ears are capable of receiving and transmitting excitations of speech, but intelligence of language is bound up with the activity of the left hemisphere only, so both acoustic nerves must enter into relation with the latter." Thus the affection we are considering, he adds, "can obviously come into existence only if the irradiations of both acoustic nerves in the left temporal region are broken through." He concludes that the union of these two tracts takes place in the cerebral hemisphere above the level of the internal capsule, and he believes it to occur somewhere in the white matter of the temporal lobe. A lesion in this situation, therefore, cutting off, as he assumes, the afferent fibres of both auditory nerves on their way to the left auditory word centre, is the explanation that Lichtheim gives of the production of "isolated speech-deafness." Déjerine's view is still less satisfactory, as he speaks only of a severance of one set of afferent fibres proceeding to this centre.[3]

In regard to these explanations it may be said that the view of Déjerine is certainly insufficient, since, as I have

[1] *Brain*, January, 1885, p. 460.
[2] *Ibid.*, January, 1885, p. 482.
[3] *Mémoires de la Société de Biologie*, 1891, p. 112.

already observed (p. 169), it is well known that the cutting off of the afferent impressions going to the brain from the right ear alone does not give rise to word-deafness. Lichtheim's explanation, on the other hand, is unsupported by any anatomical facts showing the distribution that he assumes to exist—which would really be that of a semi-decussation of the auditory nerves, since, after speaking of his imaginary arrangement on the left side of the brain, he adds in a footnote, " a similar distribution must obtain in the right hemisphere." Both Lichtheim and Déjerine seem to think that the right auditory word centre is either non-existent or possessed of no functional activity, and consequently that it is not brought into relation with the left auditory word centre. They seem also to hold similar views as to the absence of a visual word centre in the right hemisphere. If these views were correct, surely such convolutional regions in the right hemisphere ought to show some distinct diminution in size and development as compared with those of the left side. But no one as yet seems to have noticed anything of the sort.

There was no necropsy in Lichtheim's case of "isolated speech-deafness," and from the particulars given it seems clear that the patient was generally rather deaf, altogether apart from his word-deafness. Thus my interpretation of this case some time since, after a careful study of its details, was that there probably existed pretty complete deafness on the right side, cutting off the left auditory word centre from its proper incitations; whilst the slight sudden brain attack that is recorded to have occurred in June, 1882, may in some way have destroyed the commissural fibres connecting the left with the right auditory word centre. In this way we should have the isolation of the left auditory word centre brought about without the necessity for postulating the existence of any hypothetical and altogether unproved distribution of the auditory fibres within the brain. We should, in fact, have to do with a man partially deaf, capable of hearing noises only, but not able to perceive the meaning of spoken words—owing, on the one hand, to the complete isolation of the left auditory word centre from all sounds; and, on the other, to the inadequacy for this purpose of the right centre alone, because of the imperfect development of associational fibres in connexion with speech functions in this hemisphere. This view has since been rather confirmed by the record of another case by Sérieux,[1] in which it is definitely stated that the

[1] *Revue de Médecine*, 1893, p. 733. It is quoted also by Mirallié (*loc. cit.*, p. 190).

patient was absolutely deaf on the right side, as a sequence of an old otitis. The following are a few brief details concerning this case.

CASE LVIII.—The patient was a woman, aged 51 years. She had been suffering from partial word-deafness for six years. There was a complete absence of word-blindness, of motor aphasia, and of agraphia. Her hearing was normal on the left side, but absent on the right by reason of an old otitis. There was musical amnesia, the patient being unable to recognise or to sing certain airs. There was incomplete word-deafness; writing from dictation was defective, and the patient was unable to repeat words. There was slight paraphasia and paragraphia, in spontaneous speech and writing. Comprehension of written language was slightly affected.

There was progressive development of symptoms; especially subsequent to another slight cerebral attack, after which the word-deafness became complete and the power of writing from dictation was lost. Her spontaneous speech became much worse (jargon aphasia), and her writing incomprehensible. Reading aloud was possible, but most frequently without comprehension of what was read. Her general deafness became more marked.[1]

Strangely enough, it has been found that a very similar clinical combination may be produced in quite a different way, as may be seen by reference to the very remarkable case, previously cited from Pick (Case lvii.), of destruction of the auditory word centre in each hemisphere. Here the clinical condition was very similar except that there was good speech instead of jargon aphasia towards the end. There was even a certain amount of general deafness, though it was not due to middle-ear disease, nor did it seem to be definitely localised to one side. As I have already said, this case recorded by Pick seems absolutely inexplicable unless we are to suppose that the patient was a "visual" who could speak, write, read aloud, and understand what he read, mainly through the activity of the visual and the kinæsthetic word centres—at all events, without the coöperation of the auditory word centres, seeing that these were destroyed. The case shows, moreover, that general deafness does not result from destruction of these

[1] Whilst this work has been passing through the press I have seen the account of the necropsy of this patient by Déjerine and Sérieux (*Compt. Rend. de la Soc. de Biolog.*, December 18, 1897), a reprint having been kindly sent to me by M. Déjerine. As in Pick's case, both temporal lobes were affected, other parts of the brain being healthy. There were, however, notable differences; there was no trace of a focal lesion on either side. The disease (of eight years' duration) is described as a chronic poliencephalitis leading to atrophy with induration of the parts affected. The atrophy was most marked in the upper temporal convolutions, and on each side it gradually diminished from the apex of the lobes towards the angular gyrus. On the other hand, in Pick's case the softening on the left side was situated in the "posterior half of the upper temporal convolution and the supra-marginal gyrus."

special word centres even in both hemispheres. The patient's general hearing power was defective, though far from lost.

These are the only three cases of " isolated word-deafness " known to me. It will be seen from what I have said that this rare group of symptoms may be capable of being induced in two very different modes—just as we shall find subsequently that what Déjerine has called "pure word-blindness" may also be produced in two different modes, and that the modes are analogous in the two cases. Pick was fully aware of the very unusual nature of his case. He recognised that it could not be a case of sub-cortical word-deafness of Lichtheim's and Déjerine's type, though he was not prepared to advance any other interpretation. Mirallié, however, has been so very uncritical as to cite it (*loc. cit.*, p. 171) as an example of Déjerine's type, notwithstanding the fact of the existence of the double cortical lesions occupying the sites of the auditory word centres.

Before leaving this subject it will be well to refer to a very interesting case of word-deafness recorded by Byrom Bramwell[1] which presented some most unusual features. Like the cases of isolated word-deafness to which we have been referring, the patient, though word-deaf, could talk pretty well and with only comparatively few mistakes. The unusual and interesting points are, however, that the patient was able to repeat words in an echo-like fashion, and after thus uttering them could gather their meaning; also that she was able to write from dictation words that she did not understand till the writing of them had been accomplished, the patient having recovered from a temporary word-blindness.[2] The following is a very brief abstract of this important case.

CASE LIX.—A woman, aged 26 years, had a sudden cerebral attack twelve days after child-birth, which was followed by absolute deafness to all sounds for sixteen days, as well as by temporary aphasia, word-blindness, and absolute word-deafness for four weeks.

There was rapid recovery of speech, as well as partial recovery from the word-blindness, but very slow and imperfect recovery from the word-deafness.

Concerning this patient the following statements are made:—" The patient could repeat correctly words and sentences which she did not understand when spoken; and could write from dictation long sentences which she was quite unable to understand by the ear." Thus it is said: " For the past four or five weeks she has been able to repeat with more or less correctness

[1] *Lancet*, 1897, i., p. 1256.
[2] The cases of Fränkel and Abercrombie referred to on p. 158 have also a bearing upon this case.

the words of a sentence spoken slowly by another, and to take up its meaning when finished." And, again, further on it is said: "When asked a question she often echoed the last part of it, and then after repeating some of the words seemed to understand their meaning (the question here arises whether she did not get information as to the meaning of the words by the aid of the ingoing kinæsthetic impressions which result from the vocalisation and motor movements attending the production of the words)."

This last suggestion made by Byrom Bramwell does, I think, probably account for the meaning of the words being understood after she had repeated them—kinæsthetic impressions would thus come to the aid of the auditory impressions, with the result that the meaning of the words could be gathered. It would tend to show that the commissure between the auditory word centre and Broca's centre (Fig. 5, c) is also double, as I have indicated, and that the understanding of words spoken, by aid of kinæsthetic impressions resulting from speech, is a phenomenon of the same order as the understanding of writing by a word-blind patient through the aid of kinæsthetic impressions, of which several examples will be found in the next chapter (pp. 191 and 197).

As to the interpretation of the ability to repeat the words heard but not understood, Bramwell says: "The most likely supposition seems to be that the repeated (mere echo) speech passed through the right auditory speech centre (instead of through the left auditory speech centre, the usual channel) to Broca's centre in the left hemisphere." I do not, however, myself take this view of the case—in which there was no necropsy to help to clear away the difficulties besetting its interpretation. I am rather inclined to look upon it as a case of partial isolation of the left auditory word centre in a "visual" in which, however, the afferent fibres going to the centre as well as the audito-kinæsthetic commissure, in both its components, were still intact. A supposition of this kind seems to me to afford a more probable explanation of the peculiarities of the case, especially if we bear in mind the initial total deafness and the possibility, therefore, of some damage having occurred to the right auditory word centre, as Bramwell himself suggests.

I would also point out that there are certain analogies between this case and another highly interesting and unusual one that has been recorded by Déjerine (see Case lxiii.), which, I am inclined to think, is one of partial isolation of the visual word centre—its afferent fibres and the visuo-kinæsthetic commissure remaining intact. The patient in question, though word-blind, was yet able to read aloud without comprehending what she read.

CHAPTER IX.

STRUCTURAL DISEASE IN THE AUDITORY AND THE VISUAL WORD CENTRES CONCLUDED.

5. DEFECTS RESULTING FROM ABNORMAL CONDITIONS OF THE LEFT VISUAL WORD CENTRE.

THERE is not the same variety in the defects of speech caused by disease of the visual word centre that is to be met with as a result of disabilities in the auditory word centre. This is apparently due to the fact that in the great majority of persons voluntary revival of words occurs primarily in the auditory word centre, while the visual word centre is called into activity mainly by stronger stimuli—either by those coming to it through associational fibres or by still stronger incitations that come directly from without. It is only in comparatively rare cases, as we have seen, that the visual word centre takes the place occupied in the great majority of individuals by the auditory word centre, as the seat in which words are primarily revived in silent or in spoken thought. Consequently the speech defect known as " amnesia verbalis " occurs almost only as a result of a lowered functional activity in the auditory word centre, and only in extremely rare cases (where the individual is a very strong " visual ") as a result of mere lowered activity in the visual word centre.

Since a volitional stimulus is weaker than that which comes through an associational channel from a centre strongly aroused, and weaker still than that which comes to the centre directly from without; and since a lowered nutrition, or diminished molecular mobility of a centre from any cause, might lead only to a failure in the centre to respond to the weakest stimuli, it is to be expected that such results in relation to language would show themselves only in connection with the centre accustomed to respond to such stimuli (namely, the auditory word centre), and would be almost absent in connection with the visual word centre.

Consequently, we know next to nothing about the effects of mere functional degradation in the visual word centre—that is, of slight disabilities such as, when occurring in the auditory word centre, are productive of the various degrees of amnesia verbalis already described. All that can be said is that when such functional degradation of the visual word centre is present it would, (a) in the rare event of the patient being an extremely strong "visual," of itself tend to produce amnesia verbalis, as was seen in a patient under the care of Charcot, whose case is referred to by Ballet;[1] or (b) even in an "auditive" tend greatly to aggravate any disability that may have been caused by a co-existing defect in the auditory word centre, and also to hinder recovery therefrom; and finally, (c) where the functional defect involves both word centres, the amnesia ought to be more than usually bad and probably associated with more or less of paraphasia.

It happens occasionally that the visual word centre is either congenitally weak, or else has undergone some damage in early childhood, in consequence of which its activity is lowered. This shows itself by an extreme difficulty in teaching the child to read; and later on, whilst some of this difficulty continues, the child also has great difficulty in remembering what he reads. Having a poor verbal memory, there is inability to learn lessons by heart, and inability, when writing, to spell even the simplest words correctly. A patient suffering in this way was brought to me in April, 1897, by his medical attendant, concerning whom I will quote the following details.

CASE LX.—A youth, aged 18 years, well grown and healthy looking. He belongs to an intellectual family, and has himself always been a great favourite with his tutors and companions—excelling in all school sports.

He is the second of five children. Has had no severe illness except whooping cough when a child.

He was extremely slow in learning to read—his father says they almost despaired of his doing so. When writing to me about him his doctor said: "He has strong common sense, and tries his best to do his school work, but his spelling is very bad, and he has great difficulty in reading and in remembering what he reads. Algebra and arithmetic give him little trouble. He works out his problems in a way of his own (which his masters describe as working backwards) though he gets his results all right. He finds Euclid easy, unless he has to write out a problem." His father said: "He has a very defective memory for words, but not for things which have happened, or for what people have said. He has an utter inability to see through a Latin sentence, even after going over it several times."

I found that he spoke freely and naturally—though he is said occasionally to bring out wrong words. Two letters to his parents were shown to me in which many words were spelt wrongly. I have noted that he had written "hear" for *here*, that he had three times written "too" instead of *to*, and that once he had written "toothack" for *toothache*.

[1] *Loc. cit.*, p. 101.

He read to me a passage from one of Anthony Hope's novels in a very slow, hesitating manner, making occasional mistakes; for instance, saying "on" for *no*, and "now" for *when*; while he stopped altogether at the word "moustache," though it was followed by the words "on his upper lip." He made also, after hesitating long, only a very poor attempt to pronounce the word "straddled."

A case very much of this type, though with more marked disease of the visual word centre, has been described by W. Pringle Morgan[1] as one of "congenital word-blindness." The greatest difficulty was experienced in teaching this boy his letters—so that his parents thought he never would learn them. This same kind of difficulty continued up to the date when his case was reported—at the age of 14 years. Thus the writer, says: "He has been at school or under tutors since he was seven years old, and the greatest efforts have been made to teach him to read, but in spite of this laborious and persistent training he can only with difficulty spell out words of one syllable. . . . . . Words written or printed seem to convey no impression to his mind, and it is only after laboriously spelling them that he is able by the sound of the letters to discover their import."[2]

Effects produced by the Destruction of the left Visual Word Centre.—We have to pass, then, at once to a consideration of the effects produced by destruction of the left visual word centre, which is now generally supposed to be situated in the angular, and possibly in part of the supra-marginal, gyrus. These effects will be found to vary in different individuals, just as variation was found to occur when we had to do with the results of lesions in the left auditory word centre.

One important difference between the effects of lesions of the visual word centre as compared with those following upon lesions of the auditory is that speech is very little, if at all, interfered with. There is no well-marked aphasia or paraphasia, only a very slight amount of paraphasia in some of the cases.[3] That there should at times be such slight disturbance of speech is not to be wondered at if we bear in mind the close functional relationships of the visual and the

[1] *Brit. Med. Jour.*, November 7, 1896.

[2] Seeing that this boy, who was only able to read letters and short words in the tedious manner indicated, could nevertheless write from dictation (though very badly, and with all except monosyllables utterly misspelt), his case appears to be one of "pure word-blindness," as described on p. 122.

[3] For instance, in Nos. 44, 45, 47, 49, 50, and 52 in Mirallié's list of cases included under the heading "Cécité Verbale" (pp. 152-157).

auditory word centres, and how easily, therefore, the functions of the latter may be disturbed by a lesion in the former.

It is often found that word-blindness is associated with right-sided homonymous hemianopsia, that is, on the same side as paresis of the limbs—for there is often no complete hemiplegia in these cases. The hemianopsia is generally due to destruction of the optic radiations of Gratiolet; but where the lesion is quite limited to the cortex these fibres will not be involved, so that hemianopsia is not a necessary accompaniment of word-blindness.

Agraphia also is a symptom sometimes present and sometimes absent in cases of word-blindness. The presence or absence of this symptom is, however, of much more fundamental importance than the presence or absence of hemianopsia. It is best, in fact, to divide cases of word-blindness due to destruction of the left visual word centre into two categories, namely, (a) cases in which the word-blindness is associated with agraphia—these being cases of ordinary "cortical word-blindness;" and (b) cases in which there is no agraphia—these representing one form of what is known as "pure word-blindness." In the latter group it is found that the individuals can write as well with the eyes closed as when they are open, and further that they are, after a brief interval, unable to read what they themselves have written.

(a) *Cases of Disease of the Left Visual Word Centre causing Word-Blindness and Agraphia. (Ordinary Cortical Word-Blindness.)*

In the cases where agraphia is present it is commonly found that such patients are capable of writing their names correctly, although they can write nothing else. The signature is an emblem which may, by reason of its familiarity, be executed by the cheiro-kinæsthetic centre with the smallest amount of prompting; and it would appear that this prompting may come from the common visual centre in cases where the left visual word centre is destroyed. Similarly a word-blind patient may be able to recognise his own name when he sees it, though apart from this there may be complete alexia. As Déjerine says:[1] "He recognises it by its general form, by its physiognomy, and not by the assemblage of letters of which it is composed"—just, in fact, as he would recognise a geometrical figure or any other drawing. Differences of degree are, however, met with in word-blindness. In some cases, though the

[1] *Comptes Rendus de la Société de Biologie*, 1891, p. 200.

patients are unable to recognise single words (word-blindness), they can recognise individual letters; whilst in others not even letters can be recognised (letter-blindness). These two forms were distinguished by Kussmaul, though he sought to make a more radical distinction between them—believing that they represented differences in kind rather than of degree. All intermediate forms, however, may be met with. Sometimes patients will recognise certain letters and not others; thus, in a case recorded by Batterham [1] vowels only were recognised. Byrom Bramwell,[2] moreover, has recorded a rare instance in which, though individual letters could not be recognised, many words could be read.

In illustration of this first group of cases, where word-blindness is associated with agraphia, I will quote one of a very typical character that has been reported by Déjerine.[3]

CASE LXI.—A man, aged 63 years, had seven years previously a slight attack of right hemiplegia, not accompanied by speech troubles, which, at the time of his entering the Bicêtre (February 12, 1890), had almost completely disappeared.

Soon after his admission he found one morning that he could no longer read his newspaper. On examination it was ascertained that there was no trace of word-deafness; he easily understood all questions. He was unable to understand either print or handwriting, and could not even name a single letter of the alphabet. On the other hand, he could recognise his name and could name all objects shown to him. Right lateral hemianopsia was believed to exist. There was paraphasia both in his spontaneous speech and in his repetition of words, so that it was sometimes difficult to understand what he said. When asked to write he held the pen most awkwardly and could write nothing but his name, and that so badly as to be scarcely recognisable; and whether he attempted to write spontaneously, from dictation, or to copy, he wrote only his own name, "Séjalon," in an indistinct fashion.

On March 10 the initial paraphasia had almost completely disappeared, so that he could readily give particulars concerning his former life and his present condition. His writing was as bad as before, but he held the pen much better. He could not write a single word spontaneously or from dictation, nor could he copy—he made mere meaningless strokes only; but he could write from dictation a numeral or a number composed of not more than two figures. There was complete alexia, as before; he recognised his own name, but nothing else. On November 5, 1890, the word-blindness was slightly less. He could then recognise "c" and "g" of the alphabet, but no other letters. He could not make out any word, but he recognised and pronounced numerals and numbers of not more than two figures. The agraphia was as complete as before, both for spontaneous writing, dictation, and copying.

Death occurred on November 20, 1890, and at the necropsy a focus of softening having "the diameter of a five-franc piece" was found occupying the inferior three-fourths of the angular gyrus, all the rest of the cortex,

[1] *Brain*, 1888, p. 488.
[2] *Edin. Med. Journ.*, 1887, p. 241.
[3] *Loc. cit.*, p. 197.

including the foot of the third and also that of the second frontal convolution, being absolutely intact. Section of this left hemisphere showed that the softening extended inwards in a wedge shape as far as the posterior cornu, destroying the greater part of the optic radiations of Gratiolet. In the right hemisphere two small foci of softening of the size of a small nut were found—one in the putamen, the other in the anterior part of the thalamus, which destroyed by their union the genu of the internal capsule. (Figures of the lesion in the left hemisphere are given.)

The absolute nature of the agraphia in this case is a noteworthy feature, and may perhaps be accounted for, in part, by the fact that writing was a comparatively unfamiliar exercise to this man, who was a mere day labourer. Damage to the visual word centre under such circumstances might be expected to produce its maximum results, while a similar lesion in the case of an educated man much accustomed to write might lead either to no agraphia or to a much less pronounced form of this defect. Another point of interest is the temporary nature of the paraphasia that was at first set up as a consequence of this lesion strictly limited to the angular gyrus. We can only conclude, therefore, that it was due to a sympathetic disorder established for a time in the contiguous auditory word centre. A third point of interest is to be found in the strict localisation of the lesion to the angular gyrus and its sufficing to produce complete alexia with agraphia.

Another very similar case, and almost equally valuable, has been recorded by Sérieux,[1] of which the following is an abstract.

CASE LXII.—A woman, aged 73 years, entered the Asylum of Villejuif on September 29, 1891. At the first examination it was ascertained that the patient heard and comprehended all questions well, but that there was some amount of amnesia verbalis. Reading and writing were impossible, although vision was good and she could recognise surrounding objects. After a time, being troubled by her inability to write, she practised much with her pen. At first she could only copy the letters whose outline is most simple, such as m, n, u, o, a. Her alphabet was almost composed of these letters, and they were sometimes grouped so as to give the appearance of words—such as um, aa, monon, mono, muosi.

In the month of November the following observations were made: "Intelligence normal. No motor paralysis. No appreciable word-deafness. Slight paraphasia. No object-blindness. Word-blindness is almost complete: she recognises a certain number of letters only. Hemianopsia absent. Agraphia is also complete for spontaneous writing, as well as for dictation and copying." On November 23, the patient had an abrupt apoplectic attack, she became comatose, and died.

At the necropsy, in addition to a recent hæmorrhage into the right hemisphere, with extravasation into the ventricles, there was found one old lesion in the left hemisphere. This was a focus of yellow softening rather larger than a five-franc piece, which involved the whole of the angular gyrus

[1] *Mémoires de la Société de Biologie*, 1892, p. 13.

and the greater part of the supra-marginal globule. The contiguous convolutions of the superior parietal lobule as well as the posterior extremities of the first and second temporal convolutions were slightly yellowish and atrophied. All the other convolutions of this and of the other hemisphere were healthy.

Here, although the lesion was more extensive than in the last case, we find both the word-blindness and the agraphia rather less complete. The slight paraphasia, which in this case was persistent, may have been kept up by the slight degeneration of the posterior extremities of the two upper temporal convolutions. It is expressly stated that the patient did not suffer from hemianopsia, so we must conclude that the lesion was superficial and did not extend deeply enough to involve the optic radiations. Another case in which word-blindness and agraphia were produced by a patch of softening limited to the left angular gyrus has been published by Berkhan,[1] though under a very misleading title.

A case of a remarkable kind to some extent intermediate between the cases of this first group and those of the second has been recorded by Déjerine,[2] of which the following are the essential details.

CASE LXIII.—A woman, 37 years old, was admitted into hospital suffering from partial right hemiplegia with word-blindness and partial loss of speech—she being unable to pronounce a whole series of words.

"She could read aloud with facility, though without understanding what she read. The written words read aloud by her aroused in her no idea; she read, so to speak, in a reflex fashion. . . . . . When a sentence was dictated to her she wrote it correctly, then read it, but without comprehending it any more than when one made her read a sentence taken from some book."

This combination of symptoms gradually changed, so that fifteen days after her admission she had become completely aphasic, and was suffering from a well-marked right hemiplegia of flaccid type, though both these symptoms varied in degree somewhat strangely from day to day.

She died comatose a month later, and at the necropsy a gliomatous tumour about the size of a Mandarin orange was found occupying the inferior parietal lobule—the third frontal convolution and other parts of the brain being healthy.

The fact of a word-blind patient being able to read aloud in the ordinary way, though without comprehension, was recognised by Déjerine as a very rare symptom. The aphasia he considered due to some indirect pressure upon Broca's region. It seems to me that in the early stage of this woman's illness, the functional disturbances must have been due to a partial isolation of the visual word centre, while the afferent

[1] *Archiv für Psychologie und Nervenkrankheiten*, 1891, p. 558.
[2] *Progrès Médical*, 1880, p. 629.

fibres, in part at least, and the visuo-kinæsthetic commissure remained intact.[1] Another possible interpretation is, however, referred to for a somewhat similar case on p. 96.

(b) *Cases of Disease of the Left Visual Word Centre Causing Word-Blindness without Agraphia. (One form of Pure Word-Blindness.)*

In the second group of cases word-blindness is not associated with agraphia, the individuals being able to write, and as well with the eyes closed as open. As I have already said, a precisely similar group of symptoms may be produced in a totally different manner, quite independently of destruction of the left visual word centre. These cases will be fully considered in the next section.

The effects of destruction of the left visual word centre are liable to vary much in different individuals in accordance with their different sensorial aptitudes and different degrees of education, just as we have found the results of destruction of the left auditory word centre to present marked differences in different cases. The results already described under group (a) are those which most frequently follow when the left visual word centre is destroyed in ordinary individuals. But suppose a similar lesion to occur in a person who is a strong "auditive" but a weak "visual," and at the same time, perhaps, an educated person who has previously been in the habit of writing much. It may happen in such a person, after the stage of learning to write has well passed, that the activity of the visual word centre whilst writing may be reduced to a minimum.

We have previously seen reason to believe that the words about to be written become nascent first in the auditory word centre, and it seems probable that in individuals with the endowments above mentioned this centre, rather than the visual word centre, may be the one which acts upon and coöperates with the cheiro-kinæsthetic centre, just as we have seen that in certain persons who are strong "visuals" the visual word centre may be capable of coöperating directly with Broca's centre for the production of articulate speech.[2]

---

[1] It is interesting to compare this case with Case lix., in which, I think, there was partial isolation of the auditory word centre.

[2] I find that I made a suggestion of this kind very many years ago (*Brit. and For. Med. Chir. Rev.*, April, 1869), in explanation of two early cases of this type which I then referred to in the following terms :—" Dr. Osborn alludes to one recorded in 'Ephemerides Curiosæ,' in which an elderly man, after an attack of apoplexy, forgot how to read, or even to distinguish one word or letter from another; but if a name or phrase was mentioned to him he could write it immediately and accurately. He was, however,

I contend, therefore, that the preservation of ability to write in cases where there has been word-blindness and where the visual word centre has subsequently been found to be destroyed is to be explained in this way—that is, by supposing that the auditory word centre, instead of acting, as it usually does, by rousing the visual word centre to conjoint action, in these cases acts directly upon the cheiro-kinæsthetic centre by way of the commissure ff (Fig. 7).[1] The writing thus produced may be fairly good and without mistakes, though that of other patients may show defects of a paragraphic type. The handwriting itself, too, is usually larger than that which was previously customary to the patient. There is likewise the peculiarity that these patients can write as well with the eyes closed as when they are open, and the still further peculiarity that they cannot subsequently read what they have written except by a manœuvre which causes a stimulus to pass in the reverse way—that is, from the cheiro-kinæsthetic centre back to the auditory word centre in the direction $f'f'$ (Fig. 7). This is brought about by passing the tip of the finger over the outlines of the letters and so reading off the result. The possibility of this mode of reading by the "tip of the finger" seems to have been first noticed in a word-blind patient by Westphal.[2]

There are only two cases with necropsies that I can bring forward in illustration of this group, and unfortunately they are not very well defined cases, being rather complicated, not only clinically, but also by reason of the lesions found at the necropsy. The first of them is also one of the earliest recorded instances of any such defect. It is a remarkable case that was published long ago by Broadbent.[3] The real significance of the case was not, I believe, recognised by him at the time, nor was it by myself some years afterwards when I first

incapable of reading what he had written, and if asked what a letter was or how the letters were combined, it became evident that the writing had been performed automatically, without any exercise of the reflection or judgment. In this case none of the means that were employed were successful in restoring the knowledge of letters to his mind. Marcé also records a case of an attorney who was quite unable to read, who could not express his ideas at all in writing, but who wrote without difficulty any word just pronounced before him, although at the same time he remained quite ignorant of its meaning."

[1] It is worthy of note that both Kussmaul and Spamer seemed to think that this was the ordinary way in which writing is effected (see Kussmaul, loc. cit., p. 777).

[2] *Zeitschrift für Ethnologie*, 1874.

[3] *Transactions of the Royal Medical and Chirurgical Society*, 1872, p. 162.

endeavoured to interpret it.[1] Now I think its leading peculiarities may be explained by supposing the existence of a weak condition, with lowered excitability of the auditory word centre, together with a destruction of the visual word centre. The case is well recorded with full details, but the essential particulars are these.

CASE LXIV.—The patient was a man, aged 59 years, and of considerable energy and intelligence. His illness commenced early in 1870 with sickness, vomiting, and pains in the head. He was restless and delirious, and did not know his wife for a fortnight. He gradually improved, but after this attack remained unable to read either print or writing, though he could write quite well. Twelve months later he was no longer able to write voluntarily though he could still write correctly from dictation. Soon afterwards when he came under observation he was found to be unable to read. He could see the words, but could not understand them; he could not even recognise or name single letters; the only exception to this being that he recognised his own name, whether printed or written.

His voluntary speech was free except that he was amnesic and much at a loss for names. If a hand or an article of clothing, or any other familiar object, were shown him, he was quite unable to name it; while if a name came up in conversation he spoke it without hesitation. He also immediately recognised the names of common objects when he heard them. Thus, asked the colour of a card he could not give it. "Is it blue?" "No." "Green?" "No." "Red?" "Well, that's more like it." "Orange?" "Yes, orange." A square and a circle were drawn and he was asked to name either. He could not do it, but when the circle was called a square he said, "No, but that is," pointing to the proper figure. He wrote from dictation and took notes of instructions as to the day and hour of subsequent visits (which were seen to be correct), saying he was very forgetful and might make mistakes. He would be utterly unable shortly after to read his own writing, but relied on his wife making out the notes he took. He was a vestryman, and still always attended the meetings of the vestry. He was not able to take part in the discussions as he had formerly done, but it amused him, and he added, "But I dare say I am as useful as many of them still."

Broadbent adds: "He was a remarkably intelligent man, cheerful and energetic, even under his affliction. He employed in conversation a very extensive vocabulary, usually speaking fluently, sometimes stopping for want of a word (usually a name), rarely using wrong words (for which I was on the look out), the only ones I have notes of being 'soup' for 'supper,' 'nephew' for 'grandson.'" He remained under observation from May 8 to June 1, 1879.

On June 21 he had a severe apoplectic attack, due to ventricular hæmorrhage, and died within forty-eight hours. The recent extravasation had torn its way into the left temporo-sphenoidal lobe, in which there was a focus of softening, though its extent could not be ascertained in consequence of this irruption of blood. Of principal significance, however, were two old blood-clots, the size and situation of which are thus described: "One about the size and shape of an almond was closely embedded in the inframarginal gyrus between the deep parallel sulcus on one side and the secondary small gyri on the lower wall of the fissure of Sylvius on the other, about opposite the junction of the upper third with the lower two-

---

[1] "Brain as an Organ of Mind," 1880, p. 645.

thirds of the descending cornu. The other was further back and on a higher level, almost exactly corresponding in situation with the posterior end of the fissure of Sylvius externally and with the junction of the descending cornu with the body of the ventricle internally; in fact, it occupied the thickness of brain substance separating the extremity of the fissure from the ventricle. It was about the size of a bean, yellow, and very tough.

It seems pretty clear that the second of these blood-clots must have been situated in the angular gyrus, whilst the first of them was situated somewhere near the middle of the upper temporal convolution, and therefore that the patient's alexia was due to destruction of the visual word centre. Notwithstanding the damage to the upper temporal convolution and the existence of an ill-defined amount of softening in the white substance of the temporal lobe, there was obviously nothing approaching to word-deafness in this patient. The only means, therefore, of explaining his ability to write with the right hand is by supposing that his cheiro-kinæsthetic centre must have received its stimulations from the left auditory word centre when he wrote from dictation (Fig. 7, ff), even though the trifling verbal amnesia showed that this centre was slightly weakened.

Another case that should, as it seems to me, find its place in this group, has been recorded by Osler,[1] of which the most important details are these.

CASE LXV.—A man, aged 72 years, applied at the Philadelphia Infirmary for Disease of the Nervous System on November 14, 1888, complaining of uneasy sensations in his head. He was a healthy, vigorous-looking man, perfectly intelligent, and spoke well and clearly. It was not thought at first that there was anything the matter with him beyond slight headache; but it was noticed that he had occasional difficulty in getting the word he wished, and this circumstance led to a more careful examination.

For some time past he had not felt so well as usual. On November 1, while at his supper in a restaurant, he found that he could not read the daily paper. He was sure that this came on quickly, and it had been his chief annoyance, as he was an ardent politician. He had no definite headache, but complained of a diffuse uneasy sensation. Though he spoke clearly and intelligently, uttering some sentences without interruption, replying promptly and fluently to questions and evidently understanding everything, there was very distinct speech disturbance. Thus, for some time he could not give the address of his residence. He said he knew where it was, but could not utter the words. He gave the first name of the man with whom he lived, but could not give the second. He could not name his own occupation, but said "Keep, keep, keep; Oh, you say it for me." When told "bookkeeper," he repeated it distinctly. He occasionally misplaced words. In referring to a wetting which he had spoken of he said, "Deliberate attacks of wet dress." When a printed or written page was presented to him he did not appear to comprehend the words. The word "Philadelphia" at the head of a hospital blank he read "P, r, i, n, g, r, e, k." When told that it was Philadelphia he replied, "Oh, certainly it is, I've known it for sixty-five years." His age, "72," written on a slip of paper,

[1] *American Journal of Medical Science*, March, 1891, p. 219.

he read "213." He did not recognise the words "Cleveland" and "Harrison" at the top of a newspaper column, but when they were read to him he said, "I know all about them" and began making some very shrewd observations. He could write his name, but said that since his failure to see distinctly he did so with difficulty. He wrote as well with his eyes shut as when they were open, but did so with hesitation. He wrote the name of the hospital, and the words "Philadelphia Record." He could not read the words of his name after he had written them. He named objects held before him quite readily. On examination, right lateral homonymous hemianopsia was found.

For the first two weeks of his stay in the hospital there was no special change. On one occasion he wrote the word "record" when told to, but after he had written it he spelled it "freedom." On December 8 he talked less freely. He spoke intelligently and plainly at first, but after a few minutes it was difficult to understand what he stated. There were no additional ocular changes. The grip in the hands was equal. He walked with a somewhat tottering gait, though there was no actual paralysis.

Early in January, 1889, he became distinctly weaker, duller mentally and more restless, and on the morning of January 15 he was found in a semi-comatose condition, completely paralysed in the right arm, and incompletely in the right leg. He died on the following day.

At the necropsy softening was found of the left supra-marginal lobule and of the white substance of the lower part of the angular gyrus; also of the white matter of the posterior parts of the first and second temporal convolutions. There was likewise complete softening of the white matter between these convolutions and the lateral ventricle (posterior horn); while another area of softening two inches in thickness and an inch in breadth was found in the white substance of the temporal lobe, which externally touched the grey matter of the third and the base of the second temporal convolutions. Concerning the focus of softening in the substance of the left hemisphere, Osler says: " Posteriorly the softening did not extend behind a line drawn across the level of the parieto-occipital fissure. The white matter of the occipital lobe was firm, and the grey matter of the cuneus was uninvolved." The right hemisphere showed no lesions.

It is, of course, impossible to say how much these areas of softening may have extended during the last four or five weeks of this patient's life, when he seems to have become gradually worse; or, in other words, we cannot say how much more limited the lesion may have been during the first week that he was under observation, to which period the above detailed clinical record pertains. It seems clear, however, that the clinical symptoms were of the type now under consideration— that is to say, that there was no word-deafness, but that the patient was affected with word-blindness without agraphia. On the other hand, it is equally clear that even at the end, when the pathological lesions had attained their maximum development, the softening had not extended into the occipital lobe. We seem, therefore, driven to the conclusion that the word-blindness must have been due to the softening of the angular and the supra-marginal gyri.

What I take to be another example probably belonging to

this group is to be found in a patient under the care of Charcot, whose case has been recorded in great detail by Bernard.[1] It is an extremely interesting case, and here, as in that of Sérieux, previously quoted, the destruction of the visual word centre does not seem to have been complete, as the patient could still recognise individual letters. The following are some of the principal details.

CASE LXVI.—A man, aged 35 years, after unusual excitement, was suddenly seized with right hemiplegia and loss of consciousness. The next day, on recovering consciousness, he stammered and substituted one word for another. At the end of three weeks the disturbance in speech had almost entirely disappeared; the hemiplegia had been gradually subsiding, so that he could now hold a pen and write very legibly. About this time, wishing to give some order concerning his business, he took a pen and wrote what he wished to have done; then, thinking he had forgotten something, he asked to see the letter that he had just written, but found he was unable to read it. He had been able to write, but could not read what he had written just before. He was also unable to read printed matter.

He was seen five months after the onset of his illness by Charcot, who ascertained that there was no longer any hemiplegia or aphasia of motor type, but that in addition to the alexia there was a forgetfulness of a certain number of nouns and proper names. It was observed that he could write a long letter without any notable mistakes in orthography. "I write," he said, "as if I had my eyes shut; I do not read what I write." In fact he wrote quite as well with his eyes closed. Having written his own name he was asked to read it. "I know very well," said he, "that it is my name that I have written, but I can no longer read it." Nevertheless, when it was insisted that he should read some written words, it was observed, while he was making efforts to do so, that with the tip of his right index finger he traced one by one the letters that constituted the word, and he was thus able with much trouble to say what the word was. Or else when he was striving to read he traced in the air with his finger the letters placed before him. It was found that he had much more difficulty in reading print after this fashion than ordinary writing, apparently because of his lack of practice in tracing the former kind of characters. Again, if when his eyes were closed a pen were put into his hand and the latter guided so as to make it write certain words, he was able to say immediately and without hesitation what those words were.

He knew and recognised separately all the letters of the alphabet, except q, r, s, t, and especially x, y, z. Though he could not make out these latter letters when they were isolated, he nevertheless wrote them easily when they formed part of a word. Numerals he recognised well, and was able easily to perform certain simple operations; but when the multiplications were a little complex he made mistakes. This patient suffered also from right lateral hemianopsia.

In the absence of a necropsy, it cannot be definitely said that this case belongs to the group now under consideration rather than to that which is next to be considered. Looking to the fact, however, that a hemiplegic condition existed at first, that there was no hemianæsthesia or hemiachromatopsia, and to the very limited nature of the clinical defects subse-

[1] "De l'Aphasie," 1885, pp. 77-90.

quently remaining, it seems to me distinctly less probable that the patient's condition was due to lesion in the occipital region than to one further forward which also involved the visual word centre itself. If that were the case, the reading by means of kinæsthetic impressions would mean that the excitations of the cheiro-kinæsthetic centre, produced by the finger tracing successively the outlines of letters, must have been conveyed backwards along the commissure $f'f'$ (Fig. 7) to the auditory word centre—there arousing the corresponding auditory word-memories, and subsequently passing over to Broca's centre preliminary to the articulation of the words.

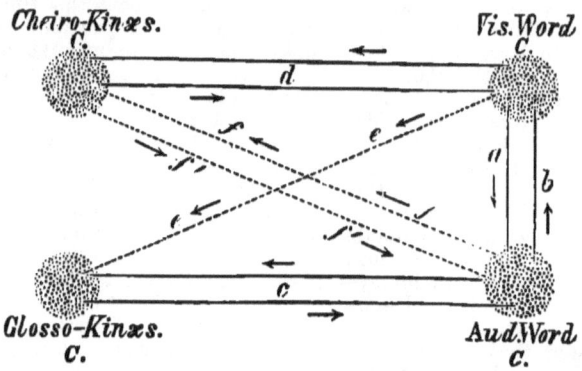

FIG. 7.—A Diagram illustrating the relative positions of the different Word Centres and the mode in which they are connected by Commissures.

The connections represented by dotted lines indicate possible but less habitual routes for the passage of stimuli.

It will be observed that in each of these three cases, in which word-blindness has been present without agraphia, the patients have been more or less educated persons well accustomed to write, and that in the first two, at all events, the incitations could not have come during their illness from the damaged visual word centre. Therefore it seems only open to us to suppose that these patients were able to write spontaneously or from dictation, simply because they could call up the familiar activity of the cheiro-kinæsthetic centre at the direct instigation of the auditory word centre. It is well known that many voluntary movements, during the period that they are being learned and sometimes for long afterwards, require for their execution the active coöperation of the visual centre, though at a later period the guidance of the kinæsthetic centres may suffice without any aid from the visual centre.

6. Defects Resulting from Isolation of the Left Visual Word Centre.

Isolation of the visual word centre is a term employed here in the same sense that isolation of the auditory word centre was previously spoken of—it is not a complete isolation, but merely a cutting of the centre off from all its own afferent fibres, as well as from all communication with the corresponding centre of the opposite hemisphere. The clinical condition itself was first referred to by Wernicke [1] as a companion to the state described by Lichtheim as subcortical sensory aphasia or isolated speech-deafness. Wernicke named this condition "subcortical alexia," while it has since been re-christened by Déjerine (who is followed by Mirallié) as "pure word-blindness." The question as to which nomenclature is to be adopted is therefore a little embarrassing. Seeing, however, that, as I have shown, each of these clinical groups of symptoms may be produced in a double way—that is, either by cortical or by subcortical lesions—it seems clear that the term " subcortical " as applied to these conditions by Lichtheim and by Wernicke would, if preserved, lead to great confusion. It seems best, therefore, to adhere to Déjerine's nomenclature and speak of "pure word-deafness" and "pure word-blindness" respectively, but with the understanding that these defects are not necessarily due to subcortical lesions.

The cases of pure word-blindness with which we are now concerned, in which the visual word centre remains intact, are comparatively few in number. Three complete cases have been recorded with necropsies, whilst of two others the clinical details only have been published. Although the two latter cases probably belong to this type, there is, of course, the possibility that they may rather belong to the second group of cases included in the last section. I shall first give some brief account of the three complete cases, and subsequently discuss the mode in which the lesions found may be considered to be capable of giving rise to so remarkable an assemblage of symptoms.

The first of these cases to be recorded which was followed by a necropsy is one of great value; it was most carefully observed, and has been published in very full detail by Déjerine.[2] The following is the abstract that he gives by way of preface to the fuller details, which will, however, well repay a careful study.

---

[1] *Fortschritte der Medicin*, 1886, p. 463.
[2] *Mémoires de la Société de Biologie*, Février 27, 1892.

CASE LXVII.—This patient was a man, aged 68 years, of more than ordinary intelligence and culture. Word-blindness was complete, both verbal and literal. Musical blindness was also complete. There was preservation of ability to read figures and also to calculate. There was no trace of word-deafness, no trace of disturbance in articulate speech, and internal language was preserved. There was no object-blindness or optic aphasia; mimicry was perfect and very expressive; there was perfect preservation of spontaneous writing and of writing from dictation—the patient could thus write whole pages correctly. Writing from a copy was very difficult and defective. There was partial right lateral hemianopsia with complete hemiachromatopsia. There was integrity of motility, and of sensibility general and special, as well as of the muscular sense. The same symptoms persisted for four years.

Sudden death occurred after the patient had presented for ten days total agraphia with paraphasia, without any trace of word-deafness. There had been perfect preservation of mimicry and intelligence. At the necropsy recent lesions were found in the left hemisphere in the form of red softening in the inferior parietal lobule and in the angular gyrus. There were old lesions (*plaques jaunes*) situated in the lingual and the fusiform lobules, in the cuneus, in the white substance of the occipital lobe, and in the posterior extremity of the corpus callosum. The optic radiations were markedly atrophied. The right hemisphere was intact.

The notable features of this case are the absence of any hemiplegia; the combination of complete right lateral hemiachromatopsia with partial right hemianopsia; the absence of word-deafness or speech defect of any kind; the absolute word-blindness coupled with ability to write freely both spontaneously and from dictation, whilst ability to copy writing was very defective;[1] and the length of time that these symptoms lasted without change. The subsequent sudden supervention of complete agraphia associated with paraphasic speech, followed by the discovery of a recent lesion in the angular gyrus and adjacent parts, were also very significant features of the case. Comments upon the old lesions found had better be reserved till some account has been given of the other two cases.

The second case was recorded by Wyllie[2] two years later, some of the principal details being as follows.

CASE LXVIII.—A man aged 72 years, formerly a clerk of the works, but then a collector of accounts for mercantile men, came as an out-patient to the Royal Infirmary of Edinburgh on December 12, 1889. About six weeks previously, when in the street, he had a very severe attack of vertigo, and ten days later a feeling of coldness and heaviness throughout the right half of the body. At this time also he discovered that he was word-blind. He could not read even his own name. At first he was also letter-blind, but before coming to the infirmary he had recovered sufficiently to be able to recognise individual letters, though he could read no single word without spelling it.

[1] There ought to be in these cases ability to form words with separate letters such as a child uses (*écriture typographique*).

[2] "The Disorders of Speech," 1894, p. 340.

On examination no right hemianæsthesia was found, but typical right hemianopsia, with marked contraction of the remaining left halves of the fields of vision. The contraction was so marked that he saw things as if looking at them through a tube. There was no agraphia; he could write well and made very few mistakes in spelling; such as he did make being evidently due not to his cerebral condition but to defective education. Although he expressed himself in writing with ease and fluency, he could read the writing only word for word, after spelling each word letter by letter. As to spoken speech, the patient expressed himself with great liveliness and volubility, being never at a loss for a word except in the case of proper names, which he was apt to forget. There was no forgetfulness of the names of things and no paraphasia. His intelligence seemed very good, but he complained of a sense of confusion in his head. This, together with his word-blindness, had made him quite unable to continue his occupation of collecting accounts.

He died four years later (January 23, 1893), of hepatic disease, with jaundice, and at the necropsy the following lesions were found. In the left hemisphere there was atrophy of the inferior surface of the occipital lobe, with a shrivelled condition of the whole lobe. Subsequent examination of sections showed softening of the white substance forming the floor of the posterior cornu of the lateral ventricle, the cornu itself being extremely dilated. The grey matter of the occipital convolutions was not involved in the softening, but the white substance was atrophied from the tip of the occipital lobe as far forwards as the cerebral peduncles. The convolutions affected were the lingual and the fusiform lobules, together with the posterior third of the gyrus hippocampus. The angular gyrus and Broca's convolution were intact.

The third case was reported in the same year by Redlich,[1] of which the following is an abstract.

CASE LXIX.—A man, aged 65 years, a copyist, had some slight cerebral symptoms for one month only in 1891. A year later he had an apoplectic attack, and on examination two days afterwards his intelligence was found to be slightly enfeebled and right lateral hemianopsia was detected. There was also slight paresis of the right arm and the right side of the face, together with slight diminution of sensibility throughout the right side of the body. There was no word-deafness, and spontaneous speech was unaffected. He had difficulty in naming objects shown to him. He could recognise the number of syllables in words. He sang perfectly both the words and the airs of popular songs. There was complete word-blindness, literal and verbal, both for print and manuscript. He could read numerals slightly. He could only write the first syllable of his name, and could not copy at all.

At a second examination, one month later, the following facts were noted. Spontaneous speech was perfect, except that he confused certain words, and had sometimes a little difficulty in finding the right word. The word-blindness was still complete. He then wrote his name very correctly; he also wrote a great number of words slowly and with difficulty, both spontaneously and from dictation, sometimes forgetting a letter. The writing was tremulous, but correct. He copied with difficulty, letter by letter.

Death occurred one year and a half after the date of this attack, and at the necropsy the following were the lesions in the left hemisphere. A focus

[1] *Jahrbuch für Psychiatrie*, 1894, p. 242.

of softening was found involving a great part of the calcarine fissure, and the lingual and fusiform lobules, and extending to the posterior part of the occipito-parietal fissure; the cornu ammonis was atrophied; the cortical softening invaded the white substance of the occipital lobe and of the fusiform lobule, as well as the posterior part of the thalamus, the tail of the caudate nucleus, and the median portion of the corona radiata. Sections in series of the occipital lobe were made and examined microscopically, with the following results: At the tip of the lobe both the white and the grey substance were intact. The focus of softening corresponded with the fusiform lobule, and the inferior half of the lingual lobule. The forceps major was degenerated only in its median portion; its supero-external portion being intact. The tapetum was in part preserved. The inferior longitudinal convolution as well as the optic radiations were degenerated. The hippocampal gyrus was much altered, as well as a part of the third temporal.[1]

A case probably belonging to this same category has been published by Hughes Bennett,[2] which, though there was no necropsy, and though some important details are wanting, is, however, in other respects very interesting.

CASE LXX.—A shipwright, 52 years old. Health good till twenty months previous to examination, "when one day he suddenly fell down unconscious and remained so for about an hour." Soon after this attack he regained his former condition, returned to work and so continued till five months before coming under observation, when he gradually became incapacitated owing to giddiness and increasing inability to read.

He came complaining of right-sided hemianopsia and inability to read. There was no right hemiplegia. His speech was good; articulation being distinct and natural. He was a very intelligent man, and his memory seemed to be good, although he complained of it. His choice of words was that of a man of education. Though quite word-blind he could name all individual letters correctly. The following details are given:—

"Although he names every letter without mistake he is unable to tell or understand a single word, no matter how short, unless he spells it out aloud like a child learning its first lesson. For example, when shown the word 'cat,' he cannot read it, and does not understand what it means. When, however, he spells it out aloud, he at once says 'cat,' and recognises the meaning of the word. The same occurs with all words of two, three, or even four letters. With longer words when they are spelt out, he is doubtful and uncertain, and frequently makes mistakes in naming them. With very long words such as 'Constantinople' and 'hippopotamus,' although all the letters are read in turn, the patient gets confused, and either miscalls them, or altogether asserts his inability to name them. It may also be mentioned that he cannot even read his own name. This difficulty in reading, applies equally to writing and print."

"The patient although he cannot read a single word, can write a letter

[1] Concerning some of the less familiar, though for us very important, structures mentioned above, it may be well to refresh the reader's memory by quoting what Schäfer says in Quain's "Anatomy" (tenth edition, 1893, p. 129), when speaking of the posterior extremity of the corpus callosum. "The fibres," he says, "from the body and upper part of the splenium which curve over the lateral ventricle form the *tapetum*, whilst a large mass of fibres from the splenium proper curves round into each occipital lobe and is known as the *forceps major*."

[2] *Brit. Med. Jrnl.*, 1888, i., p. 340.

like any educated man, and his handwriting is excellent. . . . . . I asked him, for example, to write out for me an account of his case. He did so in a very intelligent manner, giving a fair history and description of his complaint, and did not make a single mistake. On showing him his own letter he is unable to read a single word of it, and can only make out the shorter words by spelling them over as he does with print. When asked to copy writing or print he does so with perfect accuracy, but very slowly, letter by letter. Printed letters he transcribes into written characters. He does not understand what he is writing unless he spells the word out loud. His writing to dictation is without fault."

"He can read figures correctly and his powers of arithmetic are seemingly intact."

There are two points of special importance in this case—one of them so important that it induces me to quote it here, although it lacks the confirmation that might have been derived from a necropsy—I allude to the ease with which the patient was able to execute transfer copying, which must be taken as very strong evidence that the visual word centre was not destroyed, but only isolated. Then again, although this intelligent man was not tested as to his ability to read words by making him trace their outlines over with a pencil, so as to revive the kinæsthetic impressions associated with the word in question, the fact that he was not letter-blind enabled him to read short words in another way, namely, by spelling them, letter by letter, and thus rousing the auditory word centre as well as Broca's centre.

To return, however, now to the study of the three cases in which there was a necropsy; it will be observed that in the cases recorded by Wyllie and Redlich, as in that of Déjerine, there was no motor paralysis, and only a very slight amount of right-sided hemianæsthesia. There was right hemianopsia in each, but no mention is made of hemiachromatopsia in the cases of Wyllie and Redlich, or of the ability to read words by means of kinæsthetic impressions, though both were present in the case of Déjerine.[1] It seems highly probable, however, that one if not both of these characteristics might have been met with had they been specially looked for. They both existed in one of the two incomplete cases before referred to—namely, in one observed by Gaucher, and recorded by Mirallié;[2] while in the second of these cases, recorded by Batterham,[3] though there is no mention of hemiachromatopsia, the ability to read words by aid of kinæsthetic

---

[1] Déjerine is inclined to think that achromatopsia may be produced by lesions in the lingual and fusiform lobules. If the hemianopsia were complete, however, from destruction of the "optic radiations," there could be no recognition of achromatopsia.

[2] *Loc. cit.*, p. 191.   [3] *Brain*, 1888, p. 488.

impressions was present. Word-blindness was not complete in this case, and Batterham says: "When asked to spell out a word written in the 'round hand' of the copy books, she failed with several letters; but on being told to copy the unrecognised signs, or to run her pencil over them as if writing them, she in most cases recognised their name and signifi-

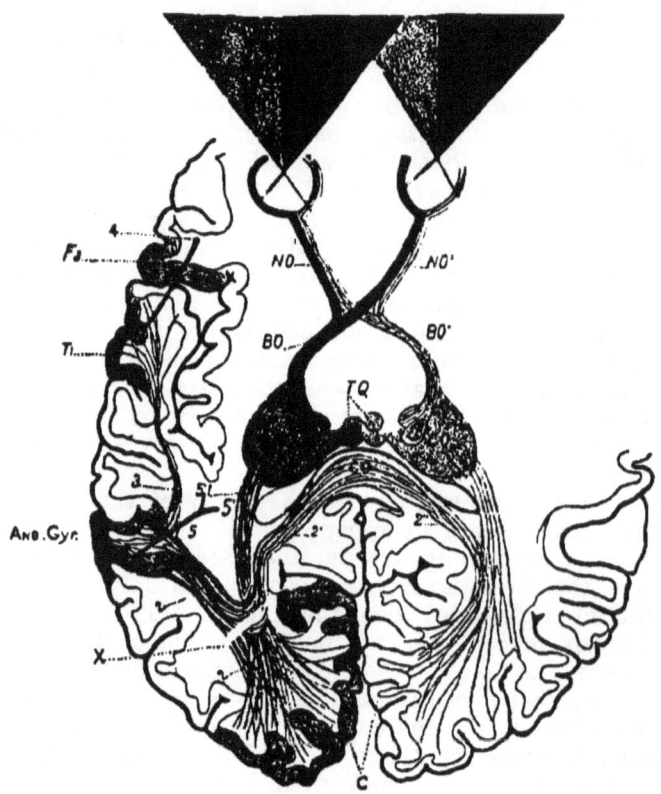

Fig. 8.—Diagram by Déjerine illustrating the site and nature of the lesion in "pure word-blindness."

cance. This experiment was repeated several times, and the patient was delighted to find that she could 'jog her memory' of letters in this way." Both these cases are well reported, and are worthy of careful study, although there is no record as to the pathological causes of the clinical condition.

The lesions found in each of the three cases where a necropsy was made have shown a striking similarity. In each there was softening and atrophy of the white substance of the occipital lobe, together with more or less damage to certain convolutions—that is, to the lingual and fusiform lobules as well as the hippocampal gyrus and the cuneus, or, speaking more generally, to some of the convolutions on the under and inner surface of the occipital lobe. It seems quite probable, however, that the lesions of the convolutions may be of little significance so long as there is the presence of extensive destruction of the white substance of the occipital lobe. Nothing more definite can be said on this subject at present. Déjerine attaches most importance to the destruction of portions of the white matter—namely, of that in which would be included the " optic radiations" of Gratiolet, the

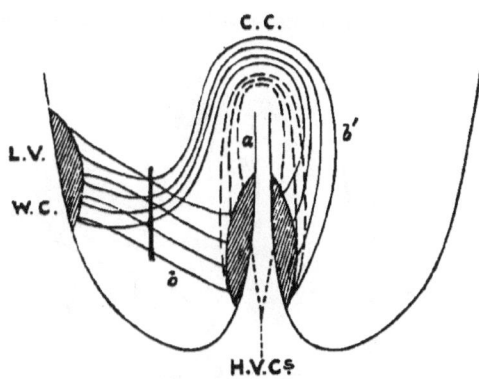

FIG. 9.—A simplified diagram representing Déjerine's view as to the mode of production of "pure word-blindness." H. V. C., Half-vision centres. L. V. W. C., Left visual word centre. C. C., Posterior extremity of corpus callosum, containing commissural fibres (a) connecting the half-vision centres, and also fibres (b') from the right half-vision centre to the left visual word centre. (The "optic radiations" have been omitted.) (b) Fibres from the left half-vision centre to the left visual word centre. The dark line indicates the site of a lesion which would cut off the left visual word centre from the half-vision centre of each side.

fibres proceeding from the left half-vision centre to the left visual word centre, as well as the fibres from the right half-vision centre to this same word centre. This he indicates in an interesting diagrammatic figure of some complexity (Fig. 8).

It is clear from the explanation that Déjerine advances of Case lxvii.,[1] and also from what he says elsewhere, that he does not believe in the existence of auditory and visual word centres in the right hemisphere in ordinary right-handed persons. On this subject he takes much the same view as

[1] *Loc. cit.*, pp. 87-89.

Lichtheim, and now explains "pure word-blindness" in a fashion analogous to that by which the latter explained "pure word-deafness"—namely, by supposing the severance from the left visual word centre of the associational fibres connecting it with the general visual centre in each hemisphere. This I have ventured to represent in a much simplified diagram (Fig. 9) in order to be better able to compare Déjerine's view with my own.

I, however, believe in the existence of a visual word centre in each hemisphere (though unequally developed), and that the two are brought into functional relation with one another by means of commissural fibres in the posterior part of the corpus callosum (Fig. 10). I believe, also, that each of these visual word centres would be in relation, by other associational fibres, with the half-vision centre of its own side (which both of

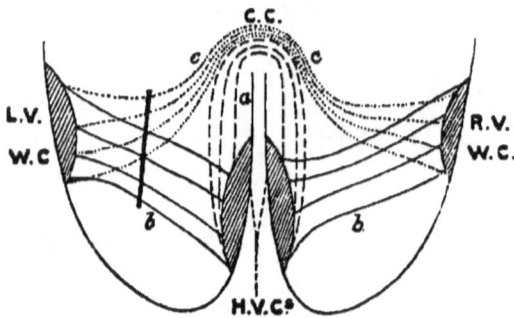

Fig. 10.—A diagram representing my view as to the mode of production of "pure word-blindness." c. c., Posterior extremity of corpus callosum. c, c, Commissural fibres connecting the two visual word centres. b, b, Fibres connecting each half-vision centre with the visual word centre of the same side.

us suppose to be commissurally connected with its fellow); and that Déjerine's form of pure word-blindness may be produced by severance of the associational fibres between the left visual word centre and its corresponding half-vision centre, together with a lesion of the commissure between the two visual word centres, in some part of its course (as shown in Fig. 10). The isolation of the left visual word centre would thus be complete; and brought about, moreover, in the same sort of way that I have postulated for the isolation of the auditory word centre in the condition that Déjerine calls "pure word-deafness."

The commissure between the two word centres may be damaged, in accordance with my interpretation, by the lesion in the white substance of the occipital lobe extending far enough forwards to involve them, just as Déjerine supposes those from the right half-vision centre to have been destroyed; or else by a separate lesion in the posterior part of the corpus callosum, such as was actually found in Déjerine's case, or in the part of it known as the tapetum.[1]

When we consider how one eye suffices for perfect vision, although only one-half of the proper amount of visual fibres goes to each half-vision centre, we may see all the more fully how complete and intimate must be the co-activity of these centres in vision, even when one eye only exists. That being so, it would be in vain, with our present imperfect knowledge, to attempt to define the actual mode of association of the half-vision centres with their outlying portions which we name visual word centres. All I can say is that such an arrangement as I have previously indicated seems to me to be much more in accordance with all known probabilities than that of Déjerine, who assumes that the right angular gyrus has practically no visual functions,[2] just as he assumes that the posterior half of the right upper temporal convolution has no auditory functions. What I have said towards the end of Chap. i. (p. 35), as well as what will be said later on as to the modes in which recovery takes place in various forms of speech defect, may be considered to lend support to my view.

As I have already indicated, the two modes of producing pure word-blindness are pretty strictly comparable, *cæteris paribus*, with two modes of production of pure word-deafness. One of these forms, which for the sake of distinction may be termed the "parietal type of pure word-blindness," is brought about by destruction of the left visual word centre in persons by whom writing can be executed under the instigation and guidance of the left auditory word centre; while the other (Déjerine's form), which might be called "the occipital type," is due to isolation of the left visual word centre from the half-vision centre of its own side as well as from the opposite visual word centre.

---

[1] The *tapetum* may perhaps include the commissural fibres between the two visual word centres, and it is specially stated in Redlich's case that these fibres were in part degenerated, as also was the *forceps major*, which probably includes the fibres connecting the two half-vision centres with one another.

[2] One result of this supposition is that, in endeavouring to account for the writing of an aphasic person with the left hand, he assumes that the right hand and arm centres are stimulated from the one and only visual word centre, the left—in fact, that the left hand and arm are in part under the direction of the left hemisphere for writing movements, while for all other kinds of movements of the left hand and arm the right visual centre would, as usual, coöperate with the kinæsthetic centres of the same side. I must confess that this does not seem to me to be very probable.

At present there seems no very definite means of diagnosing these two types of pure word-blindness from one another during life. The following considerations may, however, afford some help:—

Parietal Type of Pure Word-blindness.	Occipital Type of Pure Word-blindness.
Possibly some right-sided paresis.	Probably no right-sided paresis but possibly some slight right hemianæsthesia.
Cannot do "transfer copying."	May be able to do "transfer copying."
Speech often slightly paraphasic.	Speech not affected, or slightly amnesic only.
May be no hemianopsia or hemiachromatopsia.	Hemianopsia always, and when incomplete probably also hemiachromatopsia.

These seem to me to be the only approximations to differential characteristics that can be suggested at present, some of them being based upon the fact that destruction of the visual word centre may be associated with a lesion in the parietal region, and consequently may be associated with some amount of right-sided hemiparesis; while in the cases of occipital type, if the lesion of the occipital lobe extends sufficiently far forwards, we may get more or less marked hemianæsthesia without any motor paralysis. The association of alexia without agraphia in a person whose speech is not appreciably interfered with, and who has slight hemianæsthesia without paralysis, would, in fact, afford strong presumptive evidence that the pure word-blindness was of the occipital type, and this would be still further confirmed if the patient were able to do transfer copying, as in Case lxx.

7. Defects Resulting from Combined Lesions in the Left Auditory and Visual Word Centres.

It will have been seen from the cases already recorded that destruction of the left auditory word centre or of the left visual word centre alone (especially the latter case), may be quite compatible with the preservation of a fair amount of intelligence. The result is, however, altogether different when both these centres are badly damaged at the same time. The unfortunate individuals thus affected are generally reduced to a most deplorable condition, seeing that they can usually neither speak nor write intelligibly, and that they are unable to understand the speech which they hear or the language they may see, either in writing or in print. They can mostly communicate with others, and be communicated with, only by means of signs and gestures; and at the same time they must necessarily suffer a very distinct amount of mental impairment, owing to the blotting out of the principal linguistic symbols by means of which all but their most elementary thinking processes are carried on.

Although this is the kind of condition that has been met with in the great majority of these cases, yet a careful examination of their published records shows that, as in cases where there is destruction of the left auditory word centre, so here, with combined lesions of the auditory and the visual word centres, a considerable variation is met with in different cases. These variations depend partly upon differences in the relative completeness of the lesion in one or other of these centres; and in cases where one of the centres is incompletely destroyed, probably also, to a considerable extent, upon individual variations in original endowment—that is, upon the question whether the patients are "auditives" or "visuals," as well as in part upon their degree of education. Where the destruction of both the left word centres has been complete, moreover, variations in symptoms may depend upon the degree of development of the corresponding centres in the opposite hemisphere.

Taking the lists given by Amidon and Mirallié, and including one case by West[1] not contained therein, and two of my own, I find seventeen cases in all of this double lesion; of these, it seems desirable to exclude two—one by Chauffard, as the patient only lived three days, and one by Shaw, seeing that the patient was demented and that the clinical record is very incomplete.[2] Of the remaining fifteen cases, in three the lesion was complete in the visual and partial in the auditory word centre (in Amidon's No. 4, Mirallie's No. 38, and in West's case); in two it was complete in the auditory and partial in the visual word centres (in Amidon's Nos. 1 and 2); in four cases it was incomplete in both word centres (Mirallié's Nos. 8, 23, 24, and 48); whilst in six cases the lesion was pretty complete and equal in both word centres (Mirallié's Nos. 5, 16, 26, and 29, and my own two cases).

Looked at from another point of view—that is, as to the kind of speech defect presented by these fifteen cases—I find that speech was more or less good in two (Amidon's Nos. 1 and 2); that there was more or less marked paraphasia or actual jargon speech in seven (Amidon's Nos. 5, 16, 23, 29, 38, and 48, and West's case); and more or less complete speechlessness in six (Amidon's No. 4, Mirallié's Nos. 8, 24, and 26, and my own two cases).

[1] *Brit. Med. Jrnl.*, vol. i., 1896, p. 1242.
[2] These are Nos. 5 and 31 of Mirallié's list.

These very different results may at first seem very astonishing to others, as they did to me. But the more one thinks of cases presenting this double lesion, in which speech was only slightly paraphasic or fairly good, the more it seems necessary to suppose that the auditory word centre of the right hemisphere must have been able to act upon and with the left glosso-kinæsthetic centre. These cases cannot be explained by supposing that speech was produced by the right auditory word centre acting with the right glosso-kinæsthetic centre, because all present knowledge goes to show that this could only be brought about after a long interval, during which these centres were educated to act together. In the cases in question, on the other hand, there was evidently no such interval, seeing that the modified speech was initiated in each case just after the brain lesion occurred. It is worthy of note, also, that in the two cases in which speech was best preserved (Amidon's Nos. 1 and 2) the visual word centre was only partially damaged, so that some help may also have come from its coöperation.

A few cases may now be given in illustration of the different degrees of speech defect associated with this double lesion with which we are now concerned. The first is one that was published nearly twenty years ago by Broadbent in which there was well-marked jargon-speech.[1] It is given here only in abstract, and is No. 5 of Mirallié's list.

CASE LXXI.—A man, aged 60 years, previously very intelligent and able to read and write well, but of intemperate habits, had had some sort of fit two weeks before admission to St. Mary's Hospital. From that time he had kept his bed and had been unable to speak intelligibly.

When seen there was no hemiplegia, but slight paresis of the right side of the face and some amount of right hemianæsthesia. His speech was an inarticulate jargon and preserved the same character throughout. When questions were asked he attempted to reply, but as a rule nothing resembling a word could be detected in what he said. The voice was inflected and he appeared quite unconscious that his speech was mere gibberish. He seemed to have some idea in his mind, and to think that he was giving expression to it. Besides speaking after questions, he would go on talking, addressing himself to one or another of those present, the whole of what he said being an incomprehensible ramble, though at times a distinct word or phrase would slip out. When excited the phrase, "If you please," was several times heard distinctly. He seemed to understand nothing that was said to him, and when told to shut his eyes or give his hand there was never any attempt at compliance.

He was similarly quite unable to read writing. [Nothing is said about his ability to read print or his capability of writing, but he was probably incapable of doing either.] He would sit up in bed and appear to watch

[1] *Transactions of the Royal Medical and Chirurgical Society*, 1878, p. 147.

with interest what was going on in the ward. When the meals were brought in he looked for his portion, and ate his food naturally. When the bowels were about to act he called the attention of the nurse by knocking on his locker and then pointing to the commode. There was never anything extraordinary in his behaviour.

He died about three weeks after his admission to the hospital. At the necropsy softening of the posterior part of the left hemisphere was found in the region supplied by the third and fourth branches of the Sylvian artery. The convolutions softened were the supra-marginal lobule and the angular gyrus throughout their whole extent, and also the posterior half of the first temporo-sphenoidal gyrus, together with some portions of adjacent occipital and posterior parietal convolutions. Broca's region and the anterior half of the brain as a whole were unaffected.

This is a fairly typical case, the destruction of both the word centres being complete and speech reduced to an unintelligible jargon. In the next case (No. 29 of Mirallié's list) the lesion was equally complete, though the derangement of speech was less marked—a very bad form of paraphasia being present rather than jargon-aphasia. It has been described at length by Déjerine,[1] but I give it here only in abstract.

CASE LXXII.—A man, aged 63 years, was admitted to the Bicêtre on July 3, 1890, having had an apoplectic seizure on the previous evening. The patient was of vigorous appearance and intelligent face, and was said to have been able to read and write previously to his illness. When examined on July 4 he was found to be in a semi-comatose condition, from which he could only be roused momentarily with difficulty. He remained in this condition for four days, and then gradually recovered consciousness. On July 20 he walked about the ward and showed no appreciable hemiplegia. He began to speak two or three days before, but his speech was much altered and he understood nothing of what was said to him. On saying to him, "Comment vous appelez-vous?" he replied, "Je suis et, surtout c'est-à-dire, c'est-à-dire, non, je ne peux pas po pa." "Quel métier faisiez-vous?" "Mon père se nommait, non, peux pas." The only words that he pronounced in a suitable manner were "bonjour" when one approached his bed, and "merci" when one gave him something to eat or drink.

When a newspaper or some manuscript was shown to him he looked at the paper and then at the person who had given it to him, and it was evident that they had no meaning for him. When a pen was given him he held it correctly as if he were going to write, but made only meaningless marks on the paper, either when he was left to himself or when a phrase was dictated to him in a loud voice. If one gave him a phrase in manuscript to copy he copied the letters one after the other, but very badly and the words were illegible. Nevertheless, the general form of the letters was preserved, while in his spontaneous writing or that from dictation he was incapable of tracing even the rough outline of a letter. The patient recognised, however, quite well all the objects and persons around him. His hearing was also intact, and the slightest sound behind him would cause him to turn his head.

On December 4, 1890, the word-deafness remained about the same, but he could now recognise one question, "Comment vous appelez-vous?" and gave his surname correctly. But to any question that any one put to

---

[1] *Comptes Rendus de la Société de Biologie*, 1891, p. 167.

him immediately afterwards he still answered by giving his name. If after an interval one came back to him again he no longer replied as before with his name to all questions, but by words or phrases that had no relation to the question. For example : " D. : Qu'avez-vous fait hier ? R. : Mon père était marchand de vins.—D. : Dans quel hôpital êtes-vous ? R. : Je... vou...je voudrai......non...papapa...tou.—D. : Quel âge avez-vous ? R.: J'avais cent soixant trois.—D. : Quel métier faisiez-vous ? R. : Trois ans, six six ans, trente trois, jamais trente un ans jamais, trente, trente, trente, trente-trois ans, jamais trente, trente. trente-trois ans." The patient was still unable to read, but for the last few days he had recognised his name and been able to pronounce it aloud, though he could do this with no other word. Right lateral hemianopsia seemed to be present. The agraphia remained complete as before. The patient continued in about the same condition for another three weeks ; then his whole condition changed, he gradually sank into a semi-comatose state (owing to the establishment of a large area of softening in the right hemisphere), and about a month later he died.

At the necropsy, in addition to a large area of recent softening in the right hemisphere, there was found a large area of yellow softening completely destroying the visual and the auditory word centres of the left hemisphere, which also extended forwards into the parietal and backwards into the occipital region. Section of the hemisphere also showed that the lesion extended inwards through the white substance as far as the posterior cornu of the lateral ventricle.

It is worthy of note that there was here no speech at all for sixteen days from the onset of the seizure and for about eight days after the recovery of consciousness ; and, looking to the completeness of the destruction of the left auditory and visual word centres, it is difficult to suppose that the speech which was subsequently possible could have been effected without the aid of the right auditory word centre. Much the same thing may be said in reference to the next case, although the speech trouble was more like that met with in incomplete motor aphasia. It is No. 8 of Mirallié's list, and has been recorded by d'Heilly and Chantemesse.[1] The following is a brief abstract.

CASE LXXIII.—A woman, aged 24 years, became suddenly speechless on October 12, 1881. She entered a hospital at once. To all questions put to her she would repeat five or six times, with different intonations. "Because, because, because." Sensibility and motility seemed perfectly intact. On October 15, in reply to a question, she answered, "Thank you, sir, I am better ; " though, as Bernard points out, she was never able to repeat it. She looked very intently at the speaker, but words seemed to wake in her no image, no remembrance. When told to put her hand to her head she hesitated an instant and appeared to be trying to remember something, but she remained motionless. When the command was accompanied by a suggestive gesture she quickly obeyed. Hearing and vision were both preserved. Her power of calculation seemed preserved, as she could play écarté skilfully, making no errors as to colour or value. When asked, she called her knife, tumbler, food, wine, and plate each "du plan." When an orange was held out to her she answered all questions, "Yes, sir,"

Bulletin de la Société Anatomique, 1882, pp. 324-338.

and reached for the orange. Once when her soles were tickled she said, "Please don't, sir." She could neither copy, write from dictation, nor read.

She died on November 3, from marasmus. There was found at the necropsy a softening from thrombosis of the fourth branch of the left Sylvian artery. This softening involved the upper posterior half of the first temporal convolution, most of the inferior parietal lobule, the supra-marginal gyrus, and some of the sigmoid gyrus. The softening affected only the cortex, and on section was seen to implicate only the extreme posterior hidden part of the insula.

It seems clear, from the records of the necropsy, that the left auditory word centre was here only partially destroyed, so that perhaps we need not suppose that help came from the right auditory word centre for the production of the few short sentences that this patient was capable of uttering.

In a patient, recently under my own care, suffering from a softening of the same region of the brain, during the three weeks that elapsed before death occurred there was absolutely no speech. The following are a few notes concerning this case.

CASE LXXIV.—A woman, aged 53 years, was admitted to University College Hospital on April 28, 1896. Whilst suffering from inflamed varicose veins in the left leg she began to feel ill on April 22, complaining of headache and irritability. On the morning of April 26 these thrombosed veins were rubbed with an embrocation. She went to bed on the night of the 26th with no change in her symptoms, but the next morning she fell from the bed in an unconscious condition. She twitched her left arm and muttered indistinctly for a time, and remained in a stuporous state up to the time of admission (9 a.m. on the 28th). She was found to be semi-conscious and paralysed on the right side, showing some irritability on examination, throwing her left arm and leg about, and occasionally uttering some inarticulate sounds. There was no cardiac bruit; the respirations were natural. Her temperature was 100·8° F.

She remained in much the same condition, semi-conscious, yawning frequently, rather restless at times, till May 5, when she became more conscious, but made no attempt to speak or even utter any sound. She could not be induced to protrude her tongue. She did not attempt to respond or show that she understood any simple request, and appeared to be perfectly word-deaf. She took no notice of written sentences held before her. The right arm was rather rigid.

On May 11 she seemed more conscious than she had been but did not speak, nor did she seem to understand anything. One-fourth of albumin was found in the urine. On May 14 she began to suffer from sickness and some diarrhœa (the albumin in the urine persisting), together with lung complications, and she died about midnight on May 18.

At the necropsy the third frontal and the lower part of the ascending frontal convolution were found to be healthy, both externally and on section; but the lower part of the ascending parietal, together with the whole of the supra-marginal and angular gyri and the two upper temporal convolutions, were completely softened throughout. The softening extended to, and partly involved, the outer part of the thalamus and the corpus striatum; it also extended backwards slightly into the occipital lobe. The right hemisphere showed no focal lesion of any kind.

Here, then, there was complete loss of speech as a result of destruction of the auditory and the visual word centres in the left hemisphere, together with some adjacent tracts of brain tissue, which lasted continuously for just over three weeks. We have seen that Déjerine's patient (Case lxxii.) also remained speechless for sixteen days after the onset of his seizure, and then began to talk in a badly paraphasic fashion. It seems probable, therefore, that my patient might also have developed some sort of power of speaking had she lived longer.

My second case belonging to this group is one of a very extraordinary nature in many respects, the patient having lived over eighteen years after his seizure, and his speech defects having remained constant throughout almost the whole of this period. His spontaneous speech was limited to a few words, and, though he was neither word-deaf nor word-blind, both the auditory and the visual word centres of the left hemisphere were completely destroyed. The complete record of this remarkable case, together with illustrations of the brain, have been published in the *Transactions of the Royal Medical and Chirurgical Society* for 1897, and an abstract of the case will be found on p. 255.

In our study of the effects of separate lesions in the auditory and the visual word centres respectively, we found that speech was much more gravely affected, as a rule, in the former than in the latter group of cases. In cases also in which partial word-deafness is associated with complete word-blindness it will almost always be found that the patient's inability to speak is proportional to the degree of word-deafness. I have recently been consulted in reference to three such cases. In two of these (one at Kingston and one at Norwich), whilst the word-blindness was complete there was no real word-deafness, only a very low activity of the auditory word centre, and each of these patients was able to articulate three or four consecutive words suitably and could repeat words correctly—so that their speech contrasted notably with the unintelligible or jargon-like utterance often met with where both the auditory and the visual word centres are badly damaged, and which was actually met with in the third case. From some details of this case given below it will be seen that there was complete word-deafness at the first, with jargon-speech, but that as the power to comprehend words was regained, so did the patient's ability to speak improve, although the word-blindness remained as before.[1]

CASE LXXV.—A gentleman, aged 50, was brought to me on December 7, 1894, by Dr. G——. I was told that in the preceding spring he had begun to suffer, off and on, from pains in the left side of the head, which were ascribed to "gouty neuralgia." Early in September, while suffering from this pain, he had a strange sort of attack coming on suddenly, in which he had a frightened manner, spoke unintelligibly, and could not recognise his relatives or the doctor, whom he knew well, and who came to him about three hours after his illness commenced. Dr. G—— says there were then no intelligible words in his mumbling speech; that he not only did not recognise him, but did not even know his own son. He understood nothing that was said to him. He had had no convulsion, and there was no paralysis.

This altered mental condition, with inability to recognise familiar persons, and to understand speech, continued with little abatement for about ten days. Then he began to recognise persons, and his speech became rather more distinct.

On December 7, I found that his grip was R. 58, L. 61; that his right knee-jerk was distinctly freer than the left, and made the following notes concerning his mental condition and speech. "Does not seem to understand anything that is said to him. Told to shut his eyes, put out his tongue, or give me his right hand, he does neither of these things, but goes on talking in a rambling manner, though what he wished to say was unintelligible. He also seemed unable to read similar simple requests when they were shown to him on a card. He neither read nor did what the card requested him to do, and did not seem at all to comprehend what was printed on it. Subsequently he did name correctly two or three letters but not a single word. He could not name objects or pictures of animals, though he seemed to recognise the latter."

"On putting a pen into his hand he wrote his name quite freely and distinctly, and then went on to make a curious combination of about ten letters (not composing any definite word or words), which letters he could not a moment or two afterwards name correctly."

It was impossible certainly to make out whether he had right hemianopia or not, but my impression was that it did not exist. Before he left he seemed to understand something of one or two simple statements I made to him in reference to his prospects of improvement.

I saw him again on June 27, 1895, and found the word-deafness to be then very much less, and his speech distinctly better. He shut his eyes, and put out his tongue at once when bidden, but when I asked him "Can you read now?" he looked puzzled and did not seem to understand. When I bade him say "Garlick," he repeated again and again "Furwich," and afterwards "Furgick." He said, "I know that I am not foolish in any way, but I can't say the words. I feel right, but I am all wrong in the words. I know what to do." And then in reply to a question, he said, "Did the old gentleman, or rather the young gentleman, he is not old." He still could not read and made many mistakes in naming short words and letters—soon becoming confused. Asked to write his own name "James," after long hesitation he wrote "year," but when asked to copy his name after I had written it, he did so at once and correctly.

[1] A somewhat similar case of word-deafness and word-blindness has been recorded by Ross (*loc. cit.*, p. 19), in which the speech was unintelligible at first, but improved as the word-deafness grew less.

OBJECT-BLINDNESS.

This case is interesting also because, in addition to the patient's other defects, there was what has been known as "mind-blindness,"[1] or what is, I think, better termed "object-blindness" (Ballet and Wyllie). This latter term is not only more descriptive of the nature of the defect, it is also a better companion term for "word-blindness." The defect in question is met with not unfrequently in association with lesions in the occipital lobe, and occurs therefore at times, as in this case, in conjunction with word-blindness. Our patient for some time did not know his own son or the doctor with whom he was perfectly familiar. No other details were given concerning this side of his defect, which was probably much more extensive; but in other cases it has been found that common objects are not recognised, and that their proper uses seem to be altogether forgotten. This defect is probably due to a partial isolation of the common visual centre—the cutting across not of its afferent fibres but of the fibres by which it is brought into association with other sensory centres—so that objects can no longer be perceived and recognised.

An admirable example of these combined defects has been recorded by Bernheim[2] to which there was probably no destruction of the left visual word centre, but rather a lesion in the occipital lobe of the kind met with in Déjerine's case of pure word-blindness. At the date that it was recorded by Bernheim the situation of the half-vision centres in the cunei of the occipital lobes had not been established, so that his interpretation of the nature of the lesion is different from that which may be given now.

The following are a few details concerning this very interesting and well-observed case of combined word-blindness and object-blindness, which are all the more worthy of attention from the fact that this man was left-handed and that the lesion evidently occurred in the right hemisphere.

CASE LXXVI.—B. Nicholas, a gardener, aged 63, always left-handed, on May 4, 1893, on resting from his work felt giddy, and the next morning he found himself partially paralysed in the left arm and leg, and with some pain in his head.

He was admitted into hospital on May 12, and on examination the next day, he was found to be suffering from an incomplete left hemiplegia, together with well defined left hemianæsthesia. There was also hemianopia and the amnesic defects presently to be described.

He improved and left the hospital about the end of July, but returned in

---

[1] Its other equivalents being "cécité psychique" or "Seelenblindheit."
[2] *Rev. de Méd.*, 1885, p. 625.

much the same condition in the middle of September, and there he remained up to the date of the publication of his case. In January, 1884, he began to have attacks of left-sided Jacksonian epilepsy, which continued at times through that year and the first half of the next. The hemianæsthesia disappeared during his first stay in the hospital, but the hemianopia and the word-blindness, with other associated defects, remained throughout, with only slight variations from time to time.

Word-blindness and Object-blindness.—The patient was found to be word-blind but able to write quite legibly. He could not read what he had written, he could not recognise his own name or even a single letter that he had written. . . . . He was equally unable to read printed characters. On the other hand, when bidden to point out such and such a letter lying before him he always succeeded in doing it, perhaps after some abortive attempts. The same thing held good for numerals or drawings. Thus, Bernheim says: "We drew upon a sheet of paper a square, a circle, a house, a pipe, a head, and he recognised only the head. Another day he recognised a cross, saying at first ' c'est le bon Dieu,' then ' c'est un croix,' but he did not recognise either of the other figures. When asked to point on the paper to the square, the circle, the house, etc., he indicated each successively without hesitation."

The patient's intelligence was preserved; he spoke and maintained a conversation quite well, only being at a loss occasionally for words. In fact, there was nothing very obvious in his mental condition to attract attention beyond a slight feebleness of memory. But when shown some object, Bernheim says, "one is astonished to find that he is not able to mention its name." Thus, shown a piece of bread and asked what it is, he says, "un salière." A glass: "c'est une barre." A knife: he replies correctly. A book: "c'est une écorce pour faire des tartes." A crucifix: "c'est une catechisme pour faire des tartes." But, as it was with words, when the various objects were placed before him whose names he was unable to give—bread, a glass, a book, a crucifix, etc.—and he was told to take up successively one or other of these objects, he did it without hesitation. As with words, so here, the visual memory of objects was aroused by the activity in the auditory centre.

So far it might be thought that we have in this case little more than an inability to name at sight letters, words, or objects, as no evidence has been given that he is unable to understand what he reads or the nature of the objects which he sees—he might be merely unable to name at sight, and paraphasic in his efforts to do so. But other details given by Bernheim show that there was in this case, as he terms it, " une cécité psychique des choses."[1]

Thus he showed the patient a bunch of keys and said, "What is this?" "C'est pour marquer." "Show me how it is used." He attempts to write with one of the keys, saying, "C'est une plume." Then he appears to realize that it is not that, but seeks in vain to recognise it. "Je sais, dit

---

[1] And it should be pointed out here that Bernheim does not favour the generally received view that there is a special region in which are registered the visual memories of words, apart from that in which the more general visual memories of things are registered. His view is that "Les mots écrits ou imprimés ne sont en réalité que des choses" (*loc. cit.*, p. 637).

il, je l'ai vu cent millions de fois, c'est pour semer du grain ; c'est une
hersé." He then made with one of the keys the gesture of opening a lock;
but still did not recognise the use or the name. But when asked, "With
what does one open a door?" then only he said, "Avec une clef," and
recognised that what he had been shown were keys.

On another occasion a brush was handed to him. He could not name it.
"Of what use is it?" "C'est pour marcher." "Show me how it is used."
Then with his hand he made the brush take steps. Later he exclaimed,
"Non, c'est pour faire des barres. Non!" He continued to puzzle over it,
and after three minutes succeeded in recognising it: "C'est pour brosser,
c'est une brosse."

Allied to object-blindness there is another defect, originally described by Freund, met with occasionally in patients suffering from one or other of the forms of "sensory aphasia," to which he gave the name of "Optic Aphasia" (l'aphasie optique). An ordinary aphasic person when an object is shown to him will recognise it and make known by signs that he understands its use, though he will be quite unable to name it; and so he will continue when the object is placed in his hand, when he smells it or tastes it, if it be capable of yielding gustatory or olfactory impressions. It is different, however, with a person suffering from optic aphasia; when a common object is shown to him he will recognise it at once and show that he does so, but he cannot name it at sight. But let him smell it, taste it or, above all, touch it, and he will at once be able to call up and utter the name of the object.

This is a defect which seems to me to be due to a partial severance of the commissural fibres connecting the left general visual centre with the left auditory word centre (the visuo-temporal commissure), while its associational relations with the tactile, the general auditory, the gustatory, and the olfactory centres, may exist uninjured. On the presentation of the object to sight, therefore, there is nothing to hinder its full perception and recognition except that its name cannot be recalled. But let the object be put into the patient's hand, or let him smell it or taste it (if that be possible), and there is nothing to prevent the impressions made upon either of these centres from radiating to the other centres, so as to wake up a full perception of the object, including its name. Optic aphasia is thus one special form of inability to name objects at sight such as was present in Case lxxxiv.[1]

---

[1] There is, of course, the possibility that in some persons, naming at sight might be brought about by means of a commissure between the general visual centre and the visual word centre, and thence to the auditory word centre.

Another interesting defect pretty closely allied to object-blindness has lately been described by R. T. Williamson [1] in an article entitled "On 'Touch Paralysis,' or the inability to recognise the Nature of Objects by Tactile Impressions;" though I find a similar case, associated with object-blindness, described by C. W. Burr,"[2] under the title "A Case of Tactile Amnesia and Mind Blindness." He also refers to a second case seen in 1892, though neither of them is so fully described as those by Williamson, who cites two very interesting cases and says a few others have been recorded, to which he gives references.[3] Neither of his patients was aphasic, and nothing is known as to the seat of the lesion causing the disability in question. Full details are given concerning the first case, of which I subjoin the following abstract.

CASE LXXVII.—A young woman, aged 26 years, with a history of syphilis, was first seen in January, 1896. In the previous June she had a fit in which she lost consciousness and fell, after which there was loss of power in the left thumb. Subsequently she continued to have frequent fits that were always of the same character. They commenced with a sensation of numbness in the left thumb, which was followed by twitching in the same part; then the arm became affected by the twitchings, and the patient lost consciousness. She had about fifty of these attacks up to August, 1895; but since that date she has only had a few attacks of faintness accompanied by twitching of the left thumb. For three months previous to coming under observation she had suffered from headache, occasional vomiting, and impaired vision.

On examination she was found to have intense double optic neuritis, vision on the left side being much impaired. There was slight tenderness to deep percussion over a small area in the right parietal region. There was slight paresis of the left side of the mouth, but no appreciable paresis or ataxia of the left arm or leg; there was no loss of muscular sense, and a minute examination showed no appreciable diminution of other modes of sensibility in the left hand. Dr. Williamson adds :

"Though the examination of the above-mentioned forms of sensation gave almost negative results, there was nevertheless marked inability to tell the nature of objects by tactile impressions alone. Objects placed in the left hand were felt distinctly; the patient could feel at once when they were placed in the hand or on the fingers, however light and small the object might be. She could also feel at once when the hand was very lightly touched with the object; but she was quite unable to tell the nature of objects placed in the left hand, even when she grasped them tightly and handled them most carefully (the eyes of course being closed). With the right hand the nature of the same object was recognised at once (the eyes

---

[1] *Brit. Med. Jrnl.*, September 25, 1897, p. 787.

[2] *The Jrnl. of Nervous and Mental Diseases*, May, 1897, p. 259.

[3] One of these cases is referred to by Allen Starr (*Brain*, vol. xii., p. 89, *footnote*).

also being closed). The following were some of the objects used in examination: Penny, halfpenny, sixpence, shilling, half-sovereign, latchkey, penknife, pencil, pen, small book, small piece of paper, watch, watch-chain, card-case, pin. All these objects were recognised exceedingly readily with the right hand and fingers (the eyes being closed), but were not recognised at all with the left hand and fingers. The patient knew that there was something in the left hand, but was quite unable to say what it was. If when examining the patient two or three fingers were placed in the palm of her left hand, she felt the tactile sensation at once, but even on tightly squeezing the fingers she was unable to say what they were. With the right hand she recognised them at once, and was able to tell readily how many fingers she was grasping."

In the absence of actual paralysis and of any loss of muscular sense, it seems to me that the fits of Jacksonian type were probably caused by a sub-cortical rather than a cortical lesion; while there is no evidence to show where the defect was situated which led to the inability to carry out an act of perception on the initiation of tactile impressions. The explanation of this disability seems to me to lie in the probable existence of an isolation of the tactile centre for the hand in the affected hemisphere—a damage, that is, to the association fibres connecting it with other kinds of sensory centres, the effect of which must be to prevent that radiation of stimuli to other sensory centres which is necessary for an act of perception, by which the nature of the object presented is recognised. The defect, as I have said, has no necessary relation to speech defects; nor has it, in my opinion, anything to do with what has been mysteriously termed "active touch," to a supposed loss of which it has been attributed by some. I have merely referred to it because, in my opinion, it is a defect which, in regard to its nature and cause, is closely comparable with "object-blindness."

In Chapters vi., viii. and ix. I have striven to show that the over-weening importance attached by many to the functions of Broca's centre is not justified by facts—that too many disabilities have been supposed to result from its destruction by some, and that the power of independent activity ascribed to it by others does not exist. I have tried to show the way in which these views have influenced others to deny the existence of a cheiro-kinæsthetic centre, and the fallacious nature of many of the arguments on which they rely. Again, while alluding to the unsatisfactory nature of the nomenclature at present in vogue, I have tried to show that the so-called "sensory aphasia" of Wernicke has no claim, as he supposed, to an independent existence, but really includes a large number of distinct conditions, each of which has to be studied separately and in detail. The great importance of the auditory word centre has been dwelt upon, as well as the variety of defects produced by functional and structural derangements of this region, and also by its isolation. A study of these results, taken in conjunction with that of the less varied defects resulting from disease and from isolation of the visual word centre, led to some novel conclusions. We have seen the enforcement of the view to which we had previously been driven as to the comparative powerlessness of Broca's centre alone; we have found evidence for the existence of an unexpected possible amount of functional substitution between the visual and the auditory word centres for the production of speech and writing, respectively; and, still more surprising, we have found reason for believing that both auditory word centres are accustomed to act upon Broca's region for the production of speech—as well as much evidence tending to favour the view that, even in comparatively simple perceptional and intellectual operations, very extensive cortical areas in both hemispheres of the brain are called into simultaneous activity.

CHAPTER X.

INCOÖRDINATE AMNESIA: PARAPHASIA AND PARAGRAPHIA.

IF we regard the amnesic defects hitherto described as paretic or paralytic in nature, those to which we are about to refer are rather of an incoördinate type. The defects in speech and writing are parallel in nature, and are commonly described under the names "paraphasia" and "paragraphia."

They may be described as occurring under three grades or degrees of severity. In the minor form of the defect (1) the patient more or less frequently uses wrong words instead of those that he intended to employ. In a more severe form (2) he collocates words in such a disorderly manner as to convey no definite meaning. While in the most severe form (3) he does not make use of words at all, but utters a mere gibberish in which no actual words are to be detected; or, in writing, he makes random collocations of letters, or perhaps mere unmeaning strokes not even representing letters.

(1) In the simplest class of cases, where the patient is at the same time almost always suffering from paralytic amnesia there is the utterance of wrong words—words other than were intended, and other than are appropriate to the occasion or for the expression of the patient's wishes. Thus, in a very interesting case of speech defect recorded many years since by Banks,[1] the patient desiring to inform one of his medical attendants "that a liniment which he had been using was nearly finished said, pointing to the bottle, 'bring the cord.' On another occasion, speaking of the pills he had been taking, he said he had taken 'potatoes.' Very frequently there was some similarity in the word used to the right one; or it could be discovered that there was some association with the idea he wished to convey; for example, giving his waistcoat to be put aside, the watch being in its pocket, he said, 'take care of the break-fall.'"

At other times when a wrong word does not slip out quickly,

---

[1] *Dublin Quart. Jrnl. of Med. Science*, February, 1865, p. 78.

and the patient is conscious of his inability to find the proper word, he will substitute some noun of general import such as the word "thing," or else make use of a periphasis, in order to express his meaning.

An allied defect is also met with in the form of a transposition of syllables of words in the same sentence, or the interposition of letters that do not belong to the word, a condition to which Kussmaul has given the name of "syllable stumbling." As an example he mentions the case of an absent-minded professor who before his class spoke of "the two great chemists, Mitschich and Liederlich," meaning Liebig and Mitscherlich." Ross also gives another instance when he says: "Those who studied medicine in the University of Aberdeen some years ago, will remember that one of our most respected professors was constantly in the habit of committing ludicrous mistakes of this kind, 'cus porcuscles,' instead of 'pus corpuscles,' being an example of which we have been recently reminded." Lordat, too, states in reference to his own case: "for 'raisin' I said 'sairin,' for 'musulman' I felt inclined to say 'smulman.'"

Patients are mostly aware when they make use of wrong words, though this is by no means always the case.

Luys[1] alludes to an instance where the person was continually in the habit of using one word for another without being conscious of his mistakes. One day he pronounced the word "jardin," wishing to say "lit," repeated it several times, and afterwards fell into a violent passion because his orders were not comprehended. He was then made to write the word he wished to make use of, and the sight of the proper written symbols soon convinced him that the word which he had actually uttered was not the one he had intended to utter.

Defects in writing of the paragraphic type are also very commonly met with, and are strictly comparable with paraphasic defects. As an instance, I may quote the following letter which was written to me by one of Sir William Jenner's patients, at a time when I was in charge of his wards, in response to a request that he should give me some account of his previous history.

UNIVERSITY COLLEGE HOSPITAL,
Ward 8.
September 6, '69.

To — BASTIAN, Esq.

DEAR SIR,—I am said of my illness. As twelfth years has not the loss of my right eye, you had a lad at once reaching of a shell, and quite an

---

[1] "Syst. Nerveux," 1865, p. 395.

accident. . . . . The left eye was just for years, and do say. . . . . If I have of fifteen years, that I write Plays, and contributr that many of the London journals and newspapers. And I write Essays, Comedies, Poem, Dramatic Criticism, &c., and a thousand. . . . . . I have twenty years I gave the appointent that " read for press " for " The Examiner." . . . . . In the good health in 1863 was the " neuralgia," and go once at once, or that bad that it be is done. . . . . . In September 1867 that blood gone by head, and I cannot by that *left eye*. . . . . . In September 1868 from eye been better, and can *write* and *read!*. . . . . . In on Good Friday, in the night, has a had a " fit ; " and the right leg, right arm, and that I cannot say or can about, and of Palalysis and the Tic Dolerens. I have very ill. . . . . . In three weeks I come to that the Hospital here. That have does better. The Hospital goes to *Eastbourne*, and goes was ill that than has as ever. His " Tic " is bad, and the Doctor than the Hospital. . . . . . His has *talk* have " counchant " and the " Tic " for rampant. . . . . . I cannot write as good to need; for it at be further. . . . . . And, dear Sir, this note that not do the " Queen's English Grammar." And if be better.

I, dear Sir,
Your obedient Service,
B. W. W.

The handwriting in the letter was very good, and there were only two trifling erasures. The letter was divided into paragraphs in the manner indicated; and the punctuation and capital letters have been preserved exactly as they were in the original.

(2) What may be considered as an instance of the second grade of severity in these cases was long ago recorded by Bouillaud.[1] A patient under his care mostly made use of actual words, though they were of such a kind and so collocated as to have no resemblance to what he ought to have said. When reading aloud, however, he often uttered nothing but mere jargon.

CASE LXXVIII.—Lefèvre, aged 54, after some extreme mental anxiety, became unable to read, write or find words to express his thoughts. His sensibility and powers of movement were unimpaired, and his general health was pretty good.

When he wished to reply to questions that were addressed to him, he made use of expressions either quite unintelligible, or else having a meaning quite different from that which he intended to convey. When questioned as to his health, he replied rightly in two or three words; then in order to say that he did not suffer at all from pain in the head, he said, " Les douleurs ordonnent un avantage," whilst in writing he replied to the same question in this way : " Je ne souffre pas de la tête." When a word such as *tambour* was pronounced, and he was asked to repeat it, he said " fromage ; " though he wrote it, on the contrary, quite correctly when asked to do so. He was requested to copy the words *feuille médicale* ; he wrote it perfectly

[1] " Traité de l'Encéphalite," 1825, p. 290.

but could never exactly read the word which he had just written; he pronounced instead, "féquicale, fénicale and fédocale." Then, when made to read the word *féquical* written by himself, he pronounced it "jardait."

He often wrote upon paper phrases which were unintelligible, either by the nature of the words employed, or by their lack of relation to one another. When he was shown different objects, he generally named them correctly; but then he was wrong at times, and during the same sitting he called "une plume, *un drap;* un crachoir, *une plume;* une main, *un tasse;* une corde, *une main;* une bague, *un crachoir.*"

This is a very difficult case to explain, though it would seem that the paraphasia was more marked than the paragraphia, and that his defects were most obvious in spontaneous speech, in reading, and in spontaneous writing. A case, not very different though more severe, was a short time since under my own care for several weeks, concerning which the following are the most important details.

CASE LXXIX.—H. P., a clerk to a gas company, aged 30, was admitted under my care into University College Hospital, on December 17, 1896, unable to speak intelligibly, so that the following particulars concerning his history were obtained from his wife.

When 12 years of age a cab-wheel passed over his head (he has a dent in the left posterior parietal region) and he was in bed for seven weeks after this accident suffering from "concussion of the brain." Two years later he fell off the roof of a low house, and again suffered from "concussion of the brain." He had a "fit" ten years ago (this said to have been the first) but there were no distinct convulsions, and no paralysis followed the fit. A second attack of fits occurred four years ago—when he was ill for two days, having slight fits at intervals, and as a result of these attacks he lost sensibility and power of movement in the right arm and leg for a short time. Another fit occurred two years ago, but was not followed by any paralysis.

Present Illness.—He was in his usual health and went to work on November 30, 1896. On returning in the evening he was taken suddenly ill with a fit—a sort of *petit mal*. He did not fall, and there was no convulsion. His wife spoke to him and found that he could not speak. In ten to fifteen minutes he regained his speech, and then spoke quite naturally. The next day he tried to go to work but was deterred by another seizure. He again did not fall or lose consciousness, but was dazed, and during the fit tried to dress himself. Since this fit on December 1, he has never been able to speak in an intelligible manner. For the first fourteen days after the attack he was kept in bed.

After he had been in the hospital for a short time his condition was found to be as follows:—General intelligence and behaviour good. There has been no fit or headache and he sleeps well. There is no appreciable weakness of the right limbs, or diminution of sensibility. The condition of the cranial nerves, and the reflexes is natural, and there is no optic neuritis.

His spontaneous speech is absolutely unintelligible, consisting of a rapid utterance of an utterly confused jumble of words, out of which no sense could ever be made, except that occasionally he says three or four words correctly, when asking for what he wants; for instance, the Sister of the ward says that on one occasion he asked her for "a toasting fork" after he had been in the hospital about a fortnight.

He can count quickly up to twenty and beyond; he can go through the alphabet with only a few mistakes, when started; and similarly he can name the days of the week, but the months of the year only very imperfectly.

He can repeat words, and sentences of three or four words, correctly. He reads the newspaper or manuscript, but how much of it he understands is doubtful. Still he understands short written instructions to perform certain acts—always doing what he is directed to do.

He reads aloud very much better than he speaks, though he makes many mistakes. He will read rapidly four or five words correctly, then will either omit, alter entirely, or slur over a word—but when his attention is called to his mistake and he is told to look at it again, he will utter it correctly if it be a short word, or after one or two trials if a longer word; then a few words correctly again, and so on.

His spontaneous writing is always a perfectly confused jumble of words, devoid of meaning, though many of the individual words are spelt correctly. Here is a letter written to me, at my request.

Dr. Bastin,—Do you have no derebt a letter note we have better do as I want you, know letter want round now only a sent you a doalet reund loat.

Youss faithfully,

H. P.

He can copy writing correctly; and he can do transfer copying correctly. He can also write pretty correctly from dictation.

This man was encouraged to read aloud daily during his stay in the hospital, and made to give more attention to what he read; but with practice during several weeks he only improved to a very small extent in his power of speaking, though he learned to read more correctly. His actions and manner seemed to indicate no appreciable mental impairment. He read quickly, and wrote quickly, however imperfectly; and seemed to understand everything that was said to him.

(3) In the very extreme forms of incoördinate amnesia, speech is reduced to a mere jabber of meaningless sounds " (gibberish-aphasia " or "jargon-aphasia "), and writing to a mere scrawl in which words and sometimes individual letters are indistinguishable. It is commonly found that the two disabilities co-exist, and also that the person who jabbers does not understand what is said to him, can read nothing, and does not even seem to know that what he says is mere gibberish.

As an instance of this type, I will quote a case recorded by Broadbent.[1]

CASE LXXX.—A painter, aged 42 years, had been subject to gout, and also to epileptiform attacks for several years. During the night of October 14, 1871, while lying on the right side, he suddenly put out the left arm and began to jabber—his right arm being quite useless. There were no convulsions, and no loss of consciousness. He was found by Dr. Felce, who was called to him, completely hemiplegic, with greatly impaired sensibility of the right side, keeping up a meaningless gabble, in which m-sounds were pre-

---

[1] *Medico-Chirurg. Transact.*, 1872, p. 170.

dominant, and showing the paralysed arm. The attack was followed by much cerebral excitement, shouting and violence. He soon regained power in the right limbs, but the speech was as imperfect as ever, and he was unable to write or copy. His general health became much deranged, and finally gangrene of the left foot came on. It was soon after this, on December 14, that he was first seen by Dr. Broadbent, who says: " He received us with a profusion of bows and smiles, with gestures expressive of welcome .... His speech was a mere jabber, in which ' Ma ' and ' Mum ' were prominent, and was accompanied with an excess of gesticulation, smiles, and facial expression. The gestures were very striking, and apparently appropriate when we had a key to their meaning. .... It was stated that he said ' Yes ' or ' No,' and ' Oh, my ' at times ; but he did not use even these simple words before us. He was unable to write his own name when his signature was before him. When urged to do so, he scribbled off rapidly something in which letters of some sort were distinguishable at first, but then tailing off into a scrawl.

" He obviously did not understand anything that was said to him ; did not squeeze my hand on repeated requests, but went on shaking it and smiling ; put out his tongue repeatedly when told to close his eyes, but instantly imitated the act after Dr. Felce. It was doubtful how far he recognised the state of his speech ; he went on chattering as if he thought he was understood, but he also made signs. .... He remained in much the same state till his death, about Christmas ; once startling some friends in conversation at his bedside by exclaiming ' Exactly ' at a very appropriate moment, but not otherwise regaining speech."

Other instances of this type have been given in the section dealing with the combined effects of lesions in the auditory word centre of each hemisphere (p. 159), and also in that dealing with cases where the left auditory and the left visual word centres were destroyed (p. 202). These, in fact, are the most common conditions co-existing with "jargon-aphasia."

The explanation of some of these cases of paraphasia and paragraphia is still very obscure, and their exact mode of production is probably subject to considerable variation in different cases.

The *first* and *second* grades of this kind of defect are probably similar in nature; the defects being produced in the same manner, though due to disabilities which vary in their degree of severity.

It seems to be very generally admitted that the slighter paraphasic defects are often dependent upon perversions in the functional activity of the left auditory word centre.

Some of these slighter defects are apt to show themselves in this way in weak and exhausted states of the system, and in cases where there is impaired nutrition of the brain either from previous disease or from old age—and especially at times when the attention of the person speaking is distracted by counter currents of thought being carried on at the same time. In all such cases wrong words are particularly apt to slip out.

Many of the conditions, indeed, that favour the occurrence of amnesia verbalis are also favourable for the production of paraphasia; so that the two states very frequently co-exist—the person who hesitates for a word often bringing out a wrong one.

Even in Case lxxix., where the paraphasia and paragraphia were so marked, though without apparent diminution of general intelligence, it seems quite conceivable that the main disability was in the left auditory word centre. This patient's defect seemed to be principally in the lack of power of correctly originating words in the auditory centre for the purposes of spontaneous speech or spontaneous writing; but when a strong stimulus came to this centre, as in reading aloud, or writing from dictation, it seemed to be capable of displaying a more correct activity. Similarly, though his spontaneous speech was so bad, he could repeat words or short sentences quite correctly; and notwithstanding the mere jumble of words and no-words represented by his spontaneous writing, he could do transfer copying quite correctly. So that a strong stimulus acting directly upon the auditory or the visual word centres enabled them to function pretty correctly; whilst a strong stimulus coming to the auditory from the visual word centre also enabled it to act fairly well, as in reading aloud.

In other cases a temporary paraphasia results from destruction or serious damage to the left visual word centre. This is easily to be understood, since the topographically and functionally related auditory word centre might easily become, as a consequence, more or less perturbed in its action.

Again, it occurs with slight lesions of the auditory word centre itself or, as we have seen, with more severe lesions producing word-deafness—I mean that it may be an occasional result of such a lesion when the individual (perhaps owing to his being a "visual") in striving to speak does it by means of incitations passing directly from the visual word centre to the glosso-kinæsthetic centre, but before properly coördinated relations have been established between them.

Wyllie seems to suppose that in these cases of word-deafness, paraphasia may result from a spontaneous and unguided action on the part of the left glosso-kinæsthetic centre, owing to the guiding influence of the auditory word centre being absent.[1] But in my opinion, unguided action on the part of the left glosso-kinæsthetic centre results in the production of gibberish aphasia—the highest grade of this incoördinate defect, of which I shall speak presently.

[1] *Loc. cit.*, p. 238.

Lichtheim's view that paraphasia is dependent upon a severance of the associational fibres between the auditory and the glosso-kinæsthetic centre his so-called ("commissural aphasia") is founded upon no sufficient evidence, and, as will be seen later (see p. 242), there is reason to believe that aphasia would occur as frequently as paraphasia as a result of such a lesion—especially in those persons in whom the visual word centre is not able to act directly upon Broca's region.

How far defects in the action of the glosso-kinæsthetic centre itself may be a cause of paraphasia—that is, when the auditory word centre is sound—seems also a point worthy of consideration. Thus some five years ago I saw at Brixton, in consultation with Dr. M. Castaneda, a German merchant, who had quite recently had a slight apoplectic attack. He was not appreciably hemiplegic, but was unable to express himself intelligibly, uttering, in fact, only a mere jumble of words. Yet when I put a pencil into his hand and told him to tell me exactly how he got from his home to his office in the city, he wrote three or four lines to the point, and without any mistakes either in the collocation or the spelling of the words. But if, as I and many others believe, the primary revival of words for spontaneous writing occurs in the auditory word centre, it would seem as if, in this gentleman, this particular centre was capable of acting well, and therefore that the very disordered speech must have been due to disturbed functional activity in the left glosso-kinæsthetic centre. On the other hand, his auditory centre might have been much damaged, but being a "visual" he might have been able to write correctly under the sole initiation of the visual word centre.

After what has been said, it would be superfluous to dwell upon the mode of production of analogous *paragraphic defects*. It must suffice to say that disabilities in the visual word centre of different degrees, or, in the case of its destruction, an imperfect or incoördinate functional relationship between the left auditory word centre and the cheiro-kinæsthetic centre may be the most important proximate causes—though an independent disability in this latter centre alone must also be considered as a possible causative condition.

If we turn now to the *third* grade of these paraphasic and paragraphic defects, I would in the first place demur to Wyllie's statement that this "gibberish aphasia" is produced by destruction of the left auditory word centre alone.[1] The case of Broadbent which he quotes in support of this view was one in which word-blindness was just as well marked as the

[1] *Loc. cit.*, pp. 269 and 289.

word-deafness; while a reference to the section dealing with the effects of lesions limited to the auditory word centre itself, will show that it is not associated with the production of "gibberish aphasia."

On the other hand, it may be seen that in whatever other way this extreme form of paraphasia is producible, it is certainly occasioned in some persons by lesions which destroy the auditory word centres in both hemispheres, and also in other individuals where there is simultaneous damage to the left auditory and the left visual word centres. Here, therefore, "gibberish aphasia" would seem to be the only kind of speech producible when the left glosso-kinæsthetic centre is left to act without its usual guidance.

All these cases to which I have last referred seem to show the extremely limited autonomy of the kinæsthetic centres. They are, in fact, decidedly opposed to the views of Stricker, Hughlings Jackson and others; while they harmonise with the opinions I have long since expressed—to the effect that these centres are never accustomed to act alone, but only in response to definite incitations coming to them from the auditory and the visual word centres; also that auditory and visual word images are the real counters for thought, the glosso-kinæsthetic and the cheiro-kinæsthetic word images being only revived secondarily, and during the expression of thought by speech and writing.

In all the cases of this third group to which reference has just been made we have had the defective speech and writing occurring in the same person, and also associated with word-deafness and word-blindness. But a totally different class of cases exists in which the defect is limited to one or other of these modes of expression, and in which it exists without the association of word-deafness and word-blindness.

Thus, in a remarkable case of defective speech without defect in writing or mental impairment, which was carefully investigated and recorded by Dr. Osborn,[1] the patient was able to talk only in a meaningless jargon, and on attempting to read aloud gave utterance also to a series of articulate sounds having no intelligible meaning or resemblance to those which he should have uttered. Some of the principal particulars concerning this case are subjoined.

CASE LXXXI.—A scholar of Trinity College, Dublin, aged 26 years, of very considerable literary attainments, and well versed in French, Italian,

[1] *Dub. Jrnl. of Med. and Chem. Science*, vol. iv., p. 157.

and German, while sitting at breakfast after having bathed in a neighbouring lake, suddenly had an apoplectic fit. He was reported to have become "sensible in about a fortnight," but though his intellectual powers were restored he had the mortification of finding himself deprived of speech. He spoke, but what he said was quite unintelligible, although he laboured under no paralytic affection, and uttered a variety of syllables with the greatest apparent ease. When he came to Dublin his extraordinary jargon led to his being treated as a foreigner in the hotel where he stopped and when he went to the College to see a friend he was unable to express his wish to the gate-porter, and succeeded only by pointing to the apartments which his friend had occupied.

Dr. Osborn, after frequent careful investigations, ascertained the following particulars concerning his patient :—

(1) He perfectly comprehended every word said to him.

(2) He perfectly comprehended printed language. He continued to read a newspaper every day; and when examined proved that he had a very clear recollection of all that he read. Having procured a copy of Andral's "Pathology" in French, he read it with great diligence, having lately intended to embrace the medical profession.

(3) He expressed his ideas in writing with considerable fluency; and when he failed it appeared to arise merely from confusion, and not from inability, the words being orthographically correct, but sometimes not in their proper places.

(4) His general mental power seemed unimpaired. He wrote correctly answers to historical questions; he translated Latin sentences accurately; he added and subtracted numbers of different denominations with uncommon readiness; he also played well at the game of draughts.

(5) *His power of repeating words after another person was almost confined to certain monosyllables;* and in repeating the letters of the alphabet he could never pronounce *k, q, u, v, w, x*, and *z*, although he often uttered these sounds in attempting to pronounce the other letters. The letter *i*, also, he was very seldom able to pronounce.

(6) In order to ascertain and place on record the peculiar imperfection of language which he exhibited, Dr. Osborn selected and laid before him the following sentence from the bye-laws of the College of Physicians, viz., "*It shall be in the power of the College to examine or not to examine any Licentiate previous to his admission to a Fellowship, as they shall think fit.*"

Having set him to read, he read as follows :—"*An the be what in the temother of the trothotodoo to majorum or that emidrate eni enikrastrai mestreit to ketra to tombreidei to ra fromtreido as that kekritest.*" The same passage was presented to him a few days afterwards, and he then read it as follows :—"*Be mather be in the kondreit of the compestret to samtreis amtreit emtreido andt emtreido mestreiterso to his eftreido tum bried rederiso of deid daf drit des trest.*"

He generally knew that he spoke incorrectly, although he was quite unable to remedy the defect.

In this remarkable case in which the patient's intellect and power of writing were practically unimpaired, it seems probable that the defective speech and power of reading were both of them due to some perverted mode of activity in the left glosso-kinæsthetic centre. The case is comparable with one of complete aphasia without agraphia, such as that which has been recorded by Guido Banti (p. 89), only here, instead of no speech, there was greatly disordered jargon-speech.

Broadbent, however, has recorded a puzzling case,[1] in which though the patient's speech was limited he made use of actual words, whilst in reading he uttered invariably nothing but gibberish. A similar explanation, therefore, would not be possible here, and I am rather at a loss to understand how this particular disability is to be accounted for, though in other respects the case seems to me to be very similar to one of my own (Case xciv.), in which I believe the double commissure between the auditory and the visual word centres to have been destroyed, together with slight damage to the auditory word centre itself. Broadbent's case might possibly, however, be accounted for by supposing that the disordered reading was due to imperfect direct action of the visual word centre upon Broca's centre, owing to (and necessitated by) the fact of the visuo-auditory commissure having been destroyed.

In other cases the defective power of expression occurs in writing rather than in speech. Two examples of this type have been recorded by Hughlings Jackson. One was that of a government clerk,[2] while of the other, and more severe case, the following are a few details.[3]

CASE LXXXII.—An elderly, healthy-looking woman suddenly became ill five weeks before admission. She lost the entire power of speech for a week, and was also paralysed on the right side. When seen there was no apparent hemiplegia, but she complained of weakness in the right side. She could then talk, but made mistakes. For instance, when I was trying her sense of smell, which was very defective since the paralysis, she said in answer to a question, "I can't *say* it so much," meaning she could not smell so well. She frequently made mistakes in speaking, and called her children by wrong names. This was never very evident when she came to the hospital, and might easily have been overlooked, but her friends complained much of it. She seemed to be very intelligent. Her power of expression by writing, however, was very bad, although her penmanship was pretty good, considering that she wrote with the weakened right hand. She wrote the following at the hospital. I first asked her to write her name—I do not like, for obvious reasons, to give her real name for comparison; it had not, however, the slightest resemblance to the following, in sound or spelling :—

"SUNNIL SICLAA SATRENI."

When I asked her to write her address, she wrote,—

"SUNESR NUT TS MER TINN—LAIN."

Thinking she might have been nervous when she wrote at the hospital, Dr. Jackson asked her to bring something that she had written at home.

---

[1] See under "Disease of the Commissure between the Auditory and Visual Word Centres," p. 253.
[2] *Brit. Med. Jrnl.*, 1866, ii., p. 92.
[3] *London Hospital Reports*, vol. i., p. 432.

She did so, but the specimen (a fac-simile of which he gives) was not in the least better than what she had previously written. It is a perfectly meaningless assemblage of letters, notable only for the frequent repetition of small groups of them, in a fashion which we shall also find repeated in the next case.

Unfortunately it is not stated whether this woman was able thoroughly to comprehend written or printed characters, and without knowing her condition in this respect, no safe diagnosis can be made. There was in her case an ability to form letters, but an inability to group them into proper words —and thus a complete inability to express her thoughts by writing, even though her errors in articulate speech were comparatively few.

A very peculiar case has also come under my own observation, concerning which some details may be given here.[1]

Case LXXXIII.—The patient was a sailor, who became a criminal lunatic in 1855, and was afterwards as such confined in an Asylum. He was not hemiplegic, but there was a certain amount of dementia, together with some slight impairment of speech and an utter inability to express himself in writing. It was two years after the above date that the peculiarity in his writing about to be described began to manifest itself. It was first shown in this way: he commenced the writing of each word correctly, and then in the place of the remaining letters he wrote "*f f g*." Later there was a tendency to duplication of many of the consonants, together with an almost invariable termination with the letters "*ndendd*" or at least "*endd*."

Some seven years after the commencement of this extraordinary style of writing the man came under my observation at Broadmoor Asylum, and he then gave me at different times very many sheets of his peculiar penmanship, and from these I had sixteen specimens lithographed which are here reproduced (Fig. 11). These show plainly that he either wrote with a peculiar and continual iteration of certain sets of letters, the writing being partly intelligible; or else that it was a mere succession of letters and strokes to which no meaning whatever could be attached.

This man's mode of reading agreed with his mode of speaking rather than with his peculiar style of writing; on the other hand, he spelt a word in the way in which he would write it, and then pronounced it correctly immediately afterwards.[2]

The principal peculiarities about this case are, that whilst the patient writes in this fashion he speaks in the ordinary way, so far as he can, con-

[1] This case was originally recorded in the *Brit. and For. Med. Chir. Rev.*, Jan., 1869, p. 234.

[2] Showing that the act of spelling was apparently modified by simultaneous revivals in the cheiro-kinæsthetic and in the visual word centres. In the case of the government clerk recorded by Hughlings Jackson, the mode of spelling was entirely in accordance with the mode of writing. This perhaps might have been expected. Still it was rather puzzling to hear a man when asked to spell *cat* say deliberately "*candd*," and then immediately pronounce the word as though he had spelt it "*cat*." The same kind of thing was met with in the next case (Case lxxxiv.). In Cremen's case, however, the mode of spelling did not accord with the mode of writing (see p. 151).

sidering his demented condition and somewhat incoherent state of mind. Fearing lest my memory might have failed me concerning certain points, I wrote to my friend, Dr. William Orange, in 1868, to let me know what his condition then was. He very kindly submitted him to a careful

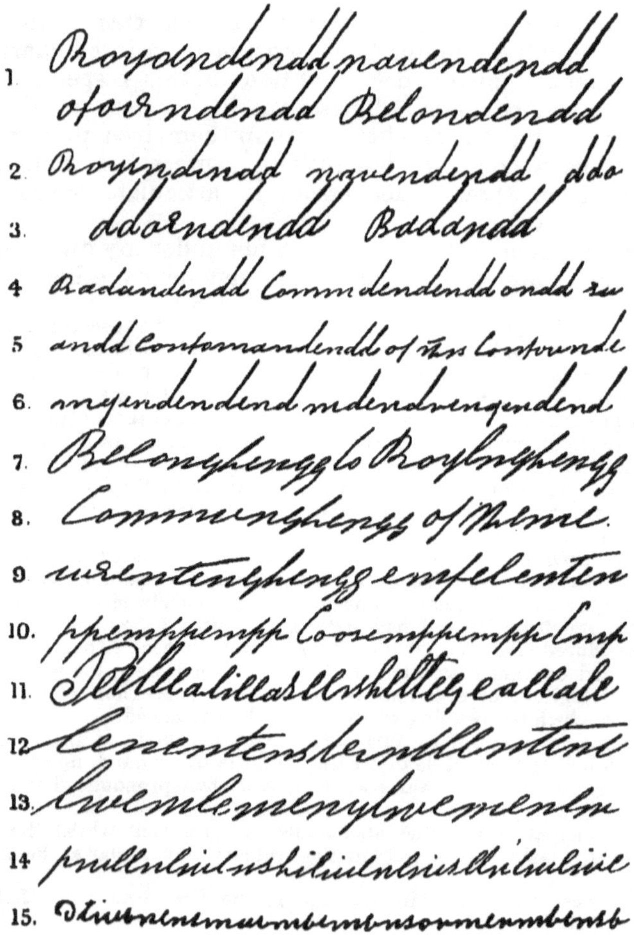

FIG. 11.—Various Kinds of Writing Employed by the same Individual at different times.

examination, and his replies to my questions are quite in accordance with my own recollections of the case, though it seems that the man had then become more demented, whilst his special defect was much less marked than it had been. The principal peculiarities observed were as follows:—

(1) He can speak fairly well for a short time, but his attention wanders,

and his voice then becomes drawling and monotonous, when he often either mispronounces a word (generally by altering its termination) or he substitutes another word or mere sound having no meaning.

(2) He can read the newspaper either to himself or aloud—but he does not seem to gather the full meaning without effort, and his power of continuous effort is small. When he reads aloud he stumbles over the difficult words, and he reads in a drawling tone, but the words which he utters, if not actually those before his eyes, are words of a somewhat similar sound and do not appear to have any obvious relation to the peculiar style of his writing.

(3) He spells a word, when asked to do so, in the way in which he would write it, and then he pronounces it correctly immediately afterwards.

(4) He can play at cards or draughts, but he seldom does play, according to his own account, because of the quarrelling; but there is no doubt the chief reason is apathy, and the mental effort which playing causes him. His memory is much impaired, especially as to recent events.

With regard to the specimens of his writing which are given in Fig. 11, No. 1 is a portion of the longest piece that was ever intelligible to me; it seemed intended for "Royal naval medical officer belonging to Admiralty." Comparison of this with the second and third specimens (written at different times), will show, in the first place, that he did frequently write the first two or three letters of words correctly —although at other times even these were quite wrong or duplicated—whilst the remaining portion of the word was replaced by *endendd* or *ndendd;* and secondly, that he spelt the same words in a similar way at different times. Having been a sailor in the navy, the words, "royal," "naval," and "doctor," appeared frequently in his writings, and these were almost invariably written in the same way. But this was not always so with other words which he did not employ so frequently; thus, the first word of No. 4 seems also intended for my name, though he usually wrote it as given in No. 3. Here it seems that the fact of the repetition of the letter *a* in the name appears to have had its influence in changing his termination from *dendd* to *dandd*. No. 6 is an example of some of the worst of his writing in this style, in which there is no attempt at the formation of words, but the mere continuous stringing together of letters. Specimens 7, 8, and 9, are examples of a style of writing which he adopted for a few weeks, and from the first "Belonging to Royal," it will be seen that *nghengg* has replaced the old termination.

Whilst he wrote in this way and in that first described he constantly made use of two words spelt rightly, *of* and *to*, and in No. 5 two consecutive words are spelt properly, though in all his writings I have never seen this occur elsewhere. When writing in the way first described he invariably wrote "*themedd*" instead of "the," and just as invariably when writing in the second manner he spelt it "*theme*" (see No. 8) simply. No. 10 is a form of writing of which I have only seen one specimen. The remaining kinds (11—16), though I have arranged them in an order of increasing simplicity, were by no means written at successive times, in the same order. He would adopt one of these forms for a time—perhaps for two or three weeks—and then generally return for a much longer period to his *ndendd* style, subsequently relinquishing this again, and adopting some other form. For some considerable time he wrote after the fashion shown in No. 12, when *entens* and *entent* perpetually recurred.

I have seen whole sheets of foolscap closely covered with markings in each of the styles 13, 14, 15, and 16; and in 1868 he was writing as shown in No. 11, which was selected from a specimen forwarded to me by Dr. Orange. The first two modes are the only kinds of writing in which words were at all recognisable, and this generally in the first few lines only; after these his manuscripts became almost entirely unintelligible.

It seems to me very interesting to compare this case with that recorded by Osborn: there the patient could write properly, though he spoke in a mere jargon, in which the same sounds frequently recurred; here the man (barring his demented condition) spoke properly, though he wrote a mere jargon of letters, with a constant repetition of certain of them.

Just as Osborn's case seemed to be due to a defect in the glosso-kinæsthetic centre, so here, although the case is by no means so simple, we may imagine the defect to be largely in the cheiro-kinæsthetic centre. That it was more pronounced there than in the visual word centre would seem to be shown by the fact that the man's reading was fairly good, and certainly showed none of the same kind of defects as were met with in his writing.

Again, in the interesting case of amnesia recorded by Cremen, an abstract of which has been given in an earlier section (Case xlviii.), a paragraphic defect existed which was even more comparable with the paraphasia in Osborn's case than was that of the sailor above referred to. It must also have been due, I think, to partial destruction of the cheiro-kinæsthetic centre.

In a case recorded by J. T. Eskridge,[1] which he speaks of as "agraphia of the motor variety," indicating disease of the left frontal region, a condition existed very closely resembling that met with in the criminal lunatic. The following is an abstract of this interesting case, which is very fully detailed by Eskridge.

[1] *The Medical News* (American), August 1, 1896.

Case LXXXIV.—A man, aged 34 years, was the subject of some slight mental derangement, and one year after this commenced (in September, 1893), he first experienced difficulty in writing.

"Before this he had been a good and rapid penman, having served as bill-clerk in the house of representatives during two sessions of the State legislature. The difficulty in writing consisted in misspelling and in transposing the letters of words. The disturbance soon became so great that it was difficult to make any sense out of what he had written. He had been before his illness fairly good in spelling, but at times was careless."

He had an attack of speechlessness in January, 1895, in which he could not utter a single articulate sound for about fifteen minutes; then speech was slowly restored, and in one hour all speech disturbance had passed away. Subsequently, on four or five occasions he experienced difficulty in speaking, which came on suddenly, but at no time was there a total inability to speak as in the first attack.

The patient was first examined by Eskridge, on April 19, 1895, and the following are some of the results: His mental action is slow. There is no form of visual or auditory defect in his speech. He repeats words readily, except that he is slow and deliberate in his articulation, and sometimes he hesitates before trying to utter some words. If he makes a mistake he at once recognises it, and usually corrects it. He has no difficulty in thinking in speech. He writes his name perfectly and copies script, and printing into script, with scarcely a mistake. In writing voluntarily he transposes the letters of a word, adds and omits letters, so that what he writes makes very little sense unless one knows what he is trying to say. On requesting him to spell words, it is found that he spells the words just as he writes them. It seems to matter but little how he spells a word, he usually pronounces it correctly. He reads aloud writing and printing quite readily, but not infrequently omits and miscalls words. He says that he has no difficulty in understanding what he reads.

Six months later (October 27, 1895), having suffered much from pains in the head in the interval, his condition was found not to have varied much. Thus, Eskridge says: "In writing simple words, such as 'dog,' 'cow,' 'hog,' at dictation, he frequently transposes some of the letters, and is as long in writing a single letter as a child is when it is first learning to write. He can write voluntarily, but it requires about two hours for him to write five or six lines. He can read much of what he has written, but miscalls words, inserts some, and leaves out others. When he makes a mistake in writing he sometimes recognises it and will try to correct it, but in trying to correct one he frequently makes a greater mistake. At other times he is either indifferent to his mistakes or does not perceive them. The effort of writing is very exhausting to him. He can copy fairly well, transcribing both printing and writing into script. On getting him to spell a word aloud, he always spells it just as he writes it, although in either case he pronounces it correctly." A letter written by this patient is given in fac-simile.

On December 5, 1895, he was operated upon and a cyst was found in what was thought to be the foot of the second left frontal convolution. The cyst was adherent to the dura mater and it caused a constricted appearance when this membrane was exposed. "On endeavouring to relieve the constriction by cutting through the dura mater at this point, which was found to be very thick, a straw-coloured fluid spurted out several inches high, apparently from the substance of the brain. The opening through which the fluid escaped was now enlarged, and about an ounce of fluid in all was found. The cyst which had contained the fluid was situated immediately below the cortex, and extended into the brain substance to the depth of half an inch or more. The globular portion of the cyst lay within the white substance of the brain, the constricted neck-like portion in the cortical substance, and its external boundary was formed by the membranes of the brain. The fluid was watery, straw-coloured, and contained no pus or tumour cells."

His general condition and ability to write is said to have distinctly improved after the immediate effects of the operation had passed away; that is, during the last two weeks of December, and through the month of January, 1896.

On February 1, 1896, he struck his head against a wooden box, the sharp corner of the box striking the head at the point where the bone had been removed. After this there were many fresh head symptoms, and in consequence of them he was operated upon again on February 13. A recent, friable, inflammatory cyst containing a quantity of blood clots was found, which extended almost completely over the frontal lobe, as well as back over the Rolandic region and the parietal lobe. When it had been removed the cortex presented a flattened appearance.

After the operation he had five or six right-sided convulsions, which began two or three days after the operation. When the patient was examined, two months after this operation, his condition as regards speech, reading, spelling, and writing, was found to have undergone very little change.

In his comments on this case, Eskridge says: "It would appear that his speech defect was an inability to spell rather than an inability to write, as he could copy both printing and writing with scarcely a mistake, transcribing the former into script. He copied much more rapidly than he could write voluntarily, except when the words were spelt for him. If a word was spelt for him he wrote it readily and correctly. Further, it should be noted that he formed his letters well, except when he was suffering severely from headache or from other causes." I agree with the recorder that in this case "the agraphia is secondary, and the fault in spelling the primary symptom." The case is, indeed, a very puzzling one, and two interpretations seem possible. The lesion, judging from what was seen at the operation, was presumed to exist at the foot of the second frontal convolution, but then nothing is known as to the condition of the visual word centre, with which the function of spelling must be primarily associated.

The disability as regards spelling might possibly have been due to an incoördinate activity on the part of the cheirokinæsthetic centre. If we look to the fact of the patient's ability to copy correctly and to write correctly from dictation when the words were spelt, it might be that under the influence of strong associational stimuli coming to it from the visual word centre, this damaged cheiro-kinæsthetic centre could act correctly—though it could not act properly under the influence of weaker volitional stimuli coming to it from the same visual word centre.[1]

[1] Still similar reasoning would apply to the possibility of its being due to some functional defect in the visual word centre itself, where

The constant repetition of letters or combinations of letters which formed such a characteristic feature about the writing in Case lxxxiii., affords an admirable example of what Gairdner [1] long ago spoke of as "intoxication" of the brain with a letter, of which he gave various examples. A similar intoxication with words is not unfrequently shown by aphasics or by others who are partially agraphic. In all such cases, however, we may mostly expect to meet with evidence of some general mental defect over and above that which may be due to the mere aphasia or agraphia. A good instance of this may be quoted from Trousseau, concerning a patient to whom reference has previously been made (p. 146).

In a man suffering from left hemiplegia, his usual "stock of words was restricted to these two, ' My faith!; ' and when he was pressed hard, he looked impatient, and uttered the oath, ' Cré nom d'un Cœur!' . . . . . I asked him what his name was, and his occupation; he looked at me and answered : ' My faith !' . . . . . I insisted, but in spite of his efforts, he only shook his head with an impatient gesture, exclaiming: ' Cré nom d'un Cœur.' As I wished to find out how many words he had at command, I said to him : ' Are you from the Haute-Loire?' He repeated like an echo, ' Haute-Loire !' ' What's your name?' — ' Haute-Loire.' ' Your profession?'—' Haute-Loire.' ' But your name is Marcou?'—' Yes, sir.' ' You are sure it is Marcou?'—' Yes.' ' What department do you come from?'—' Marcou.' ' No; that's your name.' But with an impatient gesture, he exclaimed, ' Cré nom d'un Cœur.' "

the process of spelling is undoubtedly initiated. The volitional stimuli coming from the auditory word centre (for spontaneous writing) might not be strong enough to rouse a functionally depressed visual word centre into properly coördinated activity—though the strong stimuli coming directly to the visual word centre from without (as in reading or in copying) may be adequate to enable the patient to copy readily and correctly what he sees, and also fairly correctly to read what he sees (as in Case lxxix.).

There may have been, in fact, in this case a functional defect of the visual word centre fairly comparable with the functional defect of the auditory word centre existing in cases of marked amnesia verbalis, where patients who are scarcely able to speak at all voluntarily, can nevertheless read correctly at the incitation of strong visual impressions (as in Cases xlv.-xlviii.). It seems to me perfectly clear that in writing, the spelling of words (that is, the order of the letters) is first revived in the visual word centre, and that this activity calls up a related activity, letter by letter, in the cheiro-kinæsthetic centre. The auditory word centre cannot be concerned with this process, as in ordinary writing we spell by arbitrarily arranged letters and not phonetically or by sounds.

[1] *Procced. of the Philos. Soc. of Glasgow*, 1865-68, p. 104.

CHAPTER XI.

DEFECTS OF SPEECH AND WRITING DUE TO DAMAGE TO THE COMMISSURES BETWEEN THE DIFFERENT CORTICAL WORD CENTRES.

It will be found that damage to a commissure between two word centres produces some though not all of the defects occasioned by disease of the centre from which the fibres issue on their way to the second centre, and sometimes almost exactly similar defects to those caused by a lesion of this second centre.[1]

Several examples of this will be given in the course of the present chapter.

Referring to Fig. 12, it will be seen that I have indicated the existence of the following commissures: (1) between the auditory and the glosso-kinæsthetic word centres; (2) between the visual word centre and the cheiro-kinæsthetic; and (3) between the auditory and the visual word centres. These are the commissures to the damage of which our attention will now principally be given; but there is another, not included in the diagram, concerning which something must be said, namely, (4) a commissure connecting the general visual centre with the auditory word centre.

Other commissures exist which seem to occur, or at all events, to come into use, only in some individuals, and to be mostly associated with compensatory functions. They are of interest, not because we know anything about their own injury, but because the opening up or establishment of such commissures seems necessary to enable us to account for the preservation of this or that function, when either the auditory

---

[1] It seems to me better to reserve, as I have done for some years, the word "commissure" as an appellation for the fibres which connect centres of like kind, that is, either sensory centres or motor centres; and to name "internuncial" the fibres which connect sensory with motor centres. These two orders of fibres will thus be clearly distinguished from one another, as well as from so-called "sensory" and "motor" nerve fibres, namely, those that connect the periphery with sensory centres, or those that connect motor centres in the bulb or cord with muscles.

or the visual word centre is destroyed. Of such a nature is the commissure marked (*e, e*) in Fig. 12, connecting the visual word centre and the glosso-kinæsthetic centre; and those marked (*f, f; f′, f*), between the auditory word centre and the cheiro-kinæsthetic centre. Reference has already been made several times to the probable existence of such commissures, and a few additional remarks may now be made concerning these alternative routes through which the activity of the glosso-kinæsthetic and the cheiro-kinæsthetic centres are occasionally stimulated.

As I have already remarked, there is reason to believe that in some persons the glosso-kinæsthetic centre may be stimulated directly through the visual word centre (as it is habitually in deaf-mutes who are taught to speak). Such cases as those of Hertz and Hun (quoted and commented upon in "The Brain as an Organ of Mind," pp. 623-626), in which, though the patients could not speak, they could read aloud with facility, seem only explicable in this way.

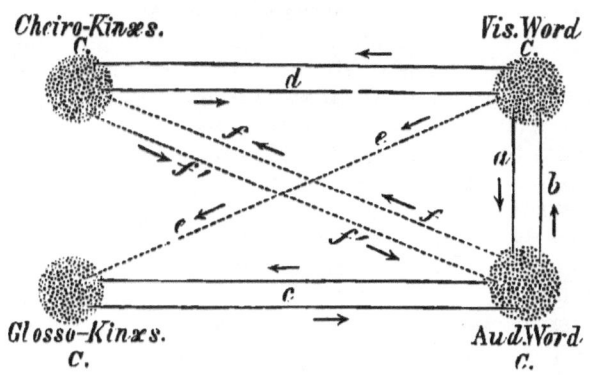

Fig. 12.—Diagram illustrating the relative positions of the different Word Centres and the mode in which they are connected by Commissures—*a*, the visuo-auditory commissure; *b*, the audito-visual commissures; *c*, the audito-kinæsthetic commissures; *d*, the visuo-kinæsthetic commissures; *e, e*, the visuo-glosso-kinæsthetic commissure; *f, f*, the audito-cheiro-kinæsthetic commissure; *f′, f′*, the cheiro-kinæsthetic and auditory commissure.

The last three Commissures represent unusual routes for the transmission of stimuli between the Word Centres.

There is also strong reason to believe that in some persons the cheiro-kinæsthetic centres may be stimulated directly from the auditory rather than from the visual centres; and when such persons become word-blind they may be able, in this manner, to write from dictation when they cannot perform the simpler act of copying writing; they write, moreover, as well with their eyes closed as with their eyes open (see Case lxvi., which is one form of " pure word-blindness"). Again, the fact that such patients are sometimes able to read writing aloud, by executing the manœuvre of tracing over the outline of the letters with pen or pencil, would seem to show that in them, at least, the cheiro-kinæsthetic centre may transmit its different impressions direct to related portions of the auditory word centres (Fig. 12, f, f'), so that the written words may then be spoken by the ordinary process.

These latter cases are very remarkable and interesting. Westphal, I believe, recorded[1] the first example in the case of a patient, of whom the following details are related:—
"He could write very well from dictation, but shortly after he was unable to read the words he had written, and he suffered in general from complete alexia. By means of a stratagem, however, as he himself very clearly explained, he succeeded in reading the word he had written from dictation upon the tablet. He passed his finger over each letter of the written word as if he were writing it again and read it while so doing. He then made a sort of calculation and counted off the sum of the separate letters." Here apparently the kinæsthetic impressions from writing movements were capable of rousing related parts of the auditory word centre so as to enable them to act through the glosso-kinæsthetic word centre, and thus evoke speech movements.

A very similar case has also more recently[2] been referred to by Charles L. Dana, who says: "A patient of mine, after a slight right-sided stroke, lost entirely the power of reading. She did not know even the letters that she saw. She could write fluently and coherently, but could not read a word of what she had written. . . . . Yet all her word-memories of the past were not destroyed, because by making her trace the letters with a pencil she could slowly read, through her muscular sense; and on reading a sentence its meaning was clear to her. But she could never be taught to read with her eyes." There is good reason for believing, therefore, that this particular commissure may be double, so as to be capable of transmitting impressions between the auditory word centre and the cheiro-kinæsthetic centre in both directions.

As I have already pointed out, however, when speaking of defects due to disabilities in the visual word centre, there is another means of explaining some of these cases. We there

---

[1] *Ziemssen's Cyclopædia*, vol. xiv., p. 776.
[2] *The Psychological Review*, November, 1894, p. 577.

saw reason for concluding that in some cases in which there is an inability to comprehend writing or print but no agraphia, we have to do with a second form of "pure word-blindness," in which the left visual word centre is not itself damaged, but where the fibres that associate it with the half-vision centre in the left occipital lobe and with the right visual word centre are cut across. In such a case there would be nothing to prevent reading at the instigation of cheiro-kinæsthetic impressions; it ought, in fact, to be more easy than in the cases previously referred to, seeing that the cheiro-kinæsthetic impressions might pass backwards to the intact visual word centre, and thence on to the auditory word centre in the ordinary way. This, however, would tend to show that commissure d, Fig. 12, is also double, and capable of transmitting impressions in both directions.

Other commissures again are invariably present, though it is only on very rare occasions that their damage, in connection with another lesion, may give rise to certain very rare forms of speech defect. I allude (a) to the commissure between the auditory word centre in each hemisphere, which, when damaged at the same time as the afferent fibres going to the left auditory word centre, gives rise to the condition known as "pure word-deafness" (p. 175); and (b) to the commissure between the visual word centres in each hemisphere, whose damage in association with that of the commissure between the left half-vision centre and the visual word centre of the same side gives rise to "pure word-blindness" (p. 200). These defects have already been considered and will not be further referred to.

A third commissure (c) is probably invariably present, namely, one between the glosso-kinæsthetic centres of the two hemispheres. The destruction of this commissure does not of itself cause any distinct form of speech defect, though it will hinder recovery in certain cases of aphemia. Its existence and functions, although already referred to (p. 67), will be more fully discussed in Chapter xii.

Other commissures still, though they do not originally exist in a perfected form, seem capable of being developed when an occasion for their existence arises, that is, as a consequence of destruction of one or more of the previously active word centres of the left hemisphere. Such, for instance, is (d) the commissure that either comes into play or seems to develop between the right auditory word centre and the left glosso-kinæsthetic centre, in cases where the left auditory and visual word centres have both been destroyed (Fig. 18, a); and

(e) that which seems to develop between the left auditory and the right visual word centre (Fig. 29), in cases where ability to write with the left hand is developed in a person whose left visual word centre or left audito-visual commissure has been destroyed (as in Case xc.).

Some of the commissures postulated by other writers are not regarded by me as having any real existence. These are the commissures going to, and coming from, an imaginary "concept centre," such as are believed to exist by Lichtheim, Wernicke, and others, and which are also figured in the schema of Charcot.[1] The reasons for my disbelief in any such commissures have been fully set forth in previous chapters, so that nothing more need now be said on this subject.

There is, however, one other commissure figured by Charcot, and the existence of which is evidently believed in by Wyllie,[2] Elder,[3] and possibly by others, to which no reference has hitherto been made by me. I allude to a supposed commissure between the left glosso-kinæsthetic and cheiro-kinæsthetic centres, which they imagine to be brought into play during the act of writing. I am not aware, however, of any facts or theoretical considerations tending to support the necessity for the existence of such a commissure. I do not believe in the autonomous activity of either of these centres. They are only secondarily, and in a subordinate manner, concerned with the web of thought; but when thought translates itself either into speech or writing, they are immediately called into full activity by strong incitations coming from the auditory or the visual word centres, with the result that there is an immediate coördinated activity upon the speech centres in the bulb, or upon the centres in the spinal cord concerned with writing movements. Nothing could be more different than the movements concerned with speech and with writing respectively (the one set of movements having to reproduce combinations of sounds, and the other arbitrary collocations of letters), and I cannot conceive any purpose which a commissure between the glosso-kinæsthetic and the cheiro-kinæsthetic centres could subserve.

The speech defects resulting from lesions of the four sets of commissures to which attention is to be especially given in this chapter will be found to be interesting and important either from the point of view of the regional localisation of aphasia and agraphia, or else because of the peculiar combination of defects which are thus produced.

---

[1] As given by Bernard (*loc. cit.*, p. 45).
[2] *Loc. cit.*, p. 317.   [3] *Loc. cit.*, pp. 83 and 84.

(1) DEFECTS DUE TO DAMAGE TO THE COMMISSURE BETWEEN THE LEFT AUDITORY WORD CENTRE AND THE GLOSSO-KINÆSTHETIC CENTRE.

Destruction of the *Audito-Kinæsthetic Commissure* in any part of its course between the auditory word centre on the confines of the occipital lobe and Broca's convolution will, in about one-half of the cases in which it occurs, give rise to an aphasia indistinguishable clinically from that which would be produced by damage to the glosso-kinæsthetic centre itself. This being true, it is of course a very important point, because it shows that it is an error to suppose that typical aphasia (as is commonly believed) can only be caused by a lesion in Broca's region.

I formerly thought, as announced in my book "The Brain as an Organ of Mind" (p. 686), that these results would always follow upon a destruction of this commissure in any part of its course; and that destruction of the auditory word centre itself would, in addition, give rise to word-deafness. The fact, however, that since that date Lichtheim and Wernicke (who have been followed by many others) have promulgated the view that destruction of the auditory word centre gives rise to paraphasia and word-deafness rather than to aphasia and word-deafness, led me, as I have previously detailed (p. 154), to carefully examine the recorded cases in which lesions have been more or less strictly confined to the left auditory word centre, with the result that neither view has been found to be correct. In rather less than half of the cases there was aphasia, and also in rather less than half of the cases there was paraphasia, whilst in the remainder voluntary speech was still less affected. This result was certainly very surprising to me, and, as I have endeavoured to show, the explanation of the apparent anomaly that aphasia was not produced in all cases is probably to be found in the fact that in some persons suffering from destruction of the left auditory word centre paraphasic speech (or, rarely, even more correct speech) may be brought about by means of incitations passing either from the left visual word centre or from the right auditory word centre to Broca's region.

This being true in regard to the effects resulting from destruction of the left auditory word centre, it must be conceded that there is some probability in favour of the view that paraphasia may occasionally result from damage to the audito-kinæsthetic commissure. Lichtheim's view (in which he has been followed by many others) is that paraphasia is always to be met with in such cases, but this is certainly an error. An examination of recorded cases would seem to show that in the majority of cases aphasia is the form of speech defect that is met with as a result of destruction of this commissure; the proportion being larger than that met with as a result of destruction of the auditory word centre itself, probably because a lesion destroying this commissure would often also destroy the visuo-glosso-kinæsthetic commissure (Fig. 12, *e, e*).

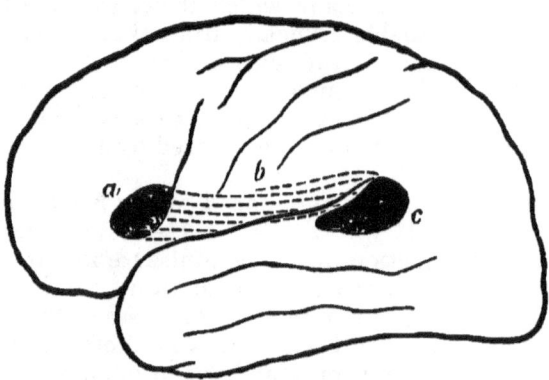

FIG. 13.—Diagram showing the possible location of an Aphasia-producing Lesion, either in the Glosso-Kinæsthetic Centre (*a*), in the Auditory Word Centre (*c*), or in some part of the commissural fibres (*b*) by which they are connected.

If we examine sections of the brain it will be seen that the commissural fibres connecting the posterior part of the upper temporal with the third frontal convolution must pass in the first place (1) not very far away from the posterior extremity or sensory division of the internal capsule, and thence (2) onwards beneath the island of Reil. But these are just the regions the damage of which may, at times, as much clinico-pathological experience has shown, be associated with aphasia.

(1) Grasset, about the same time that I dwelt upon the above-mentioned view that a section of these fibres would cause aphasia, called special attention to a fact which had not previously attracted much attention, viz., the not unfrequent association of aphasia with loss or disturbance of general sensibility (hemianæsthesia) on the right side of the body.[1] He refers to several cases illustrating this association, but gives, as I venture to think, an erroneous explanation when he attributes it to the supposed propinquity of the third frontal convolution and the sensory segment of the internal capsule

[1] "Des Localisations," 1880, 3me. édit., pp. 272-277.

(*loc. cit.*, p. 277). The posterior part of the hinder segment of the internal capsule is in reality far removed from the region of Broca, though it must be very close to the hinder extremity of the commissural fibres connecting the posterior half of the upper temporal gyrus with Broca's region.

(2) Again, the island of Reil lies just outside the direct tract which must almost certainly be taken by such commissural fibres. But, since 1868, when Meynert originally advanced the notion, and supported it by cases, that a lesion in the island of Reil might produce a typical aphasic condition, other cases of the same kind have been published, and these have been analysed by Boyer.[1] There are now over thirty of such cases on record, so that there can be little doubt as to the correctness of Meynert's position, on the understanding that a lesion of the insula would in all probability also destroy the audito-kinæsthetic commissure.

I shall not attempt to analyse these cases myself so as to ascertain, in the first place, which of them is sufficiently fully described to make the exact nature of the speech defect certain; and, secondly, how many of such cases presenting real aphasic symptoms have been due to a lesion strictly limited to this region. I have little doubt, however, that the number would be comparatively small that would strictly comply with these two conditions.

I shall content myself with recording four cases in which what I take to be true aphasia, has been produced by a lesion probably involving the audito-kinæsthetic commissure, and certainly leaving the third frontal convolution intact.

A very definite proof of this symptomatology has been given by Déjerine[2] in the following carefully investigated case, an abstract of which I subjoin.

CASE LXXXV.—A compositor, aged 20, suffering from phthisis, was admitted to the Hôtel Dieu, May 6, 1884. On July 16 at 8 a.m. the Sister, when speaking to him, noticed that he had a great difficulty in pronouncing words. Twelve hours afterwards this difficulty had disappeared. The next morning, Déjerine says: "At the hour of the visit a certain amount of motor aphasia was noticed in the patient, he pronounced with difficulty certain words and certain letters. *D*, for instance, could not be pronounced. There was also a certain degree of paraphasia, but no sensory aphasia; the patient easily understood what was said to him, and what he read. He wrote spontaneously and from dictation, without mistakes." His face was slightly paralysed on the right side.

Speech again became normal during the day, but in the evening the aphasia was found to have returned. The next morning (18th) it was yet

[1] " Etudes Topographiques sur les Lésions Corticales," Thèse de Paris, 1879, No. 115.

[2] *Rev. de Méd.*, 1885, p. 175.

more pronounced, though still only of motor type, and without agraphia. The right arm as well as the right side of the face was now weak. By the following day this was still more marked, as was also the aphasia. On July 20 there was, in addition, paresis of the right leg.

Two days afterwards, the patient being in much the same condition, Déjerine says he examined him from the point of view of the aphasia, and noted the following particulars : " The motor aphasia is complete, absolute; the patient could not pronounce a single word. He answered *bu, bu, bu*, to all questions; he can, however, say '*oui*,' and that is the only word he can utter. He cannot repeat spoken words. When asked his name, he takes his bed card and points to the name with his finger. When asked how many brothers and sisters he has, he extends one finger to the word brother, and two to the word sister. When asked his age, he counts on his fingers, and reaches twenty without any mistake. He quite understands what he reads, for if he is asked his occupation he takes his bed card and points to the word ' compositor ; ' similarly if one places on a sheet of paper several christian names, he points to his own correctly with his finger."

As Déjerine points out, there was no trace of sensory aphasia. The weakness of his right hand made it impossible to ascertain whether there was still an absence of agraphia.

On July 24 the patient had an attack of epileptiform convulsions ; many other attacks followed at short intervals during the day, and he died comatose at 2 a.m. on the following morning.

The necropsy was made with great care and was followed by a microscopical examination. The lesion was found to be a limited tubercular meningitis, involving the island of Reil and parts of the ascending frontal and parietal convolutions, while the convolution of Broca was intact throughout its whole extent. The whole of the island of Reil was covered by a fibrino-purulent exudation, containing innumerable granulations attached to the vessels which penetrated into the brain substance, so that the subjacent white fibres must also have been affected.

Déjerine holds that this case shows that a lesion of the island of Reil produces an aphasic condition impossible to be distinguished from that caused by a lesion of Broca's convolution, and believes this group of symptoms to be due to damage to the subjacent white fibres rather than to disease of the cortical grey matter itself. He considers, therefore, that this case tends to upset the view of Lichtheim as to the effects produced by destruction of this same set of fibres (the main result, according to him, being paraphasia), more especially seeing that Lichtheim's views were based upon a case[1] which is incompletely recorded clinically, and too complex from a pathological point of view to permit of any certain conclusions being based thereupon. Déjerine seems to have been quite unaware that I had previously expressed views of a precisely similar kind.

The next case, recorded long ago by H. Dogson,[2] seems to me also fairly convincing as to the reality of the aphasia, and

[1] See *Brain*, 1885, p. 445.
[2] *Lancet*, 1866, i., p. 397. This is Case xix. of Pitres.

as to the fact that the lesion must have destroyed the audito-kinæsthetic commissure, whilst Broca's region was uninjured.

CASE LXXXVI.—Mary S., aged 64, had an apoplectic attack on September 28, 1864, and remained comatose for three days. On recovering consciousness she was found to be paralysed on the right side, and speechless except for one word—"*far*." This word or syllable she subsequently made use of on all occasions up to the time of her death, and as an answer to all sorts of questions. She apparently understood well what was said to her, but on attempting to reply could only articulate the word "*far*," which she repeated three or four times in rapid succession.

The recorder says: "I have made inquiries of her attendants and cannot find, at all events during the last eight months of her life, that she ever articulated any other word. Her sister, with whom she lived for the first few months of her illness, stated to me that she had occasionally said 'yes' and 'no,' but I do not place much reliance on this statement for I have myself intentionally asked her questions requiring a simple affirmative or negative answer, but in place of that she invariably replied '*far, far, far.*'"[1]

The patient died of bronchitis about thirteen months after the commencement of her illness—on November 25, 1865.

At the necropsy an old hæmorrhagic cavity about one and a quarter inches in length was found in the white substance of the left hemisphere, external to the lateral ventricle. When its sides were opened out it was found to be about the size of a walnut, its long axis corresponding with that of the brain, and the cerebral substance around being stained of a dirty reddish brown colour. "To define its position more accurately: it was placed partly in the anterior and partly in the middle lobe of the brain, between the island of Reil on the one side, and the body and anterior horn of the lateral ventricle on the other. Its inner margin corresponded to and was opposite the corpus striatum. The original lesion being, in fact, in the outer or extra ventricular margin of that body, or between it and the convolutions of the island of Reil and of the Sylvian fissure. . . . . No lesion was found in any other part of this, or in the opposite hemisphere of the brain, which was carefully examined throughout."

Another very similar case has been recorded by Oulmont.[2]

CASE LXXXVII.—Eliza M., aged 67, entered the Salpêtrière September 16, 1874, under the care of M. Charcot, suffering from right hemiplegia and aphasia. No information could be obtained concerning her antecedents or the course of her malady.

In January, 1876, the following particulars were noted:—Right hemiplegia (face and limbs), with secondary contracture in the paralysed limbs; sensibility preserved. Complete aphasia; the patient seems to understand questions that are addressed to her, but can only reply by certain syllables that are always the same—"*teu, teu, teu.*" Sometimes she utters the word "*oui.*" When told to show her tongue, she opens her mouth widely, but the tongue remains motionless in the floor of the mouth. She follows attentively what goes on around her, and appears to comprehend what is done. The right upper extremity is motionless, with some rigidity. Lower

---

[1] We now know that there need be no grounds for these doubts, and that what the sister said and what the reporter says only tend to confirm the view that this was an undoubted case of aphasia and not one of aphemia.

[2] *Bullet. de la Soc. Anatom.*, 1877, p. 327. This is Case xl. of Pitres.

extremities both permanently flexed—the legs upon the thighs and these upon the pelvis.
March, 1877. Paralysis as before. Aphasia still complete. Death occurred on April 16, 1877.
At the necropsy the grey cortical substance seemed everywhere normal except on the left occipital lobe where there was a minute yellow patch from 2 to 3 millimetres in diameter. The lateral ventricles were dilated and filled with a yellowish coloured serum. When sections were made, the right hemisphere seemed to be everywhere perfectly healthy. In the left hemisphere was found a pale yellow focus with cellular walls, having a vertical direction, which had completely separated the anterior part of the corpus striatum from the convolutions of the island of Reil.

Another remarkable case has been recorded by Farge[1] which seems undoubtedly to be one of aphasia, and where the cause would at first seem to be a sub-cortical lesion destroying the geniculate fasciculus, which, according to my interpretation, should produce aphemia rather than aphasia. But here, as in the last two cases, it will be found that the lesion also occurs in the course of the audito-kinæsthetic commissure, the destruction of which should produce a typical aphasia—and thus would prevent the aphemia-producing lesion from revealing its existence. After a time, however, this last patient did begin to make very slight attempts at repeating words.

CASE LXXXVIII.—F. Chaslon, aged 61, a weaver, was admitted to the Hôtel Dieu at Angers, June 8, 1864, having had an apoplectic attack on the previous evening. He remained in a comatose condition for nearly two days, and in a dazed condition for two more, and was found to be paralysed on the right side.
June 11. He was more conscious and understood questions. All questions were invariably followed by the same reply, "*Ah! si....ah! oui.*" Two days later he uttered the same sound more freely and with less provocation. When stimulated by a slight prick or pinch he says, "*Ah! bon sens de Dieu.*" This was his invariable response to any slightly painful impression. He intelligently performed certain movements with the left limbs, tongue, or lips when told to do so.
June 14. Patient attempts when bidden by the doctor to repeat some words, when they are repeated over and over again before him, such as "*tisserand,*" "*médecin,*" "*bonjour.*" He is sometimes able to do it; often omits a syllable; and more frequently replaces the word demanded by "*Ah! si....Ah! bon sens de Dieu.*"
June 18. He has regained some slight power in the right arm and leg. His intelligence begins to clear. If questions are put to the patient which permit of his replying yes or no, he replies correctly, "*Ah! oui*" or "*Ah! si*" for the affirmative; "*Ah! non*" for the negative; but every insistence, every vexation or slight pain brings invariably "*Ah! bon sens de Dieu.*" Meanwhile his vocabulary increases; he can be made to repeat words, even phrases, by forcing him to repeat them. Example: D. *Chaslon, dites moi bonjour, monsieur.*—R. *Chaslon...dites moi.*—D. *Dites moi bonjour, monsieur.*—R....*moi...bon...jour, mo....* After repeating it four or five times,

[1] *Gaz. hebdom. de Méd.* 1864, p. 724.

he pronounces the phrase "*Bonjour, monsieur,*" but preceded or accompanied by "*Ah! si...Ah! oui;*" and if one insists the patient replies only, "*Ah! bon sens de Dieu.*" His pronunciation, however, is good, except for a slight thickness of speech.

June 27. Appetite bad. Is self-absorbed; replies badly and more rarely, and scarcely repeats at all. Some cough and oppression.

June 30. Patient died of pneumonia.

At the necropsy a careful examination of the surface of the brain showed nothing abnormal. The centrum ovale on the right side was healthy, while that of the left side contained a focus of softening as large as a small egg. Above, it extended to the level of the roof of the lateral ventricle. The superficial portion of the thalamus seemed slightly affected; the corpus striatum was healthy, and a careful examination of the third frontal convolution, both by the naked eye and by the miscroscope, showed that it also was healthy.

This was one of the early cases published as being out of accordance with the localisation of Broca. It is unfortunate that the exact situation of the lesion is not more fully described than by the words which I have above given. Farge only thought it necessary to show that the lesion was in the centrum ovale, and that Broca's convolution was perfectly healthy. It is to be presumed that by a small egg he meant a small fowl's egg, and if so, a lesion of this size outside the corpus striatum would almost certainly have destroyed the audito-kinæsthetic commissure. The recurring utterances were rather free from the first. If they were brought about by the right hemisphere (as is commonly supposed) this extra facility would imply the existence in this right hemisphere of more than the usual amount of organisation for special purposes, and would, therefore, account for the very early commencement of recovery, as evidenced by the beginning of attempts to repeat words even six days after the onset of his attack.

(2) DEFECTS DUE TO DAMAGE TO THE COMMISSURE BETWEEN THE VISUAL WORD CENTRE AND THE CHEIRO-KINÆSTHETIC CENTRE.

Destruction of this commissure should give rise to inability to write both spontaneously and from dictation, without any coexisting word-blindness. Even those who do not believe in the existence of a topographically separate centre for the registration of writing movements, would admit that a complete cutting across of the outgoing fibres from the visual word centre to the hand-and-arm centres would produce these results.

It is probable that this *Visuo-Kinæsthetic Commissure* is not unfrequently damaged at the same time as the audito-kinæsthetic commissure, the result being the production of a typical aphasic condition, with complete loss of the power of writing as well as of speaking—though the former defect may be obscured by the existence of paralysis of the hand and arm, rendering the common as well as the very special movements of these parts alike impossible.

It is only in very exceptional cases, therefore, that any distinct evidence can be obtained of the effects of damage to this visuo-kinæsthetic commissure.

While they would be hidden in the cases above referred to where there was coexisting paralysis of the hand, they would also be unconvincing to many if associated with aphasia but without paralysis of the hand, because of the belief held by many (though I have brought forward much evidence to show that it is erroneous) that destruction of Broca's centre of itself entails agraphia as well as aphasia.

It is only, therefore, as I have said, in altogether exceptional cases that we are at all entitled to postulate the existence of a lesion of the visuo-kinæsthetic commissure. These exceptional cases would be those in which the ability to write has been much more gravely impaired than the ability to speak, where there has been no word-blindness, and where subsequently a lesion has been found in a situation likely to have cut across these fibres.

Reference can be made to three such cases, namely, Cases xxviii., xxix., and xxx., recorded in the section dealing with the subject of agraphia, and to two others which have been recorded by Byrom Bramwell.[1]

What has been said as to the effects of lesions involving these two commissures may enable us thoroughly to admit the validity of some of the objections formerly raised against the doctrine promulgated by Broca and his immediate followers, that the posterior part of the third left frontal gyrus is *the* region always damaged in cases of aphemia. Many cases of aphasia were gradually made known where the necropsy revealed no lesion of Broca's centre. We now know that these were in part cases of aphemia, due to sub-cortical lesions involving the geniculate fasciculus, rather than of true aphasia; and still further that true aphasia may be produced by a lesion comparatively remote from the third frontal convolu-

---

[1] *Lancet*, 1897, vol. i., pp. 796 and 1404. I find in Allen Starr's paper on "The Pathology of Sensory Aphasia" (*Brain*, vol. xii., pp. 95 and 100) reference to a case of agraphia recorded by Sigaud which, from the characters he assigns to it on the latter page (Case xli.), ought also to be a case in which there was a lesion of the visuo-kinæsthetic commissure. Unfortunately, I have been unable to find the record of this case, as Allen Starr's reference to it is incorrect, and so is that of Mirallié (*loc. cit.*, p. 209). I have been unable to find any such case in *Le Progrès Médical* for 1887.

tion, so long as it cuts across the audito-kinæsthetic commissure in any part of its course.

(3) DEFECTS DUE TO DAMAGE TO THE COMMISSURES BETWEEN THE AUDITORY AND THE VISUAL WORD CENTRES.

The commissures between the auditory and the visual word centres are two in number (Fig. 12, *a*, *b*). One, the *Visuo-Auditory Commissure*, is habitually called into play in certain mental operations, such as reading aloud, when stimuli have to pass from the visual to the auditory word centre before the reading aloud can occur. The other, the *Audito-Visual Commissure*, is habitually called into play in writing from dictation, and probably in writing any spontaneous effusion, when stimuli require to pass between these two centres in an opposite direction, that is, from the auditory to the visual word centre.

It seems quite possible that one of these commissures only may, on certain occasions, be damaged by a lesion while the other escapes. I have neither seen myself nor read of a case in which the visuo-auditory commissure alone appeared to have been injured;[1] but I can quote the clinical histories of three cases in which some of the principal defects seem to have been due to destruction of the audito-visual commissure.

Subsequently I shall detail other cases in which the symptoms point to the destruction of both these commissures having occurred—resulting in each case in a significant and extremely interesting grouping of abilities and disabilities.

Damage to the Audito-Visual Commissure.—The leading peculiarity in this case would be that the individual would be able to read aloud quite well, though he would be unable to write either spontaneously or from dictation. There would be nothing, however, to prevent him from slowly copying writing, or printed capitals (as he would copy Hebrew or copy a drawing), so long as the visual word centre itself remained intact and in communication with the hand-and-arm centres, though he might not be able to do transfer-copying, or form words from block letters.

The first case of this kind that I will quote is one that was published by Dingley.[2]

CASE LXXXIX.—John C——, aged 56, a brewer's drayman, was admitted into hospital on March 27, 1884. He was rather more intelligent than men

---

[1] I formerly thought that I could refer to such a case (see "The Brain as an Organ of Mind," 1880, p. 645); but I have lately put a different and, I think, a more correct interpretation upon this important case recorded by Broadbent. It now stands as Case xliv. of this work.

[2] *Brain*, vol. viii., 1886, p. 492.

in his position usually are, and could, previous to his illness, read and write well, and keep simple accounts.

On March 25 he complained of headache; on the 27th he went to work as usual but came home at 8 a.m., "looking strange," and got into bed with his clothes and hat on. After a time he got up, and walked about the house in a delirious condition. He could speak plainly, but his right arm was weak, and he dragged his right foot slightly.

For the next two days his speech was thick and slow, his mind seemed somewhat confused and dull; he only answered questions and did not volunteer any statements. After a week his mental condition cleared up considerably, but in the course of three or four weeks he got into an almost stationary condition, the nature of which may be gathered from the following abstract of the notes given by Dingley:—

He can speak fluently and well though he is often at a loss for words. Adjectives, verbs, etc., seem to present little or no difficulty, it being almost entirely noun-substantives that are unable to be recalled.

He can read aloud with perfect ease, reading long and difficult words clearly and distinctly. He also thoroughly understands what he reads.

He cannot name common objects at sight, except a few very familiar ones. When asked the name of an object he sometimes makes blundering attempts somewhat resembling the correct one in sound; at other times he makes use of a periphrasis as, when asked to name a camel, he said, "Egypt—a long way—go a long way—carry things—hot place." All the time he is quite clear as to the word he wants, and will recognise it immediately it is spoken.[1]

He can repeat at once and easily any words that he hears.

He can copy written words quite well, and can transfer from print into written characters.

He cannot write from dictation, except his own name or such simple words as "cat" and "dog." He cannot write more difficult words; he sometimes attempts them, but only makes a scrawl; though occasionally he gets the first word right.

His power of spontaneous writing is not so much impaired, as he is able to write a few simple sentences. A good measure of the extent of his defect in this direction is given in the fact that although he can repeat the Lord's Prayer fluently he is quite unable to write it.

More than eighteen months afterwards (Oct., 1885), this patient was seen again and found to be "in very much the same condition."

The fact that this man's power of writing spontaneously was a little greater than his power of writing from dictation, makes it possible that he was to some extent a visual, and therefore that he was able to call the visual word centre to a slight extent into independent action for writing—that is, independently of prompting from the auditory word centre. In addition to the assumed defect in the audito-visual com-

---

[1] Dingley makes an interesting remark in relation to these peculiarities when he says: "The fact that he can read aloud, and yet cannot name objects, shows that there must be a visual word centre, as the auditory word centre can be stimulated by the reception of impulses from it, but cannot be roused into activity by volitional impulses originated on seeing the object." Naming at sight is supposed mostly to be brought about through the intervention of the "occipito-temporal commissure" (see p. 264).

missure, it is clear that there was also some functional degradation in the auditory word centre, leading to a certain amount of amnesia verbalis.

What I take to be another example of damage to the left audito-visual commissure is a case that has been most fully and completely recorded by Pitres as an example of agraphia due to destruction of the cheiro-kinæsthetic centre. Much discussion has taken place in regard to this case and another very similar one which was also recorded by Pitres in the same communication.[1] My reasons for dissenting from his interpretation have already been given (p. 106), and will, I think, be justified by the following details of his case.

CASE XC.—M. L——, a wine-merchant, aged 31, had led for several years a very irregular life. On July 30, 1882, whilst in a café he became ill, partly losing consciousness, and this was followed within an hour by the development of an incomplete right hemiplegia. His condition gradually became worse for several weeks, so that early in September he was in a semi-comatose condition and completely paralysed on the right side. He was then put upon an anti-syphilitic treatment, under which he rapidly improved, so that by the end of 1882 he could walk without assistance. This improvement continued during the following year, so that when he was examined by Pitres early in 1884 there only remained "some relatively slight but very interesting defects," such as are about to be described.

He complained of nothing unnatural except a slight stiffness of the right lower extremity, together with a complete inability to write with the right hand, although he could use this hand in almost all other ways.

On submitting him to a most careful examination Pitres found his condition to be as follows:—

There was no deviation of the face or tongue and a barely appreciable weakness of the right limbs—though there was slight rigidity of the leg with exaggeration of deep reflexes. Acuity of vision was normal, but there was a distinct right-sided hemianopsia.

The movements of the right hand were generally executed with precision, though a little more slowly than those of the left. Muscular sense in the right upper extremity was scarcely at all impaired. Thus Pitres found that if whilst the patient's eyes were closed he impressed upon his right hand the movements necessary for writing in the air the word "Paris" he readily recognised the word he had been made to write in this fashion.

The patient's intelligence was unimpaired; his speech was fluent, and his articulation good. He could read aloud well either printed or written matter; and he perfectly comprehended what he read.

During 1883 he had practised writing with his left hand, and had become able to write quite legibly with it.

When asked to write with the right hand the word "Bordeaux," he took up the pencil, held it naturally and without difficulty, but found it impossible to write a single letter. But he knew perfectly well of what letters the word was composed. Thus he spelt the word correctly, and he showed the several letters in a newspaper lying before him. He said, "Je sais très bien comment s'écrit le mot Bordeaux, mais quand je veux écrire de la main droite je ne sais plus rien faire."

[1] *Rev. de Méd.*, 1884. p. 855.

But with his left hand M. L—— wrote very plainly and without any mistake the word "Bordeaux." Then taking the pencil again in his right hand he was able with difficulty, and by looking each instant at the letters he was about to trace, to reproduce them with the right hand.

There were exactly similar results when he was asked to write individual letters and even numerals. He knew quite well what he was asked to write but could never succeed in tracing a single one of them with his right hand. He could write the letters or numerals quite readily, however, with the left hand; and then slowly copy what he had before him with his right hand.

He could make mental calculations without difficulty. Thus, if several numerals are placed below one another he can add them without making any mistake, though he is unable to write the total with his right hand.

Although, as above stated, this patient was able to copy writing with his right hand, he could not when asked to copy a printed word transfer the letters into writing characters, he could only slowly copy the printed characters themselves.[1]

Another most interesting point ascertained was this. When the patient was asked to draw with his right hand a circle, a triangle, and an octagon he did it at once without any notable hesitation. He could even sketch with his right hand a pretty well proportioned human figure in profile.

Seven months later this patient was found by Pitres to have undergone no appreciable change, although during the interval he had perseveringly tried to regain the power of writing with his right hand.

Although no specific mention is made in these notes in regard to the patient's power of spontaneous writing, I think from the general statement made by Pitres that he complained " d'une impossibilité absolue d'écrire de la main droite," this power must have been as completely lost as that of writing from dictation. This again would tend to show that he was a poor visual, and in this respect rather different from Dingley's patient, who could write a few words spontaneously though not from dictation, and could do transfer copying.

The fact that this patient of Pitres could draw geometrical figures with his right hand, and even sketch a human figure is not a contradiction, as these performances would have been brought about through the left general visual centre acting upon the corresponding hand-and-arm centres.

This patient, it is said, had gradually learned to write with his left hand during the year 1883—presumably, therefore, only after long practice. Now it has been noted by more than one writer that where there is word-blindness owing to destruction of the left visual word centre, such patients are mostly unable to acquire the power of writing with the left hand. This is explicable in accordance with my views, since in such a case the right visual word centre would be cut off from all relations with the left auditory word centre in which the initial process for spontaneous writing, or writing from dictation, occurs. It would, however, be equally cut off from the left auditory word centre by damage to the audito-visual commissure, which I assume to have occurred in this case.

[1] This would seem to show that he was a very poor visual.

It seems, therefore, as if, in this case, the left auditory word centre must have established direct relations with the right visual word centre by the development of a new set of commissural fibres (as represented in Fig. 29). Such a process is a possible one also in a case of word-blindness from destruction of the left visual word centre.[1]

That the defects shown by this patient are to be explained by damage to the audito-visual commissure rather than, as Pitres supposes, by a damage of the cheiro-kinæsthetic centre, seems to me clear for four more or less decisive reasons:—(*a*) Because the proved identification of writing movements made with the right hand in the air could only take place through an intact cheiro-kinæsthetic centre; (*b*) because of the patient's ability to copy with comparative ease words or letters with the right hand; (*c*) because the existence of hemianopsia is more in accordance with the presence of a lesion in the neighbourhood of the visual word centre than of one situated in the foot of the second frontal convolution or thereabouts; (*d*) because he was able to draw geometrical figures and even the outline of a man with the right hand when bidden, though he could not write letters or words—differences which are readily explicable in accordance with my interpretation, seeing that the general visual centre may have been undamaged, and also the commissures passing from it to the hand-and-arm centres.

The other case reported by Pitres in the same communication, though belonging to the same category, is not nearly so neatly defined. It was that of a Russian officer seen by Charcot.

CASE XCI.—M. X——, a Russian officer, who could speak Russian, French and German with facility. Six months after a slight threatening of cerebral disease he found to his great surprise, at an evening assembly, that when addressed in French or German he could not reply in either of these languages; though he understood perfectly what was said to him in them and could still express himself in Russian with his usual facility. Little by little he improved; regaining the power of speaking French, though that of speaking German still remained lost.

Then wishing one day to write he found that it was impossible for him to trace a single word, though his hand and arm were not at all paralysed. This led him to consult M. Charcot on April 10, 1883, who found him in the following condition:—

His intelligence was well preserved. He spoke French freely, replying to questions exactly, and related with precision the various phases of his malady.

He had a little difficulty in the movements of the fingers of the right hand, together with slight cutaneous anæsthesia, and incomplete loss of the sense of position of the fingers.

---

[1] See what has been said on p. 92 concerning left-handed persons learning to write with the right hand.

He could read aloud in either Russian, French or German. But he could not write in either of these three languages—not even in Russian. The latter fact surprised and disturbed him much; he said: "Je suis très affligé de ne pouvoir écrire en russe, bien que je le comprenne, que je le parle, et que je possède la force suffisante pour diriger la plume."

M. Charcot asked him to name and then write his address at Paris. He replied immediately: "Je demeure hôtel de Bade, boulevard des Italiens," but when he tried to write these same words, he could scarcely trace "je dem. . . From dictation he could, however, succeed in writing the rest. He could copy writing placed before him.

Asked to write M. Charcot's name in Russian, French or German, he could write it without any great difficulty in Russian; he had more difficulty in writing it in French; whilst it was impossible for him to write it in German.

A few days after this examination M. X—— died suddenly, and there was no necropsy.

Here neither the voluntary writing nor the writing from dictation were completely abolished, so that if the case really belongs to this category (as to which I freely admit there is room for doubt) we must suppose that this audito-visual commissure was badly damaged rather than destroyed.

Damage to both the Audito-Visual and the Visuo-Auditory Commissures.—Both sets of commissures may be simultaneously damaged, and this seems to have been the cause of the most notable defects met with in two of my patients whose cases will be recorded in this section.

The first of them came under observation at the National Hospital for the Paralysed and Epileptic in 1869, but nothing similar was encountered till about ten years later when the second patient came under my care. The following are a few details concerning the first case.

CASE XCII.—A middle-aged woman had an attack of right hemiplegia with pretty complete aphasia in the early part of the year 1868. In the course of some months she improved considerably, though she continued subject to "fits" at intervals. After twelve months she was able to walk about with a little assistance, though she was still incapable of using the right hand and arm.

She seemed thoroughly to understand everything that was said to her, and had in great measure regained her power of speaking. She could repeat almost any word uttered in her hearing, and this without hesitation, though she could not read even the simplest words in large type. Yet the same words could be uttered with ease immediately that she heard them pronounced. She copied the written word "London" fairly well with her left hand, but could not write "cat" or "dog" after merely hearing them pronounced, though she could spell the same words quite well. She could not even write the first letter of either of these words.

Twelve months afterwards she was found to be in much the same condition. She could not read aloud even such simple words as "and" and "for;" could point out any letters which were named with the greatest ease, but could not herself name the letters when they were pointed to. She had improved in her power of walking, and was also able to talk rather better. She could read a letter silently so as to understand it, though she did not always seem to comprehend what she read in a newspaper or a book.

When seen again, four years afterwards, this patient was found to be in much the same condition.

It is worthy of note that during the earlier stages of this woman's illness, she seemed to be suffering from ordinary aphasia with right-sided paralysis; it was only after she recovered her power of speaking that it was possible to obtain evidence of the more special defects above enumerated, which pointed, as may be seen, to a severance of functional relations between the left auditory and visual word centres. Thus she could not read aloud, neither could she write from dictation—both of them being acts which require the conjoint activity of these two centres—especially in persons not very well educated, and therefore not thoroughly habituated to the performance of these processes. Exceptions, however, may occur to this rule.[1]

A case presenting certain differences, but which I think belongs to the same category, has also been recorded by Broadbent.[2] This latter case, however, differs from mine inasmuch as the patient could read aloud, though what he read was gibberish, and this difference is possibly explicable from the fact that the patient was a well-educated young man in whom the gibberish reading may have been produced by incoördinate attempts to excite the left glosso-kinæsthetic centre directly from the visual word centre, as it could not be done by way of the commissure to, and with the aid of, the auditory word centre itself. There seems here also to have been some damage to this latter centre, seeing that the patient's spontaneous speech was very limited.

[1] See "The Brain as an Organ of Mind," 1880. p. 624. As evidence that almost if not quite from the first I put the above-mentioned interpretation upon this case, and recognised the importance of these commissures, I may say that after a brief reference to the case in my work "Paralysis from Brain Disease," 1875, p. 201, the following note was appended: "There can be little doubt that in this woman the connections between the visual and the auditory 'perceptive centres' were broken through in the left hemisphere. Speech, therefore, could not be initiated through sight impressions though it was easily done through the sense of hearing. It could not be roused through the sense of sight, simply because the impressions revived in the visual centre were unable to awake their corresponding impressions in the auditory perceptive centre. And without this latter revival the motor incitations to speech cannot occur. This, at least, is my view."

[2] *Brain*, vol. i., 1878, p. 484.

CASE XCIII.—A well-educated and intelligent young man one evening, about two weeks before his admission to the hospital, partly lost power over the right limbs, and speech, together with loss of consciousness.

It was then found that he understood all that was said to him, but could not answer—said "Yes, yes, yes," and would wave his hand. After ten days he spoke a little more, saying "Yes, yes, very much better;" and after two months there was still further improvement. When unable to express himself he sometimes said, "But I can speak very much better than I could." Still he was unable to give a connected account of anything requiring more than a few words. He mostly spoke in broken sentences, thus :—" Brother—brother—'Merica—letter—New York—two brothers in America."

He could not write spontaneously or from dictation, but could copy.

He read the newspaper regularly, "with all the marks of intelligent interest," and gave proofs that he understood it; when asked whether this was so, he replied, "Oh, yes, perfectly." But he could not name objects, and when asked to read aloud the result was mere gibberish. A specimen of this is given, concerning which Broadbent says : "It will be seen that there was no traceable relation between the passage and the reading of it, beyond a certain imperfect correspondence between the number and length of the words . . . . . it was clear that there was no regularity in the substitution of certain constant inappropriate sounds for given combinations of letters." In copying, moreover, as he wrote each letter he named it aloud, and always wrongly.

His intellectual impairment seemed to be very slight, as judged from a careful observation of his condition and manner generally.

My other case is a most remarkable one in many respects. The patient remained under observation for a period of eighteen years, and repeated examinations showed that during the whole of this period his symptoms after the first four months underwent comparatively little change. The necropsy after this long series of years unfortunately threw very little light upon the extent and site of the original lesion, though the very large area of brain atrophy discovered opened up questions of great interest as to the degree to which some of the functions of the left hemisphere had been gradually taken on by the right hemisphere. Thus, at the necropsy, we found complete destruction of the hinder two-thirds of the left upper temporal, and also of the supra-marginal and of the angular gyri, and yet from the date the patient came under observation (three months after the onset of his illness), there had never been anything like either word-blindness or word-deafness. This case has been described in detail in the *Transactions of the Medico-Chirurgical Society*, vol. lxxx., 1897, but it will be here reproduced in an abbreviated form. It will be observed that in addition to damage of the commissures between the auditory and the visual word centres, the patient's symptoms pointed to the existence of partial aphemia, and also to a lowered activity of the auditory word centre interfering with spontaneous speech.

CASE XCIV.—Thomas Andrews, aged 32, a tin-plate worker, had always been temperate, and had never suffered from syphilis. He had a slight fright about the middle of December, 1877, and two days after, as he was going out to work, about 8 a.m., he suddenly fell backwards on a chair. He was not convulsed, and did not lose consciousness, but he lost power in the limbs of the right side and "did not speak coherently after the commencement of the fit." The limbs seem to have been completely paralysed at first, and there was incontinence of urine and fæces for four or five days. "He also had headache which persisted for some time."

He was admitted under my care into University College Hospital on March 12, 1878, by which time he had recovered a good deal of power in the paralysed limbs. There was no cardiac bruit at the base or apex. The right angle of the mouth was somewhat lower, and not raised so well as the left. The tongue was protruded considerably to the right. The grasp of the right hand was extremely feeble. He could raise the right foot two feet off the bed, and could resist flexion of the right knee with considerable force. The movements at the right ankle-joint were greatly impaired, though not altogether absent. He walked with difficulty, dragging the right foot somewhat. Tactile impressions were badly appreciated, and painful impressions still more so, on the limbs of the right side. (No mention is made of trunk or face, in this respect.) Sight and hearing good.

He continued to improve slowly, and his condition as regards speech and related powers were thus described on April 2:—" He recognises common objects, but cannot name them; repudiates a false name, and recognises the real one at once when he hears it. Can never remember his own name till it is suggested to him. On being asked to repeat it, after a few trials which vary each time, he pronounces it Anstruthers or Anstrews. His first name seems to come more readily, and he can often attempt this without prompting. But, either after it has been repeated to him, or when he says it spontaneously, he pronounces it 'Touvers.' The letter l is difficult for him to utter; sometimes he pronounces it like a d, and at others like a v. He has been taught to count, and can fairly pronounce the numerals from one to twelve; after twelve he is uncertain, the articulation and order becoming rapidly worse. He is conscious when he makes a mistake, but cannot correct himself, and ends in a hopeless muddle."

"In reading from a book, the words he pronounces have no relation to the print, either in length or sound; neither does he seem to understand written characters, as he will not attempt to answer a question written on a slate, though he will at once endeavour to respond when the same question is put to him orally. He, however, recognises figures from 1 to 9 when written, and is conscious when they are not placed in regular order. He cannot name any coins, but seems to have some idea of their relative value. He indicated on his fingers that sixpence was worth six pennies—not being able from sight to utter its name."

The patient was discharged from the hospital on April 6, but was re-admitted on April 16, on account of his having had two fits on the morning of that day. The second fit left his speech rather worse than it was before, and his right limbs also became quite powerless for a time.

It was found that the patient's state in the early part of the month of May was distinctly different from what it had been in the early part of April, previous to the occurrence of the two fits on April 16. At the later date, whilst the patient's utterance in repeating words which he had heard had become more distinct, he could not even emit an unintelligible jargon in attempting to read. At the same time he had become able to understand what he read, though he still could not name a single letter at sight, nor could he write a single word from dictation—both these processes requiring for their performance the proper relation between the visual and the auditory word centres, and therefore the integrity of the commissures by which they are united. That half of the commissure which conveys stimuli from the visual to the auditory word centre (as in reading aloud) seems to have been more extensively damaged after the two fits than it was before.

His speech defects had, in fact, now taken the form which they subsequently retained throughout the remainder of his life.

But while his speech defects preserved a constant character through so many subsequent years, evidence of the extension of the original damage done to the patient's brain was furnished by the recurrence of fits from time to time, and by the increase, especially during the last two years of his life, in the amount of paralysis of the right limbs and in the degree of right-sided hemianæsthesia.

In regard to his mental condition and speech defects, I subjoin some notes which were taken in January, 1879.

Patient is quite rational, and able to play games such as draughts.

The Auditory Word Centre with its Afferent and Efferent Fibres.—He readily understands any question put to him. He can repeat short words correctly, but there is often an aphemic difficulty in utterance. Thus, he cannot properly pronounce words beginning with the labials, but those beginning with gutturals he pronounces much better. He can count quite correctly up to twenty, but beyond this gets rather confused with the repetition of the first word. When started, he can say the alphabet correctly as far as "*i*," after that very badly. His spontaneous speech was very limited, but one day before returning to the hospital he said, "Never mind, they may come to-morrow;" and on another occasion, "Mrs. Foster will come to-morrow." Otherwise his speech has been limited, as at present, to "Yes" and "No," "No, it isn't," "Good morning," and some simple expressions of this kind.

The Visual Word Centre with its Afferent and Efferent Fibres.—He understands any written question. He reads books and newspapers much. He says he understands what he reads, and there is every reason to believe that this is true. He could detect mistakes in his wife's accounts, but could never explain the error to her. He can add two rows of simple figures. Subtraction he finds more difficult, but can get a simple sum right after a short time. Over a simple sum in multiplication he requires some prompting before he gets it right. He can copy writing well with his left hand, but he cannot transfer from printed to written characters, or *vice versâ*. Can write nothing spontaneously, not even his own name.

Commissures between the Auditory and the Visual Word Centres.—
(*a*) He cannot read aloud at all, either from print or writing. He cannot read aloud the simplest words or even name single letters, except that occasionally he can name *a* or *o* correctly. (His disability in this respect is always most striking when contrasted with the rapidity with which he pronounces the same word or letter as soon as he hears it uttered.)

(*b*) He cannot write anything from dictation; but if a short word is spelt for him he can sometimes write the letter just named after a long time and with much difficulty. (His inability to write a single word from dictation always contrasts notably with the readiness with which he will copy the same word as soon as it is written.) Simple numerals, such as 7, he can write from dictation; 12 he finds more difficult, making it 31, then 10, and shortly afterwards 12. He puts on the 5 at the right place to make it 125.

Sixteen years later, that is in 1894, his condition in regard to speech defects was thus recorded:—

Understands all that is said to him. He can repeat all simple common words, but uncommon polysyllabic words he is unsuccessful with, producing mostly a mere jumble of sounds. He can repeat the numerals up to 20; and can say the alphabet when started as far as *m*, but cannot go beyond without missing many letters. Can name the days of the week, but often omits Sunday. His spontaneous speech is limited to his name (he cannot give his address), together with short affirmative or negative answers, and occasionally very short phrases such as " Never mind."

He understands what he reads. Recognises all words, letters, numerals, common objects, and pictures. He can copy numerals, letters, and words quite well, with the left hand. He can go on writing numerals when the first one is given, but letters and days of the week cannot be written in the same way. He cannot write a single letter or word spontaneously.

(*a*) Cannot name at sight words, letters, numerals, or common objects.

(*b*) He can write simple numerals from dictation. Cannot write a single letter or word from dictation.

He was seen on many occasions after this date up to within one month of his death, and his condition in regard to speech and writing was found to have undergone no change.

Death occurred on March 29, 1896, about thirty hours after the occurrence of an apoplectic attack associated with convulsions, which was subsequently found to have been caused by a thrombosis of the right middle cerebral artery.

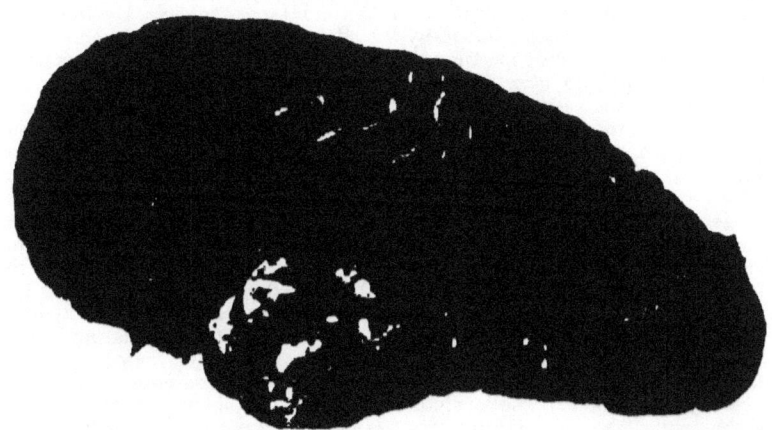

FIG. 14.—Left Hemisphere of the Brain of Thomas Andrews, after removal of the Membranes.

At the necropsy nothing unnatural was found in the skull-cap or the dura mater, but on the left side of the brain there was what appeared to be a large pseudo-cyst. The organ was handed over to Dr. Risien Russell for preservation, and he subsequently gave me the following report, the accuracy of which I can fully confirm :—

"The brain showed evidence of an old softening resulting in complete atrophy of the left hemisphere throughout the whole of the area of distribution of the middle cerebral artery, with the exception of the area usually supplied by the first cortical branch of this artery, which area was intact, and thus Broca's convolution had escaped softening and subsequent atrophy. There was, however, well-marked extension of the atrophy into the posterior portion of the second frontal convolution, corresponding to the area of distribution of a branch of the second branch of the middle cerebral artery, which supplies this area.

"An unusual amount of the upper part of the ascending frontal and parietal convolutions had escaped softening and subsequent atrophy, viz., 4½ cm. of the former, and 3½ cm. of the latter, which portions presented a healthy and unaltered appearance. The total lengths of these convolutions on the right side were, ascending frontal 10¾ cm., and ascending parietal 10 cm.

"The angular and marginal convolutions were destroyed; but almost the whole of the superior parietal convolution had escaped, the width of the intact area of cortex in this region amounting to a little more than 3 cm.

"The superior and inferior occipital convolutions were intact, but the atrophy had extended to an unusual degree into the middle occipital convolution, 3 cm. of this convolution being intact, however.

"The whole of the superior temporo-sphenoidal convolution was destroyed, with the exception of the anterior one-third, 4½ cm. in length. Of the middle temporo-sphenoidal convolution, the anterior 5 cm. were perfectly intact; but posteriorly only a narrow portion of the inferior part of this convolution remained, and that was in a discoloured and degenerated condition.

"The whole of the atrophic area that has been described (from which all trace of brain tissue had completely disappeared) was occupied by a large pseudo-cyst which was continuous with the lateral ventricle, the fluid with which it was filled being shut in by the slightly thickened pia arachnoid.

"On examination of the thrombosed branches of the left middle cerebral it was impossible to ascertain in which of these the process had commenced. The trunk of the vessel and all its branches had now become much attenuated. When the brain had been hardened and the membranes completely removed from the left hemisphere a great gap was found in the region indicated, 11 cm. in length by 3½ cm. broad, and 4 cm. in maximum depth (Fig. 14). The upper part of the descending and nearly the whole of the posterior cornu, together with the posterior half of the body of the lateral ventricle were opened up. Much of the posterior segment of the internal capsule together with the greater part of the thalamus had disappeared, the latter being represented only by a small rounded portion, having rather less than one-third of the bulk of a normal thalamus. Anteriorly the atrophy had extended into the white substance up to the corpus striatum, which was also much diminished in size. All other portions of the cortex of the left hemisphere, beyond the region above defined, presented a perfectly healthy appearance. The left half of the pons was slightly flatter than the right, whilst the left pyramid was very distinctly smaller than its fellow. The lateral lobes of the cerebellum showed no inequality.

"The right middle cerebral artery was occluded by a recent thrombus. There was practically no evidence of softening of the cortex of this hemisphere (nor were there lesions of any other kind), but there was extensive softening of the basal ganglia and internal capsule, probably accentuated by *post-mortem* changes."

The interpretation of this remarkable case is full of difficulties. It seems clear that in the first attack of 1877, there could not have been a complete occlusion of the middle cerebral artery. The transitory nature of the complete paralysis, as well as some of the other symptoms, were unfavourable to this view. It was, moreover, negatived by the *post-mortem* revelation that the third frontal convolution was intact. The occlusion of the vessel was, therefore, evidently beyond the point whence the first cortical branch is given off, which is the nutrient artery of this convolution.

It is even difficult to suppose that at first the whole of the other cortical branches of the middle cerebral were involved. Had this been so it seems clear that there must have been a more than usually free anastomosis between the terminal ramifications of these branches and those of the corresponding anterior and posterior cerebral arteries. That there was such a free anastomosis between the second and third cortical branches of the middle cerebral artery and these other vessels seems evidenced by the unusually small amount of the customary territory of these branches which underwent softening and atrophy. The same kind of evidence does not, however, tell in this direction for the fourth branch, in the territory of which there was found a full amount of atrophy.

It is the clinical evidence that makes it improbable that the whole of the territory of this fourth branch was at first cut off from its blood-supply. Had this been the case, word-deafness as well as word-blindness would have been produced. But no word-deafness seems ever to have existed in the early stages of this man's illness; and though there was word-blindness at first, this had completely cleared up early in May —that is, about five months after the onset of the hemiplegia.

At this latter date the speech defects assumed the form that they maintained during the following eighteen years.

Though this was in all probability the condition existing for an indefinite period from May, 1878, onwards, the results of the *post-mortem* examination showed that a progressive destruction must have taken place in these auditory and visual word centres. This being so, one of the most remarkable features of this case will be found to be the unvarying character of the speech defect, notwithstanding the progressive destruction of such centres.

A similarly progressive destruction of brain tissue must have extended even into parts of the hemisphere beyond the field of distribution of the terminal ramifications of this fourth cortical branch of the middle cerebral—that is, into the contiguous territory of some of the anastomosing branches of the posterior cerebral. This is shown by the amount of atrophy that had taken place in the white substance of the hemisphere, in the posterior part of the internal capsule, and in the thalamus.

The increased sensory and motor paralysis that occurred during the last two years of the patient's life was almost certainly due to the destruction of the posterior part of the internal capsule.

The difficulties in reconciling the persistent and often verified clinical condition with the *post-mortem* record are extreme. As I have said, in the absence of word-deafness and word-blindness as initial symptoms, one cannot suppose that complete softening at first existed in the auditory and the visual word centres.

The total atrophy in these regions found after a period of eighteen years seems, therefore, as above stated, irreconcilable with the fact, shown by the clinical records, that during this long series of years there had been no appreciable change in the speech defects. How, then, it must be asked, could this complete atrophy have occurred in the region of the auditory and the visual word centres without manifesting itself clinically?

We must, I think, conclude that the destruction of these parts had been gradual. Had it been sudden, all present knowledge entitles us to conclude that it must have revealed itself by the occurrence of well known symptoms, namely, word-deafness and word-blindness in combination. And assuming that the destruction had been gradual, we can only suppose that this process must have coincided with a gradual development in the functional activities of the corresponding convolutional regions of the right hemisphere.

What is altogether novel and surprising, however, is that restoration should only have taken place in such a very imperfect manner—that imperceptibly, as it were, the imperfect functioning of the left hemisphere, and no more, should have been taken on by the right hemisphere. It may, perhaps, be supposed that the low, though unequal, activity which was alone attained in the right auditory and visual word centres did not even suffice to develop the usual commissures between these two centres, and that owing to their absence the original clinical combination of speech defects was preserved—even when the cerebral activity on which they were dependent had been gradually transferred from the left to the right hemisphere.

Whilst offering these mere suggestions concerning a very difficult problem, I fully realise how much they fall short of anything like an adequate explanation of the very puzzling riddle presented by this case.

(4) DEFECTS DUE TO DAMAGE TO THE COMMISSURES BETWEEN THE GENERAL VISUAL CENTRE AND THE AUDITORY WORD CENTRE.

It seems clear that two commissures must exist between the general visual centre in the occipital lobe and the auditory word centre in the posterior part of the temporal lobe.[1] One of these, which we may call the *Occipito-Temporal Commissure*, would be called into play on "naming at sight" any object presented, other than a letter or a word. It is destruction of this commissure that gives rise to the condition which has been named "optic aphasia" (see p. 212).[2]

There is, however, the possibility that the "naming at sight" of ordinary objects might be brought about by a different process, namely, by the stimulus passing from the general visual centre to the visual word centre, and thence across the visuo-auditory commissure to the auditory word centre. But the case recorded by Dingley (Case lxxxix.) makes it probable that in this particular individual, at least, it could not have been so brought about, seeing that the patient could read aloud, although he was unable to name objects at sight—still, even here the break might have been in the connections between the general visual centre and the visual word centre.

The other of the two commissures, which we may call the *Temporo-Occipital Commissure*, conducts stimuli in the reverse direction, and would be called into play when a patient is told to point out any one from a number of objects that may happen to be before him.

---

[1] Allen Starr (*Brain*, vol. xii., 1890, p. 90) speaks of such a commissure as the "visual-auditory tract," and makes the following remarks concerning it:—"Such a tract, according to Meynert, lies beneath the cortex covering in the white matter between these areas, a region which must be termed temporo-parieto-occipital, since the gyri of all three lobes lie upon it. Hence lesions of the annectant gyri convolution, also those of the parietal lobule, are very likely, if at all deep, to involve this tract."

[2] Of course "naming at sight" would be impossible also when there is "object-blindness," as if an object seen is not recognised, it could not be named.

CHAPTER XII.

MODES OF RECOVERY FROM APHEMIA AND APHASIA. MODES OF PRODUCTION OF RECURRING AND OCCASIONAL UTTERANCES. LICHTHEIM'S TYPES OF SPEECH DEFECT.

WITH certain reservations we must suppose that both hemispheres are educated concurrently, by the advent of similar sensory impressions to each of them. Such concurrently educated sensory centres in the two hemispheres are, in all probability, connected with one another by commissural fibres forming part of the corpus callosum. This should be the case for the cortical centres for hearing, vision, smell and taste; though it would probably not be the case with the centres for tactile and kinæsthetic impressions, which are called into action independently of one another in the cerebral hemispheres, and whose impressions have in all cases to be strictly localised, not merely to one side but to particular parts of one side of the body.

Kinæsthetic centres are, however, the regions from which the appropriate stimuli issue for calling into action the different motor centres in the bulb and spinal cord. For limb movements such stimuli must pass off in cross relation from each hemisphere — from the left hemisphere for the right limbs, and *vice versâ*. In speech movements, however, although they are always varying aggregates of bilateral muscular contractions, the muscles seem to receive their incitations mainly from one hemisphere.

As we have seen, the evidence in our possession tends to show that speech incitations pass down, in most right-handed persons from the left hemisphere; while in left-handed persons there is reason to believe that they most frequently pass off from the right hemisphere.[1] Thus it happens that one hemisphere, and that the most highly organised (through the preponderance at least of tactile and kinæsthetic impressions), seems to take the lead; so that effective speech inci-

---

[1] An exception to this occurred in the case of Dr. Dickinson's patient. Case xxiii.

tations begin and continue to pass off, in the main, from it alone. This being so, aphemic difficulties of utterance and aphasia ought to be, as they are, slight or rare with lesions of the right hemisphere in right-handed persons.

It would not be correct to say that the whole body of speech incitations passes off from the one leading hemisphere. As stated in an earlier chapter, it would be more correct to say, with Kussmaul,[1] as follows : — "The main current of the centrifugal impulses of speech passes downwards through the left cerebral hemisphere. But since dysarthric troubles—usually of a trifling kind—are noticed, as a rule, in connection with left hemiplegia likewise, we must conclude that, besides the main current just referred to, a weaker accessory current is transmitted through the right hemisphere."

In relation with this subject we may now briefly consider (1) the mode in which recovery is brought about in some cases of aphemia and aphasia; and (2) the mode in which "recurring utterances" and "occasional utterances" are produced by aphasics.

(1) Mode in which Recovery is brought about in Aphemia and Aphasia.

Many years ago Broadbent started the notion that a speechless condition (aphemia) dependent upon a lesion of the pyramidal fibres starting from Broca's region (if not too close to the cortex) might be gradually recovered from owing to the establishment of a "new way out" for the motor stimuli. This was supposed to be brought about by the passage of the stimulus across from the third left frontal convolution, by way of the corpus callosum, to the third right frontal convolution, and thence downwards by the pyramidal fibres to the motor centres in the bulb.[2]

This hypothesis assumes, therefore, that the two glossokinæsthetic centres are naturally connected by means of commissural fibres (Fig. 15, a), and looking to the fact of the bilateral character of speech-movements it seems quite possible that they may be so connected, and that in this case, as well as in others, we may have exceptions to the rule above stated that the kinæsthetic centres are not bilaterally associated with one another. The possibility of recovery then, according to the mode suggested by Broadbent, would require that Broca's region should be intact, and also that the

---

[1] *Loc. cit.*, p. 677.
[2] Lichtheim suggests (*loc. cit.*, p. 480) another anatomical arrangement which seems to me much less probable.

lesion in the pyramidal fibres should be situated (as at *d*, Fig. 15), below the level of the assumed commissural fibres forming part of the corpus callosum.

But this communication between the two glosso-kinæsthetic centres would be of little avail unless we are to suppose that the centre in the right hemisphere had also become pretty fully organised and provided with its own set of efferent fibres. This difficulty, therefore, stands in the way of Broadbent's hypothesis, so long as he or others suppose such centres to be real motor centres.

For if the right third frontal centre is almost never used for speech in right-handed individuals, and if it is supposed to be a motor centre, how can we imagine that it could ever have become organised?

The difficulty is very distinctly lessened, however, in accordance with my view that these centres are really sensory

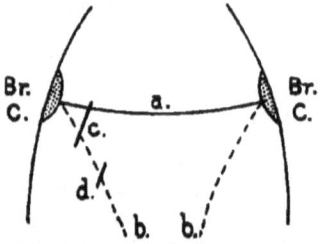

Fig. 15.—Diagram showing possible mode of recovery from Aphemia when lesion occurs in some part of the pyramidal fibres (*b*), as, for instance, in the position of line marked *d*. When lesion is in position of line *c*, recovery can only take place as in Aphasia, that is, in the same way as it must occur when Broca's Centre (Br. C.) is destroyed.

centres of kinæsthetic type, because we are bound to admit that kinæsthetic impressions, proceeding from the bilaterally symmetrical parts involved in speech-movements, must ascend equally to either hemisphere, and ought therefore to be registered in the right as well as in the left third frontal convolution.

If the right centre, however, be not habitually used for the actuation of speech-movements, one can easily understand that efferent fibres from it to the motor centres in the bulb would either be absent or scanty. For reasons above stated we are compelled to admit that a certain number of these efferent fibres do exist; and perhaps under exceptional circumstances more may become developed without great difficulty.

In this way Broadbent's hypothesis unquestionably affords a possible means of explaining the recovery in some cases of aphemia of an incomplete character; and inasmuch as recovery from this kind of defect is much more commonly met with than recovery from aphasia, we may suppose that the process by which it is effected is more easily brought about than one or other of those changes that we shall presently see must occur when we have to do with recovery from this latter condition.

In complete aphemia, however, from a lesion in the situation marked by the line *c* in Fig. 15, such a mode of recovery would not be possible, because the commissural fibres between the two third frontal convolutions would be cut across as well as the outgoing fibres from the gyrus of the left side. With a lesion in this situation recovery could only take place as it does in aphasia, due to destruction of Broca's centre.

There are two methods by which this seems possible. Either the left auditory word centre becomes more intimately

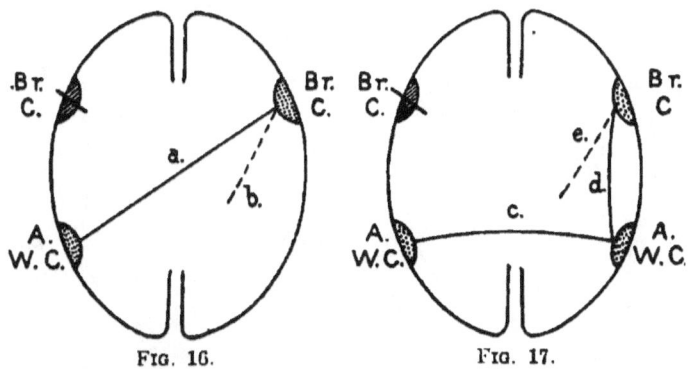

FIG. 16.   FIG. 17.

FIG. 16.—One possible mode of recovery from Aphemia or Aphasia. *a*. Commissure between the left Auditory Word Centre and Broca's Centre on the right side; *b*. Outgoing, or "Internuncial fibres" from latter Centre.

FIG 17.—Another possible mode of recovery from Aphemia, or Aphasia. *c*. Commissure between the two Auditory Word Centres; *d*. Commissure between the right Auditory Word Centre and Broca's centre on the right side; *e*. Internuncial fibres from the latter Centre.

connected with Broca's region on the right side, and the stimuli pass therefrom to the motor centres (as represented in Fig. 16); or else the left being in intimate functional relations with the right auditory word centre, a set of commissural fibres becomes developed between this latter and Broca's centre of the right side (as represented in Fig. 17), from which, as before, the outgoing stimuli pass, after gradual development with multiplication of the internuncial fibres between it and the motor centres in the bulb.

There are some undoubted cases of recovery from aphasia on record, in which a necropsy has subsequently verified the fact of the destruction of the leading third frontal convolution. Thus, Case xxiv. is an instance of aphasia occurring in a left-handed boy in whom speech was recovered, though the necropsy twelve months after the onset of his illness showed a large lesion in the right hemisphere which must have destroyed the third frontal convolution. This boy was ambi-dextrous, and that may have made the recovery more easy, owing to the opposite cerebral hemisphere, which had to take on the new functions, being rather more highly organised than usual. Certainly, the recovery was altogether unusually rapid.

In other rare cases in which word-deafness and disturbed speech is recovered from, it seems probable that this restoration of power must be brought about in ways that are very

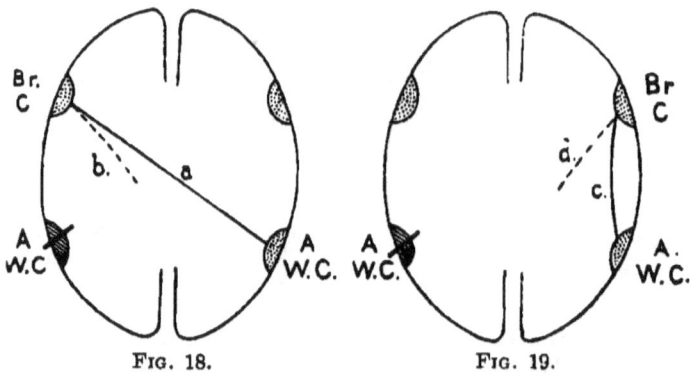

FIG. 18. FIG. 19.

FIG. 18.—Diagram showing one possible mode of recovery from Word-Deafness, due to destruction of the left Auditory Word Centre. *a.* Commissure between the right Auditory Word Centre and Broca's Centre on left side; *b.* Internuncial fibres issuing from the latter Centre.

FIG. 19.—Diagram showing another possible mode of recovery from Word-Deafness caused in the same way. *c.* Commissure between the right Auditory Word Centre and Broca's Centre of the same side; *d.* Internuncial fibres issuing from the latter Centre.

analogous. In each case in the absence of the left auditory word centre the centre of the right side must take on more and more of functional activity, and then more complete commissural connections may be opened up between it and Broca's centre of the left side (as indicated in Fig. 18); or else the commissural relations between it and Broca's centre of the right side must become more perfect (as indicated in Fig. 19), and be followed by a more complete development of this centre, as well as of the outgoing channels from it to the motor centres in the bulb. In regard to the first of these suggested modes of recovery, it will be recollected that in Chapter viii. we were compelled to conclude that a commissure between the right auditory word centre and Broca's centre of the left side must exist more or less frequently.

It has always seemed to me very difficult to understand why such complete word-deafness or word-blindness, as is usually produced, should result from destruction of the auditory or the visual word centres in the left hemisphere alone, and why in ordinary right-handed persons the destruction of the homologous centres in the right hemisphere should produce no appreciable defects of this kind, if, as we must assume, word-impressions, both auditory and visual, go equally to the two sides of the brain.[1] These facts seem, however, to point pretty definitely to the conclusion that for auditory and visual word-perceptions, as well as for speech, in the majority of persons, the right hemisphere is very distinctly less potent than the left. On the other hand, the existence of variation in the degree of this inequality in different persons may account for the surprisingly slight effects that have resulted from destruction of the left auditory word centre in some recorded cases, that is, where the amount of word-deafness has been comparatively slight, looking to the extent of the lesion found in the left upper temporal gyrus.

(2) MODE OF PRODUCTION OF RECURRING AND OCCASIONAL UTTERANCES.

The other question as to the mode in which "recurring" and "occasional utterances" are to be explained in aphasics in whom the left third frontal convolution has been destroyed, is no less difficult and obscure. It is well known that the recurring utterances, whether they consist of actual words or of mere meaningless sounds, are used in a random manner; whilst the occasional utterances, generally more apposite, are mostly produced under the influence of some excitement by a strong emotional stimulus. Neither of them, moreover, can be deliberately reproduced by the patient when he is bidden to do so. The question which now presents itself, however, is: How, if the left third frontal convolution is destroyed, and

---

[1] In left-handed persons the results are, of course, altogether different. The case of Marcou, related by Trousseau, in whom there was left hemiplegia and amnesia, and also a case recorded by Bernheim, may be taken as illustrations of this (see Cases xliii. and lxxvi., and p. 233).

ordinary speech is lost as a consequence, are we to explain the common production of certain sounds or words, or the occasional utterance of oaths, or even short sentences, with a distinct articulation?

No one has treated this side of the question with more zeal and thoroughness, or with anything like so much wealth of illustration as Hughlings Jackson;[1] and yet, as it has been done from a rather peculiar point of view, I find myself, unfortunately, as little able to agree with him here as I am in regard to the fundamental positions from which he starts. It will, therefore, I think, conduce to clearness if I, first of all, state my own view as to the mode of production of these recurring and occasional utterances, and the extent to which I can adopt the explanations advanced by Hughlings Jackson, and subsequently refer more especially to the points concerning which I feel compelled to differ from him.

What I have already said as to the necessarily simultaneous organisation to some extent of the right with the left glossokinæsthetic centre, and as to the fact, attested by Kussmaul, that a weak or smaller current of centrifugal impulses seems to be habitually transmitted from this right centre to the bulbar speech centres, are facts that seem to me of much importance in this relation.

If the recurring utterances do not consist of actual words, but of mere syllabic sounds, such as "poi-boi-ba," "baddi-baddi," "cousisi," "sapon, sapon," etc., it seems to me just possible that they may be uttered through the instrumentality of the badly damaged, but not totally destroyed, left glossokinæsthetic centre. Where, however, as is so often the case, the recurring utterances consist of actual words distinctly articulated, I agree with Hughlings Jackson in thinking that such an explanation will not suffice. He says (*loc. cit.*, vol. ii., p. 333): "As the recurring utterance is, as a series of articulations, very elaborate, and, as the syllables are plainly enunciated at any time, it is not credible that slight remains of nervous arrangements can be concerned during the utterance of them, especially when, looking at the matter on another side, we see that these remains do not serve the patient to utter any other words whatever, except, perhaps, 'yes' and 'no,' which, however, I believe to occur also during activity of the right half of the brain."

In the case, therefore, of such recurring utterances as consist of actual words, constituting short phrases, and also in the case of "occasional utterances" of similar complication, we

---

[1] See more especially his paper in *Brain*, vols. i. and ii. (1878, 1879).

cannot suppose that they are produced through the intermediation of the badly damaged left glosso-kinæsthetic centre. The only other explanation that seems possible, therefore, is the supposition that such utterances are effected through the intervention of the right cerebral hemisphere. To this latter explanation, therefore, I appeal, just as Hughlings Jackson does—though, as will be seen, for the most part on different grounds.

To a considerable extent, as Hughlings Jackson points out, such words as "yes" and "no" are the "most general, most automatic" of all statements; they are certainly, as he says, never of "high speech value." What he means by this distinction may be gathered from the following quotation:—
"We may now say that speech of high value, or superior speech, is new speech; not necessarily new words, and possibly not new combinations of words; propositions symbolising relations of images new to the speaker, as in carefully describing something novel; by inferior speech is meant utterances like 'very well,' 'I don't think so,' ready fitted to very simple and common circumstances, the nervous arrangements for them being well organised."

These common utterances are, therefore, just those that would be likely to be most perfectly represented by organisations in the right third frontal convolution, and, therefore, those which we may both of us suppose to be most easily actuated through the right hemisphere.

As to the more complex recurring utterances such as "List complete," or "I want protection," I do not profess to be able to understand how the utterance of such phrases can be brought about almost at once after an attack by the little used right hemisphere—that is, without the need of tuition and gradual practice. Nor does Hughlings Jackson himself seem to me to be able to throw any light upon this difficult point, though his supposition[1] that these utterances constitute the last propositions, or part of the last propositions, either uttered or about to be uttered at the moment of the onset of the attack by which the patient is rendered speechless, is possibly correct.

There remain for consideration the cases in which occasional utterances of some complexity, or others of a more simple though equally unusual character, are brought out by patients who have been, perhaps for weeks or months, almost speechless. Some of these latter expressions may be oaths or interjectional remarks, but in all such cases, as well

---

[1] *Brain*, vol. ii., p. 215.

as where the expressions are of a more complex character, we find that they occur under the influence of some temporary excitement or surprise. Some of the interjections and oaths are, it is true, rendered all the more easy by the fact that they are what Hughlings Jackson would term the most general and automatic expressions, like "yes" or "no," and, therefore, easily producible in the way already mentioned. It may, however, be said of all these rarely occurring utterances that they owe their origin principally to the strength of the antecedent stimulus—to the fact, that is, that they occur under the influence of emotion.

What I said many years ago[1] on this head seems to me still to hold good and to afford the best explanation of the mode of production of these unusual utterances, whether we suppose them to be brought about through the agency of the damaged left hemisphere, or by that of the undamaged, but little used, right hemisphere.

"The emotional stimulus being a much stronger one than the volitional—a thing of higher tension—it forces its way along channels and against resistance which the volitional stimulus alone is utterly unable to overcome; for though the individual may swear or utter such an expletive as "Oh, dear!" at the incitation of emotion, he is quite unable to repeat the same phrase by any volitional effort that he may make, a few moments after, when the passion has subsided. I cannot, therefore, altogether agree with Dr. Hughlings Jackson, when he says, in explanation of this peculiarity of aphasics, that their ability to swear depends upon the fact that there are two distinct kinds of language, emotional and intellectual, of which the latter only is impaired in the aphasic condition. . . . . I do not think he is correct in stating that 'swearing is, strictly speaking, not a part of language.' Seeing that oaths and expletives consist of actual articulate words, they must be called up and evolved by precisely the same physiological methods as are concerned in the enunciation of any more strictly intellectual proposition. The sole difference, in my opinion, is that in the one case the words are evolved, in spite of opposing difficulties, by the superior force and energy of an emotional stimulus; whilst in the other they often fail to be produced under the influence of the more slender and feebler stimulus of volition."[2]

Some quotations may now be given in illustration of Hughlings Jackson's other special views with which I find myself unable to agree, but which, on account of his eminence as a thinker and clinical observer, cannot be lightly passed over.

[1] *Brit. and For. Med. Chir. Rev.*, April, 1869, p. 480.

[2] In regard to this it now occurs to me that whilst a voluntary speech incitation is unilateral, where emotion comes into play we have in all probability to do with bilateral phenomena, which may be capable of sending stimuli simultaneously towards both third frontal convolutions, and this may in part account for the difference in results.

I give the quotations because I prefer to let him speak for himself, as much as possible.

He says in explanation of his own views: "It is believed that the process of verbalising and every other process is dual, but that the more automatic a process is, or becomes by repetition, the more equally and fully it is represented doubly, in each half of the brain" (*loc. cit.*, vol. ii., p. 222).

"The hypothesis is that words are in duplicate; and that the nervous arrangements for words used in speech lie chiefly in the left half of the brain; that the nervous arrangements for words used in understanding speech (and in other ways) lie in the right also" (*loc. cit.*, vol. i., p. 319).

"In healthy people every word is duplicate. . . . . The speechless patient has lost that set of words which serves in speech; he retains another set serving in other ways" (*loc. cit.*, vol. ii., p. 326).

"Every proposition in health occurring during activity of the left side of the brain is preceded by the revival of the words of it during activity of the right half. . . . . There are thus, it is supposed, two services of words by the right half of the brain; the 'reception' of words by others, and the reproduction of words which precede our own speech. These are the two ways other than speech in which words are supposed to serve. Doubtless at bottom these two are alike" (*loc. cit.*, vol. ii., p. 327).

Thus it will be seen that Hughlings Jackson supposes a similar organisation of words to occur in the third frontal convolution of each hemisphere. That in the right hemisphere is what he calls (p. 327) an "automatic service of words"—words which are revived on hearing the speech of others; so that he says (p. 204): "The most important thing showing the duality of the speechless man's condition is given very generally by saying that speechlessness does not imply wordlessness." This automatic service of words is, he says, also called up in the right hemisphere as an immediate antecedent of their revival in the left hemisphere (doubtless through the intervention of callosal fibres) by which voluntary utterance is effected. Such antecedent revival in the right hemisphere corresponds with the "conception" of the particular speech-movements; he considers it to be the rule, indeed, that "the conception of a voluntary movement" should take place only in the right hemisphere (p. 350). When the revival occurs in the left hemisphere (the second stage of the process) there is then brought about the execution of the same movement—the part of the process which is impossible in aphasia.

It is then from this basis that Hughlings Jackson explains

some of the recurring and occasional utterances of aphasics; he thinks that in some manner, very imperfectly described, they are able to make use of their "automatic service of words" in the right hemisphere, in the production of the higher forms of these utterances.

Now, by way of criticism of this series of speculations, it may be pointed out that Hughlings Jackson says absolutely nothing as to the mode in which either "service of words" has become organised. Apparently he believes that this organisation occurs in cortical motor centres—since he declares that "words are motor processes," and accepts one of the dicta of Bain, to the effect that the "materials of our recollection" in the case of words are "suppressed articulations." But I submit that there could be no organisation of words in motor centres, cortical or other, without their previous registration in guiding sensory centres. And as to the existence of any such purely sensory seats of registration of words (whether auditory or visual) Hughlings Jackson says nothing. His supposed motor registration seems to be in the left third frontal convolution, but the homologous right third frontal centre is credited with two functions which are in part exactly such as I and many others now assign to the left auditory word centre, viz., "the reception of words by others, and the reproduction of words which precede our own speech." How, therefore, the existence of this auditory "service of words," supposed by Hughlings Jackson to be registered in the right third frontal convolution, can, when the left third frontal is destroyed, be deemed to account for the production of the recurring and occasional utterances of aphasics consistently with his own views, I am quite at a loss to understand. The former sensory centre is supposed in some mysterious way to take on also the motor functions previously assigned to the destroyed third left frontal—which is the same thing as supposing, in accordance with more commonly received views, that when Broca's centre is destroyed the auditory word centre might in part take on its functions whilst retaining its own.

As to the existence of an auditory or a visual revival of words, Hughlings Jackson makes no mention, although the possibility of the auditory revival of words in aphasics affords an all-sufficient explanation of the fact that, as he insists, "though speechless they are not wordless"—and from that point of view does away with the necessity for postulating his "automatic service of words" in the right hemisphere. This postulation is further condemned by the fact that damage to the third frontal convolution in this hemisphere has never been known to produce word-deafness, as should be the case in accordance with his hypothesis that one of the uses of this automatic service of words is for the understanding of the speech of others.[1]

As a consequence of his views, Hughlings Jackson regards inability to write and inability to speak as essential constituents of an aphasic condition. He cannot believe in the existence of what I call complete aphemia, so many cases of which are now on record. Thus he says (*loc. cit.*, vol. i., p. 320): "If a patient does not talk because his brain is diseased, he cannot write (express himself in writing)." Again, he says (p. 329): "I submit that the facts that the patients do not talk and do write, and do swallow, are enough to show that there is no disease at all"—in other words, that they are either shamming or suffering from hysteria. Hughlings Jackson can scarcely say that the presence of such symptoms as one of the manifestations of hysteria (as in the numerously recorded cases of "Hysterical Mutism") is not their production by disease; but if he demurs to accept, as proof of the *bonâ fide* existence of this group of symptoms, their association with mere functional disease, there is the very conclusive case recorded by Guido Banti (Case xxii.) in which, though the posterior third of the left frontal convolution was destroyed, and the patient was speechless, he could nevertheless freely communicate his thoughts by writing; and there is the case recorded by Barlow (Case xv.) in which, even though the posterior part of both third frontal convolutions was destroyed and the patient was speechless, there was still some ability to write. Moreover, there are many cases on record in which sub-cortical lesions in the left hemisphere have given rise to a similar grouping of symptoms—that is, where, though speechless, the patients have been able to write. Examples will be found in Cases xi. and xii., recorded by Déjerine.

Recovery from Agraphia.

Then again there are cases like those of Dickinson and Wadham (Cases xxiii. and xxiv.) in which left-handed persons who had been taught to write with the right hand became aphasic owing to damage to the right side of the brain, while retaining their power of writing unimpaired. This shows that,

---

[1] Hughlings Jackson is, in fact, so oblivious of other modes of registration of words, and therefore of other forms of verbal memory, that he actually says, I suppose by way of reproach to some of us who think differently: "It is not always vividly realised that retention of memory of words can mean anything more than a retention of words, the word-'memory' in that context being really surplusage."

in them, the left cheiro-kinæsthetic centre must have been habitually stimulated either from the visual word centre of its own side (which in such persons has probably become developed on a par with the visual word centre in the leading hemisphere for speech), or else through an oblique commissure connecting it with the right visual word centre.

We have, in fact, to do here with much the same kind of functional adaptation that is bound to take place (only in the opposite hemisphere) in the case of ordinary aphasics suffering from right hemiplegia, when they learn to write with the left hand. This can only be brought about in one or other of two ways.

Déjerine suggests that in such a case the left visual word centre acts upon the right centres concerned with writing through an oblique commissure in the corpus callosum, as

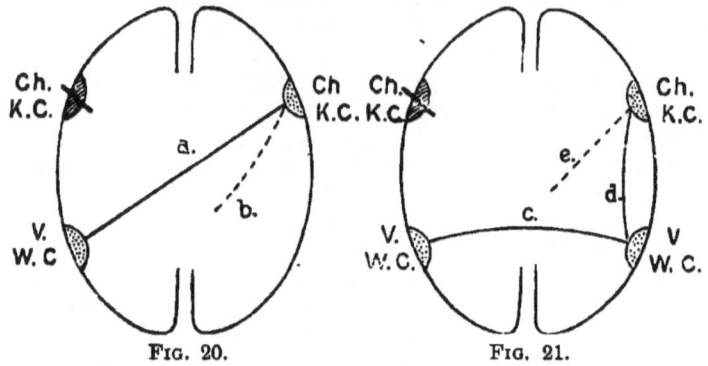

Fig. 20.

Fig. 21.

Fig. 20.—The mode of recovery from Agraphia suggested by Déjerine, when the left visual word centre is not destroyed.

Fig. 21.—The mode of recovery from Agraphia, under similar conditions, which I deem more probable.

represented in Fig. 20. This arrangement is, I think, a very unlikely one, as the effect would be that writing movements of the left hand would be primarily instigated from the left side of the brain while all its other less special movements would (following the ordinary rule) be primarily instigated from the visual centre of the right side. Such an arrangement would seem to be especially unlikely.

The other way in which it seems to me more probable that writing with the left hand may be learned by an agraphic patient paralysed on the right side, would be that indicated in Fig. 21. The left visual word centre being in relation with the right visual word centre by means of the commissure in the corpus callosum, this latter centre would (when efforts to write with the left hand were made) gradually assume more activity, and, as a result of the writing efforts, commissural fibres would gradually be formed between this centre and an embryo cheirokinæsthetic centre on the right side, which would undergo progressive organisation, and lead likewise to the perfecting of internuncial channels between it and the motor centres in the bulb. The latter part of the process would, in fact, be pretty strictly comparable with what must take place on the opposite side of the brain when the young child learns to write with the right hand.

Before dismissing this part of our subject, I may say it is believed by Déjerine and others that if in an ordinary aphasic the left visual word centre happens also to be destroyed, the power of writing with the left hand cannot be acquired. I think this is probably true for the great majority of cases, and the reason would be that the right visual word centre when cut off from the corresponding centre in the leading hemisphere cannot alone and of itself take the initiative in bringing about the development above described. I am not sure, however, that this disability would always exist provided the left auditory word centre remained intact. A consideration of Case xc. recorded by Pitres, in which, as I believe, the left audito-visual commissure was damaged, and where the patient learned to write with the left hand, although he remained quite unable to write with the non-paralysed right hand, seems to show that he was (a) either a "visual" who could call his left visual word centre into activity independently of the left auditory word centre, or else (b) that the left auditory word centre could establish commissural relations with the right visual word centre, and thus bring about the changes above referred to as necessary for writing with the left hand (see Fig. 29). But if this latter process is a possible one, then it also would be a means by which the aphasic could learn to write with his left hand even though his left visual word centre was destroyed.

Is Ordinary Aphasia a Form of Paralysis?

There is another theoretical point which may be best referred to here. Much discussion has taken place from time to time as to whether aphasia is or is not a paralytic defect. It will be seen that Ross comes to a similar conclusion to my own when he says (*loc. cit.*, p. 99) : " From these considerations it may be concluded that aphemia[1] and motor

[1] He uses this term aphemia, as I did in 1869, to designate the loss of speech pure and simple which occurs as a constituent of the aphasic state, due to destruction of Broca's centre.

agraphia are a paralysis of highly specialised movements; that in rare cases the special movements of articulation, and of the hand in writing, are paralysed without the general movements of the respective organs being interfered with; but that in most cases not only the special but the general movements also suffer to some extent, and then only is it popularly recognised that the speech disorder is accompanied by, or is a part of, hemiplegia."

After having furnished adequate evidence in establishment of this point that "motor aphasia" is "a paralysis of the special movements of articulation constituting aphemia; and of those of the right hand in writing, constituting motor agraphia," he goes on to say this (p. 106): "Some pathologists, however, believe this statement to be imperfect, and they supplement it by saying that this form of aphasia is a loss of the memory of those special movements. Now, if this theory of loss of memory be true as applied to the explanation of the inability of a patient suffering from aphemia to execute the articulatory movements necessary to the expression of words, it must be equally true as an explanation of the inability of a hemiplegic patient to execute the movements necessary to normal locomotion. But were we to say that the inability of a patient suffering, for example, from crural monoplegia from cortical disease was due to a loss of memory of the movements of locomotion on the affected side, the absurdity of the statement would be immediately recognised, and yet it is not a whit more absurd than the statement that aphemia is due to a loss of memory of the movements of articulation."

In reference to these remarks I am quite unable to agree with Ross. This second explanation that aphasia is due to a loss of the memory of special movements is, in my opinion, only another mode of stating Dr. Ross's own conclusion that the defects in question are due to a paralysis of special movements. This is obviously so from my point of view—that these defects are due to destruction not of motor but of kinæsthetic centres, in which the sensory impressions resulting from speech and writing movements are registered. The re-awakening of these sensory impressions constitutes a necessary stage for the production of voluntary movements—without which such movements become impossible. Of course this second explanation would not, however, be applicable to the cases of aphemia (as I now apply the term) due to sub-cortical lesions. In this case the kinæsthetic memories would not be lost, and we should have to do with a case of paralysis caused by an arrest in the transmission of the stimuli destined to excite the bulbar centres for the production of speech-movements.

Again, the statement which Ross says would be absurd concerning the cause of the paralysis in the case of a crural monoplegia of cortical origin would, in my opinion, be correct enough on the understanding that it was a loss of unconscious or sub-conscious memories that was postulated—as I do not believe that any motor centres exist in the cortex in close contiguity to the kinæsthetic centres, as Ross seems to imply. The lesion in the so-called "motor" region causes the paralysis solely, as I maintain, owing to destruction of the sensory centres of kinæsthetic type there situated—since the activity of such centres is essential for the production of voluntary movements. The destruction of these centres would, therefore, necessarily involve a loss of the memory of corresponding movements. Consequently I see no absurdity in saying that the paralysis of a patient suffering from a crural monoplegia of cortical origin is due "to a loss of memory of the movements" of the affected limb. It would be an unusual, though a correct, mode of expressing the exact pathology of the affection. It is obvious, however, from what he said on the following page, that Ross fell into the common error of not distinguishing between the effect of destruction of the kinæsthetic centres themselves and that of their afferent fibres; and that it was on this account, as well as for the reason previously indicated, that Ross was induced to deny a "loss of memory of words" as a cause of the inability to speak or to write in ordinary aphasia. His reason for the statement is to me most inconclusive, when he adds: "This inference being fully attested by the fact that the subjects of these disorders understand vocal or written speech, although unable to give expression to their thoughts in words." But surely the fact that the auditory and the visual memories of words are preserved, cannot be taken as evidence that the kinæsthetic memories of words are not lost.

LICHTHEIM'S TYPES OF SPEECH DEFECT.

In an earlier chapter I promised (p. 45) to show later on how the seven types of speech defect assumed to exist by Lichtheim—and which either in simple form or in combinations represent, as he thinks, the great majority, if not all, of such defects—are explicable without any necessity for having recourse to the notion of the existence of a separate "centre for concepts," such as he and several others postulate. This, however, is not the only reason for drawing attention more in detail to his views. His memoir is admittedly a very able one, and has attracted much attention in this country as well as abroad, so that for this reason also it may be useful for me to show to what extent we are in accord, and what different interpretations I am disposed to place upon some of his seven simple types of speech defect.[1]

TYPE I.—*Destruction of Broca's Centre.*—This I regard as a sensory rather than a motor centre; and if the lesion is limited to it I do not agree that there would necessarily be (*d*) loss of volitional writing, or (*e*) of writing to dictation. The amount of these defects would vary much in different persons.

TYPE II.—*Destruction of the Auditory Word Centre.*—If complete, I say there would often go with the word-deafness an aphasic loss of speech. (*b*) Understanding of written language, would, I think, often not be lost; (*e*) the faculty of reading aloud might be preserved in an educated visual; (*f*) the faculty of writing spontaneously would mostly be lost.

TYPE III.—*Destruction of the Audito-Kinæsthetic Commissure.*—(*d*) Volitional language would often be lost; (*f*) repetition of words, and (*g*) reading aloud would also be lost, except it may be in a visual. Paragraphia is certainly not a necessary consequence of such a lesion; nor do I believe that paraphasia would be, except perhaps where the commissure is merely damaged rather than actually destroyed, or where the patient is a visual and Broca's centre begins to be incited from the visual word centre. (*h*) Writing from dictation should not be interfered with.

TYPE IV.—*Destruction of a Commissure between a supposed "Concept Centre" and the Glosso-Kinæsthetic Centre.*—The case adduced in illustration of this type I regard as a typical instance of the combination of defects produced by a low excitability of the auditory word centre.[2] I agree with all the characteristics given, and can thus explain this type without any necessity for postulating the existence of a separate concept centre.

TYPE V.—*Destruction of the Internuncial Fibres in some part of their course between the left Glosso-Kinæsthetic Centre and the Bulb.*—This is the condition that I name "complete aphemia," and the real existence of which Hughlings Jackson doubts. I agree with all the characteristics given.

The five types above enumerated Lichtheim believes to be " well known varieties of aphasia; " the remaining two he regards as rare. Thus he says, " it would have been natural had I had no actual examples of them to adduce, though it

---

[1] See *Brain*, 1885, pp. 438, 439, 442, 447, 449, 453, and 460.
[2] It is quoted in abstract on p. 150 as Case xlvii.

would obviously have weakened my whole position, the full confirmation of which requires a filling up of every gap. I have, however, found the two missing types,[1] and their entire correspondence with the theory of aphasia here developed is a warrant for the accuracy of the hypothesis."

TYPE VI.—*Destruction of a Commissure between the Auditory Word Centre and the supposed "Concept Centre."*—Lichtheim's case, representative of this type, is reported in much detail, but the first one and a half pages of the report will be found to contain all the essential characteristics on which its interpretation must depend. Subsequently, there is a record of progressive improvement up to the time of the advent of a fresh cerebral attack—almost immediately after which the patient no longer continued under Lichtheim's observation.

The attack, the effects of which were recorded, was an abrupt one occurring in a man aged 60 years, and the clinical condition described is that which existed three days after its onset, when the patient had already begun to improve. In the early days of an acute attack like this the symptoms are often of a general, ill-defined, and more or less transitory character. It is stated, indeed, that on the fifth day, in this case, " the improvement was already very marked." It appears therefore, to my mind, very unsuitable to attempt to found a new type of speech defect upon the initial and transitory symptoms presented by such a case.

The case seems to me an interesting but ill-defined one, in which there was a partial resemblance to the symptoms met with in combined word-deafness and word-blindness, owing to disease involving the corresponding centres in the left hemisphere.[2] Still word-deafness was not complete, as the patient " could repeat correctly all that was spoken before him," and he could understand some things said to him. The word-blindness was, however, nearly complete, but as it is said that he " copies perfectly," this word-blindness must have been due to partial isolation rather than to destruction of the visual word centre. Much the same kind of thing may, indeed, have happened in regard to the auditory word centre.

Lichtheim's scheme requires that there should be, in this type, preservation (d) of volitional writing, and (f) of ability to read aloud ; but this patient is reported to have written only one word, and to have read only two, when he was examined on the third or fourth day. And later on, when he had gained

---

[1] One supposed case answering to each variety.

[2] In the subsequent attack, however, the right hemisphere seemed to have been affected. There was no autopsy.

more power in these respects, it could no longer be said that he had (in accordance with the requirements of the type) a loss (*a*) of understanding of spoken language, and (*b*) of written language.

Clearly, if the case is to be regarded as one that is in agreement with the scheme, it must be that all its requirements are satisfied at some one time—and not some at one time and some at another, as in this progressively improving case.

Lichtheim refers to two other cases (*loc. cit.*, p. 460), as appearing to him to belong to the same category—one published by Schmidt,[1] and one by Broadbent.[2] The first of these, recorded in abstract as Case lxiv., was one of nearly complete word-deafness together with word-blindness to a lesser extent. And the disorder of speech was of a kind closely resembling what is most common with partial combined defects of this type—to which category, with Kussmaul, I am inclined to refer it. The record of Broadbent's case (an abstract of which has been given as Case lxiv.) does not seem to have been seen by Lichtheim in the original, otherwise I feel sure he would never have cited it as belonging to his Type vi. It is really not at all in accordance therewith, seeing that there was no loss of (*a*) understanding of spoken language, and no paraphasia; that (*d*) volitional writing was lost; as also was his power (*f*) of reading aloud.

Neither here, therefore, nor in his Type iv. is any good evidence to be found in support of the existence of a separate "centre for concepts" standing apart from the perceptive area.

TYPE VII.—*Destruction of the Afferent Fibres going to the Left Auditory Word Centre.* — This type, Lichtheim says, "does not really belong to the subject of aphasia, for the faculty of speech remains perfect." But he adds, "it is necessary to consider it here because its symptoms cannot be properly understood except in connection with the present subject."

The mental condition is defective only as in a deaf person, owing to (*a*) inability to understand spoken language; (*b*) to repeat words; and (*c*) to write from dictation. This state is assumed by Lichtheim to be due to destruction of the afferent fibres, in some part of their course, which proceed from each ear to the left auditory word centre.

Lichtheim says (*loc. cit.*, p. 433) this affection "can obviously come into existence only if the irradiations of both acoustic nerves in the left temporal region are broken

---

[1] *Allg. Zeitschr. f. Psych.*, 1871, Bd. xxvii., s. 304; and Kussmaul, *loc. cit.*, p. 772.

[2] *Med. Chir. Trans.*, 1872, p. 162.

through." But I believe there is no evidence to show that the fibres from the left auditory nerve ever go to this centre; and that the right auditory word centre is only brought into relation with the left through the corpus callosum. Lichtheim would seem to think that the right auditory word centre is possessed of no functional activity, and, moreover, that it is not in relation with the left centre through the intervention of the corpus callosum.

This is the kind of defect now commonly known as "pure word-deafness," which has already been fully described and illustrated.

CHAPTER XIII.

THE READING AND WRITING OF NUMERALS. PRESERVATION OR LOSS OF MUSICAL FACULTY. AMIMIA. MIRROR WRITING.

CERTAIN apparent inconsistencies in connection with the powers of aphasic and amnesic patients have long been recognised. One which at first excited much surprise was that some persons who are speechless could, nevertheless, freely express their thoughts by writing. The reason for this apparent anomaly has been dealt with, and the conditions under which it arises set forth. There remain, however, two other apparent but well known inconsistencies which, though they have been mentioned in the narration of some of the cases, have not yet received any special attention. I allude to the ability possessed by some patients to read and write numerals when they are quite unable to read and write letters and words; and the ability possessed by others to comprehend music which they hear, to read musical characters, to write musical notes, and even to sing, when they are quite unable to comprehend ordinary speech, to read, to write, or to speak ordinary words.

These peculiarities will be found to be explicable in a manner very analogous to that by which the occasional persistence of ability to write is preserved while articulate speech is impossible. In each case we probably have to do, in the main, with different kinds of memory, having cerebral seats more or less distinct from one another—some of which may be damaged whilst others are spared. Though our knowledge is still rather ill-defined in regard to these subjects they are sufficiently interesting to deserve a brief consideration.

APPARENT ANOMALIES IN REGARD TO NUMERALS.

It cannot be said that aphasics have any more power of uttering the names of numerals than of uttering other words; nor do I think that word-deaf patients have any more power of comprehending the names of numerals when they hear them pronounced than of comprehending other names. Word-blind patients can, however, often recognise numerals, and even write them when they are quite unable to do the same with words or letters.[1] The preservation of both these abilities probably depends on one common cause—and this is, that the visual impressions derived from numerals are registered in a brain region which, owing to its being slightly removed from the visual word centre, has remained undamaged.

Destruction of the visual word centre is the common cause of word-blindness and also of an agraphia dependent thereupon. If there should be a separate seat for the registration of numerals it is suggested that its escape from damage would account in all probability for the preservation both of the ability to read and the ability to write numerals.

That this is the true explanation seems all the more probable, because on rare occasions it has been found that loss of ability to read and comprehend numerals exists in the absence of word-blindness. One such case has been recorded by Trousseau[2], and another by de Capdeville,[3] in which there was a transitory defect of this kind. At other times, and much more frequently, these two defects are combined.

The difficulty is to understand why the seat of registration of these impressions should be separate from that for letters and words. This, however, is the same kind of difficulty that had originally to be met in regard to words. *A priori* it might have been thought little likely that words should be registered in a cerebral region apart from other visual impressions, yet clinical and pathological facts teach us that this is so. And now, it would appear, we must go even further in this direction and postulate a separate localisation of visual impressions.

In regard to this point Déjerine[4] holds that numerals (and also a patient's name) can be read by word-blind patients at times because they are like drawings of objects rather than letters, indicating therefore that their seat of registration may be in a part of the general visual centre rather than in a part

---

[1] Instances of this have been mentioned in regard to several of the cases previously quoted (Cases xxvi., lxi., lxvi., xciv., etc.), and very numerous examples might be cited. I will, however, only refer to three cases recorded by Bernard (*loc. cit.*, pp. 85, 94 and 108) in which this difference existed to a well-marked degree; and to six cases cited by Jas. Hinshelwood in *Lancet*, December 21, 1895.

[2] *Bullet. Acad. Imp. de Méd.*, 1865, t. xxx., p. 653.

[3] *Marseille Médical*, 1880.

[4] *Compt. Rend. de la Soc. de Biologie*, 1891, p. 200.

of the visual word centre. Francis Galton [1] seems also to hold a view which favours this opinion since he says: "Persons who are imaginative almost invariably think of numerals in some form of mental imagery. If the idea of six occurs to them, the word 'six' does not sound in their mental ear, but the figure 6 in a written or printed form rises before their mental eye."

Another explanation may also be possibly found in a direction favoured by Broadbent, namely, that numerals are signs of concepts, and that therefore they may be registered in a region apart. He would say in a special "concept centre," whilst my interpretation from this point of view would be that their seat of registration should be looked for in a mere extension, or annex, of the visual word centre.

In either case the supposition is that the seat of registration of numerals is slightly different from that of letters and words, and that therefore in different cases of "sensory aphasia" it might occasionally happen that the former centre had either escaped or been less damaged than the latter.

As additional reasons for the exceptional preservation of ability to read and to write numerals, occasionally met with, there are two others deserving of some consideration:—(1) that numerals are so few as compared with letters; and (2) that more attention is generally given to individual numerals than to individual letters. Words are our counters of thought rather than letters; but the simple numerals with which amnesics show themselves familiar are also counters for thought—counters which are few in number, whilst verbal counters are very much more numerous.

Defects in Musical Faculty.

"Musical aphasia" and "amusia" are terms that have of late been used as generic designations for the different forms of defect of musical capacity resulting from cerebral disease. From a clinical point of view it has long been known that in some cases (a) the musical faculty equally with the faculty of speech may be destroyed, completely or partially, by brain lesions, so as to give rise to different forms of amusia; that in other cases (b) disorders of speech may occur alone, leaving the musical faculty more or less intact; while in another very limited group of cases (c) there may be one or more forms of amusia existing apart from any defects of speech.

The defects that have existed in the musical faculty as

---

[1] "Enquiry into Human Faculty," p. 114.

a result of cerebral disease or injury have only been noted and recorded in some persons. Some forms of this defect can, in fact, only be diagnosed in patients who have had more or less of a musical education, and others again only in those who are musical by nature.

Edgren[1] has been able to collect fifty-three cases in which the condition of the musical faculty has been noted, and these he divides into three groups:—(a) cases of aphasia without amusia, 24; (b) cases of aphasia complicated by amusia, 22; and (c) cases of amusia without aphasia, 6.

It has been ascertained that the different forms of amusia are closely analogous to the different forms of aphasia—using these terms in their most general sense. Edgren, therefore, as well as others, has come to the conclusion that the different forms of amusia are dependent upon damage to special anatomical centres and commissures which are adjacent to, though not identical with, those the damage to which gives rise to the corresponding forms of speech defect. This view is probably correct, so that it may be assumed that auditory centres for tones will correspond with the auditory word centres, a kinæsthetic tone centre with the kinæsthetic speech centre, and visual centres for notes with the visual word centres—the latter being called into play both for the reading and for the writing of music. In addition, portions of the general kinæsthetic centres concerned with the production of instrumental music of different kinds must be postulated, together with commissures connecting these several centres with one another and with true motor centres in the bulb and spinal cord.

The integrity of such centres and commissures would seem to be necessary and sufficient for the perception and production of musical tones for singing, for whistling, and for instrumental music of different kinds.

Knoblauch[2] has, however, made an elaborate attempt to distinguish many different kinds of disorder of the musical faculty, parallel with the disorders of speech described by Lichtheim, whose system he adopts as the basis for his comparisons. Among these, therefore, he includes disorders supposed to arise from damage to the altogether imaginary "centre for concepts," and to the commissures that are supposed to connect it with the auditory centre on the one side and Broca's centre on the other. As I do not agree with these points in

---

[1] *Brit. Med. Journ.*, December 22, 1894, p. 1441, and *Deutsch. Zeitsch. f. Nervenheilk.*, vol. vi., 1895, pp. 1-64.

[2] *Brain*, 1890, vol. xii., pp. 317-340.

the classification of Lichtheim, it would be useless for me to attempt to follow Knoblauch further than to say that he proposes the useful terms " tone-deafness " and " note-blindness " as names for the musical defects corresponding with word-deafness and word-blindness respectively; while for the so-called motor disorders corresponding with aphasia (in the more limited sense of the term), he proposes to restrict the term "amusia"—that is, to make it connote disorders of the faculty of musical expression.

In regard to all this it must be clearly understood that we have as yet no definite anatomical knowledge of the subject; all that we know is based upon clinical observation, supplemented by hypotheses which seem capable of giving some sort of explanation of the several defects in musical faculty, whether they exist alone or in combination with speech disorders, as well as of the preservation of this or that musical faculty when the allied power in regard to speech is lost.

Some of the combinations of these two kinds of defect and ability are comparatively common and have been long known, others are less common, while others still are very rare.

One of the most common and best known of the combinations consists in the *preservation* of an ability to correctly sing or whistle airs, with the aid of a few sounds or syllables, on the part of patients who are comparatively speechless. This is met with occasionally in patients suffering from true aphasia, and must be supposed to depend upon the kinæsthetic tone centre having escaped damage, when Broca's centre has been destroyed.

Falret,[1] however, long ago noticed that there was also a very rare class of cases, apparently similar except that the patients, besides being able to sing airs, could also sing the actual words of a song, although they could not, either before or afterwards, pronounce the words alone voluntarily.

The first class of cases is sufficiently simple and well understood, and needs no further illustration; but this is by no means the case in regard to the rare second class of cases. Of this class Knoblauch gives an illustration of his own, and refers to several that have been cited by others. He attempts to explain these cases on the supposition that they have been examples of true aphasia due to destruction of Broca's centre, though with preservation of the allied tone centre. Wyllie seems to be much of the same opinion, since he says[2] in refer-

---

[1] " Dict. Encyclopæd. des Sciences Méd.," 1866, t. v., p. 620, Art. " Aphasie."

[2] *Loc. cit.*, p. 274.

ence to the two classes above-mentioned: "Most of these cases, no doubt, have been cases of motor aphasia." And in regard to those who could sing songs correctly with the actual words he suggests that this power may have been brought about by aid of the "uneducated speech centres" of the opposite hemisphere—a suggestion which had previously been made by Gowers concerning a case that came under his own observation, but which is so briefly referred to as to make it impossible to feel sure as to the real nature of the speech defect.[1] This is one of the examples referred to by Knoblauch. Three of the others are of still less value—one of them being merely referred to in two or three lines by Brown-Séquard[2] upon the basis of information supplied by the relative of a patient; while another is merely mentioned by Hallopeau[3] as a patient who, "though completely aphasic," sang the words as well as the air of the "Marseillaise;" and the third is the case of an officer referred to by Grasset,[4] concerning whom Knoblauch speaks in the following terms (*loc. cit.*, p. 323):— "An officer who could only utter "pardi" and "b......," and was quite unable to pronounce the words "enfant" and "patrie," could nevertheless sing correctly both as to words and music the first verse of the "Marseillaise." In addition to his own case the only other to which Knoblauch refers is one that has been recorded by Bernard.

Some details concerning these two cases will be given presently, but meanwhile I may say that I think it extremely doubtful whether these patients have been real aphasics. All that we know of aphasia due to destruction of Broca's region seems to me to tell against it. I believe that these patients have rather been amnesics whose spontaneous speech has been more or less completely abolished by reason of a lowered functional activity in the auditory word centre, such as we found in Cases xlv.-xlviii.

It has been shown that such patients, though speechless, or nearly so, can often read fluently enough—that is, though the auditory word centre is not sufficiently sensitive to react to voluntary incitations, it will nevertheless respond to strong impressions coming to it from the visual word centre. Similarly, in these cases now under consideration, it may be that the undamaged auditory tone centre has been capable

---

[1] See "Diagnos. of Dis. of the Brain," 2nd ed., 1887, p. 133.

[2] *Compt. Rend. de la Soc. de Biologie*, Ap. 19, 1884, p. 256.

[3] "Patholog. Générale," 3rd ed., 1890, p. 665.

[4] *Montpel. Medical*, 1878, t. xl. I have been unable to obtain a sight of this periodical in London.

of rousing the auditory word centre so that patients otherwise almost speechless, were able to sing the words of songs quite correctly.

Unfortunately, owing to other coexisting defects, or to the insufficient way in which the cases have been recorded, it is not quite easy to settle this matter in regard to some of them. And even if we take the case recorded by Knoblauch[1] (an abstract of which is subjoined), it will be found to be of a rather doubtful nature.

CASE XCV.—The patient was a girl, aged six years, who could neither read nor write. After recovering from an attack of scarlet fever followed by nephritis, she was seized with general convulsions on December 21, 1886.

"On December 26 consciousness slowly returned, but there remained a condition of right hemiplegia with aphasia. The child could not speak at all at first. Later on she said 'Mamma,' and apparently *repeated* a few words. She could sing the song 'Weisst Du wie viel Sternlein stehen,' etc., but she could not recite the text of the song, or speak voluntarily single words of the same."

Soon after she improved in general health, but on February 8, 1887, she was admitted into the Clinical Hospital at Heidelberg on account of the hemiplegia and the speech defects. In regard to the latter the following details are given:—"Mentally, as far as one can judge, she is very well developed. As she is aphasic she has to make herself understood by gestures; spontaneously she only utters 'Mamma.' She is *able to repeat a few words*, but very imperfectly. If one commences the song 'Weisst Du wie viel Sternlein stehen,' she sings it with the right melody in an automatic way, being unable either to continue or to begin afresh when she once stops. All the words of the text which she is unable to pronounce spontaneously are, while she sings them, articulated perfectly. The comprehension of spoken language is quite normal. The patient has not yet learnt to read or write."

After this date she improved remarkably under treatment, so that by February 21 "she was able to *repeat most words correctly*, with considerable trouble it is true. She could count up to three if some one started her with 'one.' In the beginning of March she was able to sing the song 'Weisst Du wie viel Sternlein stehen' quite alone, and certainly with a much purer intonation than at the beginning of the treatment. On March 8, she succeeded for the first time in reciting the text of the song without singing the melody. In the beginning of April the patient had acquired a considerable vocabulary, and she even attempted to form small sentences. In the middle of the same month she could utter almost all words, but could not yet form connected sentences, though she managed to make herself perfectly understood."

I cannot think that this was a case, as Knoblauch terms it, of "pure motor aphasia." The patient was able to repeat a few words even from the first, though with difficulty, and she steadily improved in this direction before she regained the power of voluntary speech. It may therefore, I think,

[1] *Loc. cit.*, p. 318.

well have been a case of partial damage to the auditory word centre, causing the centre to be unable to respond to voluntary incitations, and leading to imperfect utterance at first, unless the auditory word centre was under stimulation from the auditory tone centre. Unfortunately, as the child was unable to read, no further evidence in favour of this view is forthcoming such as would have existed had it been shown that she was also able to read aloud. Still, I contend that no really aphasic patient attempts to repeat words and succeeds as this child did.

Much the same sort of criticism is applicable to the case that has been recorded by Bernard,[1] except that it was even more clearly, when seen, not a case of Broca's aphasia.

CASE XCVI.—A widow, aged 45 years, formerly a teacher of the piano, entered the Salpêtrière on December 12, 1881. Three years previously (in June, 1878), she awoke one morning finding herself paralysed on the right side, and unable to pronounce a word. Speech returned little by little, but having exhausted her small resources, she was obliged to take refuge in the hospital.

When examined on June 1, 1883, marked contracture with exaggeration of deep reflexes was found in the paralysed limbs of the right side, and the following particulars are given concerning her speech.

"H—— expresses herself with difficulty. She is sometimes obliged to make an effort for a tolerably long time in order to utter the terms of her replies, whenever they are a little complicated. 'Je sais, repète-t-elle, je ne puis pas dire.' She can tell the name of the majority of the objects that are shown to her. If, her silence being prolonged, one utters names that do not belong to the object, she protests, while she repeats immediately, with signs of satisfaction, the right words with which one furnishes her. She even articulates them then several times. The various details that we possess concerning her previous life, the date and the progress of her affection, have been given us by herself, and their correctness has been verified, since there have been no contradictions in the accounts given at different times."

"H—— is known in the ward by the name *la Dame blanche*, because she is fond of singing the celebrated air:—

'La dame blanche vous regarde,
'La dame blanche vous entend.'

"The words are distinct, and the air exactly repeated. At the same time that she sings, H—— accompanies herself with her left hand, whose fingers move in accord on the bed-clothes."

"She showed no trace of word-deafness, and she could repeat correctly any word that she heard. She was partially word-blind—her power of reading being rather limited. She was completely note-blind, not being able to read a single note of music. When pressed she would say, 'Je ne sais pas, je ne sais pas.'"

From August 20, 1883, onwards, she had several attacks of Jacksonian epilepsy, the twitchings beginning in the face and ending with a prolonged coma. On recovery from various of these attacks, the nature of her speech defects is said to have undergone no change.

[1] "De l'Aphasie," 1885, p. 119.

She died from pneumonia on May 11, 1884. A large "plaque jaune" was found on the external surface of the left hemisphere, involving all the upper half of the insula, the middle and posterior portion of the three frontal convolutions, together with the lower fourth of the ascending frontal. A prolongation extended backwards under the operculum to the posterior extremity of the Sylvian fissure, where it broadened out over each border slightly. At the inferior angle of the superior parietal lobule there was another small plaque. On section of the hemisphere the lenticular nucleus was found to be laid bare, and the whole of the internal capsule, except the sensory region, had a grey tint. There was marked wasting of the left peduncle, the left side of the pons, and the left anterior pyramid.

This case is certainly a very remarkable one in many respects. In the first place it is clear that the clinical details with which we are furnished of the patient's condition at the expiration of five years from the date of her original attack are not those of a case of true aphasia, but rather one showing a partial disability of the left auditory word centre. Although her spontaneous speech was limited, yet she could repeat any word when bidden, and could name objects shown to her. The account of the necropsy when compared with these symptoms gives occasion for the greatest surprise. The two records seem so incompatible as to suggest a doubt whether in reality the patient's condition as regards speech could have remained the same from June 1, 1883, up to the time of her death in May, 1884. All that Bernard says on this subject is that after some of the epileptiform attacks (which commenced on August 20), when the insensibility had disappeared, "les divers examens que nous pratiquâmes sur elle, nous montrèrent que l'état de son aphasie ne variait pas;" and the date of the last examination yielding such results is unfortunately not given. I make these remarks because it seems just possible that a considerable portion of the lesions actually found in the left hemisphere may have been produced as a sequence of further vascular occlusions occurring at the time of the epileptiform attacks above referred to, and consequently that no such extensive lesions as were found at the necropsy existed in June, 1883, when the above clinical condition was recorded.

If this possible explanation will not hold, it seems certain (in view of the complete destruction of the posterior portion of the third left frontal convolution), that in the interval between the patient's first attack in 1878 and the date of her clinical examination in June, 1883, the right hemisphere must gradually have taken on some of the functions which the left hemisphere had become unable to perform.[1] All we are told, from the

[1] So that speech may have been produced in the manner indicated in Fig. 16.

patient's own account, is that "La parole revint peu à peu." The case in this respect has some important resemblances to that of Thomas Andrews, previously recorded (Case xciv.), in whom at the time of death lesions of such extent were found in the left hemisphere as to make it seem almost certain that even the limited power of speaking possessed by the patient must have been effected in the main through the intervention of the right hemisphere.

It so happens that I have at present under my care in the National Hospital for the Paralysed and Epileptic an admirable example of this combination of almost complete loss of voluntary speech, together with a marked ability to sing not only airs but the actual words of hymns and ballads with wonderful fluency and correctness. When admitted, this patient's condition was totally different; she was then absolutely speechless, word-deaf, and word-blind, apparently as a result of thrombosis of the left middle cerebral artery. The following is an abstract of her case.

CASE XCVII.—Sarah A. B., aged 40 years, was admitted under my care on October 1, 1897. Is married and has had nine children, the last having been born seven weeks since. One month ago she had an attack of left pleurisy followed by right pleurisy. Before leaving her bed after this last illness she one day lost the use of her arm; twenty minutes later she entirely lost her speech, and soon after the right side of her face and her right leg became weak. There was no loss of consciousness and no convulsions.

On admission she was found to be dull and apathetic; occasionally emotional, but unable to utter any articulate sound. She did not even make any attempt when bidden to repeat a word. Word-deafness nearly complete. Word-blindness quite complete. With left hand cannot write a single letter or numeral, though she can copy the printed letter A. Right hemianopsia probable. Slight facial weakness on right side and deviation of the tongue to the right. Absolute flaccid paralysis of the right arm and leg. Some hemianæsthesia on the right side. Incontinence of urine and fæces. No cardiac bruit. Some albumen but no casts in urine.

October 16. Can now say "no" but not "yes." The incontinence has ceased.

November 4. She can now sing, and articulate distinctly when singing. She has been heard to repeat the alphabet to another patient. She now smiles, nods, and shakes her head in answer to questions, the word-deafness having passed away, though the word-blindness continues.

November 25. Examination by the House Physician (Dr. J. S. Collier). No word-deafness now. She corrects me directly when I make a mistake in the multiplication table. The only words she uses voluntarily are "no," which she uses correctly, and "Bull," the name of the patient next to her. If the alphabet be repeated slowly to her she joins in and will continue to repeat it alone correctly. Sometimes, however, she makes a mistake, shakes her head and says "no," and cannot continue until she is started afresh. When started by counting aloud, she can count up to twenty alone, with some defects of articulation, such as "en" for ten, "fixteen" for sixteen, "tenty" for twenty. She cannot say the easier part of the multiplication table. She cannot repeat a single word after me. She was made to say "eighteen, nineteen, twenty," about a dozen times by leading up with "sixteen, seventeen" repeated by me aloud, and then when I asked her to say "twenty," she did so once, but could not repeat the performance.

She can sing a tune to order. She commences humming and then joins in with the words, many of them perfectly articulated, some of them badly articulated, and in the place of others mere lalling. The following is a specimen of her singing of the hymn "Hark, hark, my soul," her mistakes being printed in italics.

"Hark, hark, my soul, angelic songs are swelling
O'er earth's green *cas* (seas), and ocean's *nave mint ore* (wave beat shore),
How sweet the truth those blessed strains are *selling* (telling)
Of that new life where sin shall be no more."

She sang three verses of this hymn. She also sang to order verses of the following hymns:—"Onward Christian soldiers;" "Jesu meek and gentle;" "Awake, my soul;" "At even ere the sun was set;" and others, as well as some popular ballads, such as "Belle Mahone," "Cherry ripe," etc.

She can start singing these herself. She can, moreover, repeat the above mentioned verses without singing if she is started by my beginning them aloud, but she cannot say them without being first put upon the track. Her articulation of quite difficult words in the singing is often very good, but in repeating poetry her articulation is not so good as when she sings.

She cannot repeat a single word dictated to her.

She is still absolutely word-blind. She names letters but quite wrongly. When shown a letter upside down, she at once placed it right side up. When shown her own name she evidently did not recognise it; she spelt it out, but did not get a single letter right, thus—

Sarah Brown
iptea eavr no

December 11. She is still completely word-blind; she cannot pick out a single letter, or recognise her own name spelt with capitals. She has said a few more words spontaneously, such as "oranges" and "fish." She still cannot name any object that is shown to her. She can now repeat words a little such as "father," "paper," "nice,"—has done so about a dozen times in all. She understands complicated orders at once, and obeys correctly.[1]

In this patient there may have been a "motor aphasia" at first; or, on the other hand, Broca's centre may have escaped damage, owing to the thrombosis only involving the second, third and fourth cortical branches of the left middle cerebral. In this latter case, the absolutely speechless condition of the patient during the first week or two after her attack may have been due to the cutting off of the blood supply from the auditory and the visual word centres. After a time, however (when the establishment of a collateral circulation had gradually in part restored the blood supply of the auditory word centre), the word-deafness disappeared, and the patient began to utter two words voluntarily, and to repeat many more whilst sing-

[1] Since this date she has greatly improved. She can now (Feb. 26, 1898) speak more words spontaneously, and can repeat almost any words when bidden.

ing. That she did not after a time regain more in the way of voluntary speech, was, I believe, due to the continuance of a lowered functional condition of the auditory word centre, similar to that which existed in Cases xlv.-xlviii. Unfortunately, we could not test this view further by ascertaining whether she was able to read aloud (as such patients usually are when there is no coexisting defect in the visual word centre), owing to her continued absolute word-blindness. But the escape of the auditory tone centre and its stimulating influence may be supposed to have been the mode of excitation of the auditory word centre, aided by the same kind of facility for speech derived from the existence of a strong associational nexus between the words of the song, as exists between numerals from 1 to 20, or between the several letters of the alphabet. The ability to repeat these, and the good power of articulation possessed by the patient, seems to make it quite clear that the defect was in the left auditory word centre, and not in Broca's region, although, at first, her marked inability to repeat single words when bidden was an altogether unusual accompaniment of such a condition.

Defects in musical expression somewhat akin to paraphasia have been met with. Though I am not quite clear whether it might not be better to class some of them rather with incomplete aphemia. This doubt seems to me to hold in regard to a case of "Broca's Aphasia" recorded by Kast,[1] of which the following details are given by Knoblauch (*loc. cit.*, p. 333).

CASE XCVIII.—An aphasic patient who was before his illness a "prominent member" of a musical society in his native place, concerning whom Kast gives the following details:—

"As a matter of fact the way in which the patient performed the first exercises which I gave him to sing or repeat, a simple melody ('Heil Dir im Siegerkranz'), showed clearly that his musical capacity, at least at present, was far from being such as would satisfy even the modest claims of a rural musical society. I did not fare any better in trying him with hymns, even the most familiar and simple. Yet he always showed that he could get the rhythm of the correct melody; and he also knew the value of every note, though he sang throughout false tones, and with false intervals, and that in spite of the fact that he was evidently conscious of his defective musical capacity, and consequently not easily induced to lend himself to further experiments."

"An attempt was now made to get him to repeat tones, and here, too, considerable disturbance was apparent, although he recognised the falseness of the tones which he produced and manifested considerable anger thereat. When, however, the patient was got to give me tones which I had to reproduce he remarked the smallest error, corrected me with a loud 'Nein, nein,' and was only satisfied when the correct sound was given. He recognised

[1] *Münch. Méd. Wochensch.*, 1885, No. 44, p. 62.

all the melodies and songs which were sung to him, and expressed his dissatisfaction if songs he knew were sung incorrectly."

"In the knowledge of writing notes the patient was not sufficiently trained to stand a thorough examination, nor was he able to play any musical instrument."

This case is in its leading features the reverse of those previously referred to—the patient losing his power of musical expression but retaining speech, instead of losing speech but retaining the power of musical expression.

It is common enough for both these powers to be lost in the same patient; but perhaps not very common, under these circumstances, for the patient to exhibit no tone-deafness or tone-blindness. Yet Proust[1] briefly refers to such a case. The patient was a lady with aphasia, who had been very musical before her illness came on. She could read and write notes, even compose and recognise melodies sung to her, but she herself was unable to sing the melody.

Again, it sometimes happens that an ability to write musical notes is retained in patients who are completely aphasic and agraphic for ordinary words. Thus, many years ago, Lasègue had a patient who was able to write with facility the music of melodies which he had heard, although in other respects he suffered from complete aphasia and agraphia. Again, one of Bouillaud's patients, suffering from similar defects, could compose and write in musical manuscript without mistakes, and sing the air afterwards while playing his own accompaniment.[2]

It may be well in this connection to mention again the remarkable case referred to by Charcot, in one of his lectures, of a trombone player who had, as an isolated defect, lost the memory of the associated movements of the mouth and of the hand necessary for playing upon his instrument.[3]

Note-Blindness.—Many cases are on record in which with word-blindness there has been the association of note-blindness.

Note-blindness in the absence of word-blindness is, however, a much rarer defect. Charcot in his lectures on aphasia in 1883, referred to the case of one of his colleagues who suffered from right hemiplegia and aphasia, and whose symptoms commenced with note-blindness. One day he placed himself at the piano, opened a piece of music, and could not decipher a single note, though he could play correctly and with facility. A patient of Finkelburg, again, could play from memory and

---

[1] *Arch. Gén. de Méd.*, 1872, p. 310.
[2] *Archiv. Imp. de Méd.*, 1865.　　　[3] See p. 112.

play the airs which he had heard sung or played by another, though he was quite note-blind.[1]

Proust[2] refers to a patient who could compose and write musical notes, though he was quite unable to read them—a defect comparable with "pure word-blindness."

Tone-Deafness.—The existence of tone-deafness as an isolated and inherent defect is probably far from rare. The case recorded by Grant Allen[3] of the young man who was incapable of discriminating the two notes of an octave, and had attempted in vain to take music lessons, is to be regarded merely as one of a fairly numerous class. I have known several persons in this unfortunate condition, who, notwithstanding the most frequent opportunities of hearing music, remained quite incapable of distinguishing one tune from another. As Elder[4] says: "Probably almost every one hears music as music, but there are enormous differences in the powers which people possess, not only of appreciating it in the sense of its giving pleasure, but in the power they possess of recognising its pitch and other qualities." Again, Wyllie[5] points out, in reference to the existence of what is commonly known as a "musical ear": "The development of this special faculty seems to bear no intimate relationship to the development of the other faculties of the mind. It is known that many men of genius have been utterly destitute of a musical ear; and it is further known that in idiots and imbeciles the musical faculty is often well developed,[6] the proportion of individuals with a musical ear being among imbeciles almost the same as it is among the general population."

These facts show that cases of disease in which tone-deafness is found to exist cannot safely be regarded as of any importance unless it is known that the individual had an ear for music previous to the onset of the cerebral disease or injury in question. When a musical faculty formerly possessed becomes blotted out, however, such a defect is a matter of considerable interest. This occurs most frequently in association with word-deafness, and many instances of it are on record. An example will be found in Case cv.

The same kind of thing may also occur in cases of "pure

---

[1] Referred to by Bernard, *loc. cit.*, p. 119.
[2] *Archiv. Gén. de Méd.*, 1866.
[3] *Mind*, April, 1878.   [4] *Loc. cit.*, p. 239.
[5] "Disorders of Speech," p. 114.
[6] He gives good instances of this on pp. 125 and 127..

word-deafness" as in the first case recorded by Lichtheim.[1] This patient had formerly been musical, but he lost the power of recognising even common melodies; and sometimes asked his children to stop singing quartettes " as they made too much noise." In a very similar case observed by Wernicke there was also, with word-deafness, a partial impairment of the musical faculty, the patient having lost the power of appreciating certain notes of a high pitch. See also Cases lvii. and lviii.

On the other hand, it happens occasionally that while there is word-deafness there is no tone-deafness. This was the case with one of Elder's patients, of whom it is said that though he "had a considerable amount of word-deafness, his hearing for music was quite unaffected. He sat for hours listening to a piano, and the slightest defect in a note was at once detected." See also Case civ.

Edgren[2] has recorded a remarkable case in which, after a slight injury, tone-deafness existed almost alone for a period of three years; and he has also referred to a few other cases in which either complete or incomplete amusia existed as transitory troubles independently of speech defects.

His own case is not only interesting but important, as the following details will show.

CASE XCIX.—A man, aged 34, the day after knocking his head against a lamp-post, complained of severe headache, difficulty in seeing, and mental confusion; he vomited, had illusions of taste, and eighteen days after the injury developed tone-deafness as well as aphasic symptoms. The patient, an intelligent and musical man, came home and told his wife that he had been unable to make out the melodies played at a place of public amusement. He had gone to several, to test himself. He could only hear indistinct sounds, and was unable to make out any melody; he could not even distinguish between a march and a valse or polka. When his wife spoke to him he did not understand her. The following two days the patient did not speak. His word-deafness and other aphasic symptoms disappeared within a month, but his tone-deafness remained unchanged till his death from purpura three years after the accident.

The chief *post-mortem* changes were the following: In the left hemisphere the anterior two-thirds of the superior temporal gyrus and the anterior half of the middle temporal gyrus were destroyed, exposing to view the island of Reil. At the junction of the posterior and middle thirds of the superior temporal gyrus the surface was sclerotic and adherent to the thickened pia. Under this sclerotic patch, and for about 1 cm. behind it in the first temporal gyrus, the softening was continued in the grey matter, but with decreasing depth. The surrounding parts were all healthy.

In the right hemisphere there was a somewhat similar defect, surrounding the horizontal limb of the fissure of Sylvius. The defect here included the superior and external surface of the superior temporal gyrus in its posterior half and the lower edge of the supra-marginal gyrus for a similar

[1] *Brain*, 1885, p. 461.   [2] *Loc. cit.*, p. 1442.

distance. The destruction extended deeply down to the bottom of the fissure of Sylvius, and had spread to the white matter which was softened to a much greater extent than the grey. The central ganglia and the internal capsule were healthy.

Relying in the main upon this case, Edgren supposes that the auditory tone centre may be situated in the first and second temporal convolutions in front of the area, the destruction of which causes word-deafness.[1] This is the only piece of evidence we at present possess (whatever it may be worth) bearing upon the localisation of this or of any other of the musical defects above referred to.[2] The fact, however, of their frequent existence apart from their related speech defects, and *vice versâ*, leads us reasonably enough to believe that the centres upon the functional activity of which musical endowments depend must be topographically distinct. On the other hand, the frequency with which the allied forms of these two varieties of defect are involved together, as well as certain general considerations, entitle us to believe that the allied centres for speech and music are probably contiguous to one another.

It seems probable that on the receptive side the centres connected with music (that is, the auditory tone centre and the visual note centre) must be more or less equally developed in both hemispheres, and yet we see from Edgren's case that a lesion in the anterior half of the temporal convolutions of the left side was alone sufficient to cause tone-deafness. This superior importance of the centre on the left side must be supposed to be dependent upon the fact that the expressive centre (that is, the kinæsthetic tone centre) is like that for ordinary speech all powerful on this same left side of the brain. Of course the functional activity of these two expressive centres during singing will be so intimately associated that the combined centres must be capable of acting like a single centre.

There would also be the same preponderating action of the receptive and expressive note centres on the left side of the brain during the act of writing music with the right hand.

There are, however, also the instrumental modes of ex-

[1] A case by Bruns, of which an abstract is given by Mirallié (*loc. cit.*, p. 151), is, however, rather opposed to Edgren's supposition.

[2] In connection with this case it is interesting to note that though the posterior half of the right upper temporal was badly damaged, there was no word-deafness. It tends to confirm, therefore, the notion commonly held as to the much greater importance of this region in the left hemisphere for the production of any such defect.

pression, often necessitating the coördinated action of both hands (as in the playing of the piano and the organ more especially), and these would seem to be scarcely possible without a fairly equal development of auditory tone and visual note centres in both hemispheres.

Amimia.

Amimia is the term that has been given to loss of the power of producing and of understanding what has been termed "natural language" as opposed to "artificial or acquired language"—that is, the loss of the power of producing and of understanding pantomime, as conveyed by gestures, alterations in facial expression, and by inflections of voice.[1]

This subject may be dismissed in a few words, as it is found that these defects vary in the same ratio as the intelligence of the individuals affected.

Where we have to do with sub-cortical lesions, intelligence is not diminished, and the power of the sufferers to express their thoughts by gesture or to comprehend the pantomime of another remains unabated. With lesions in Broca's region these powers may be only slightly impaired. But where the auditory word centre is also involved, and still more with a coexisting involvement of the visual word centre, there is more and more of mental degradation, and a corresponding diminution in the individual's power to make use of and to comprehend pantomime. This would always be specially marked where object-blindness existed, and also whenever we have to do with important lesions in both cerebral hemispheres.

Mirror Writing.

Mirror writing is that kind of writing which is executed from right to left, and with the normal position of the letters reversed—so that it can only be easily read with the aid of a mirror.

This is another subject which, though it has been commonly referred to in connection with aphasia, has no very special relation therewith, and may be dismissed in a comparatively few words—although the subject is interesting in itself and very obscure.

Until quite recently its sole connection with the subject

[1] Bateman (*loc. cit.*, pp. 165-174) has brought forward some interesting contributions bearing on gesture language.

of aphasia was due to the fact that some aphasics, being paralysed in the right hand, find when they attempt to make use of the left hand in order to write that they execute mirror writing. Three such cases were described by Buchwald,[1] and he it was who gave the designation "Spiegel-Schrift" to this form of writing.

The tendency to write in this fashion has, however, no necessary connection with the existence of an aphasic speech defect, although Elder[2] has recently endeavoured to show that it has a bearing upon some much controverted points in connection with agraphia.

He has made an important independent investigation concerning mirror writing, having tested or caused to be tested 451 persons of different ages and sexes, as to the mode in which they write when they are first bidden to do so with the left hand. He found that twenty-three of these (or 5·1 per cent.) were mirror writers. Elder, in fact, attempts to show that the existence of mirror writing can be most easily explained on the supposition that only one special centre for writing movements exists, and that in the left hemisphere—instead of supposing, as I do, that such a centre may become developed in either hemisphere, for writing with the opposite hand.

I cannot say that Elder's reasoning convinces me, and to some of the stages by which he attempts to establish his position I am quite opposed. Still, certain of the facts he adduces in support of his views are deserving of attentive consideration. Thus he says that, in accordance with his hypothesis, if the left graphic centre were suddenly called upon to direct the movements of the left hand the writing would not be after the usual fashion, but in the peculiar form known as mirror writing, and in support of this position he adduces the following interesting facts:—"If the mirror writing is held before a mirror or read through the paper so as to compare it with the same individual's right-hand writing, it will be seen that there is no mistaking that the handwriting is the same. Every detail of the handwriting is reproduced in the mirror writing, and the resemblance is so great that one must admit that such resemblance must be brought about by the two hands being guided from the same centre. I believe that the centre that guides the two hands is a graphic centre in the left cerebral cortex, and I cannot possibly imagine that

[1] *Berlin. Klinische Wochensch.*, 1878.
[2] *Loc. cit.*, p. 214.

that centre is the visual centre, seeing that the natural tendency of the visual centre would be to reproduce the writing in the usual shape and not in the mirror fashion."

But without entering into any detailed criticism of these views and statements, we may ask: How does Dr. Elder explain the results of his own extensive observations? The fact that only about five per cent. of those who were called upon to write with the left hand proved to be mirror writers can scarcely be said to lend any very striking support to his theoretical considerations, even when all reasonable allowances are made. They do not tend to show that only one writing centre exists, and that this centre in the left hemisphere controls the writing movements of the left as well as those of the right hand. I believe that the numerical results of his observations would rather go to support the view that writing movements of the left hand are controlled by the conjoint activity of visual and kinæsthetic centres in the right hemisphere, just as the writing movements of the right hand are controlled or coördinated by similar centres in the left cerebral hemisphere.

CHAPTER XIV.

THE ETIOLOGY OF APHASIA AND OTHER SPEECH DEFECTS.

LITTLE requires to be said here on the subject of the etiology of the speech defects we have been studying; and even that must partake somewhat of the nature of a recapitulation, since reference has already been made in preceding chapters to the extremely varied causes that may be operative in their production.

It will have been seen that these causes are very roughly divisible into two classes—the one class giving rise to speech disorders due to so-called "functional defects," and the other to the speech disorders due to "organic lesions." This is confessedly a most arbitrary and uncertain division. Of course when we say that certain speech disorders are due to functional defects all that is intended to be implied is that there is probably an absence of a gross lesion, of anything that could be detected by the naked eye or even by the microscope—though there may be a cessation or perversion of those normal molecular actions that are accustomed to take place in particular regions of the brain, in association with a healthy state of nutrition and normal physiological action.

But the demarcations or boundaries of this class of functional speech defects are extremely uncertain. Its representatives as a rule give rise to more or less temporary disorders—in other words, the conditions are curable, and may ultimately completely disappear. Still many of the causes of temporary speech disorder may not be restricted to mere molecular alterations in the nervous system; they may be represented by other conditions which would more rightly find a place in the category of organic lesions, such as temporary spasms or congestions of vessels in particular regions of the cortex, or small thromboses or embolisms which either soon disappear or are speedily compensated by the establishment of a collateral circulation. To decide, in any given case, between the existence of such conditions and those which would be more rightfully included as causes of mere functional defects, is often impossible.

Functional defects of speech have already been fully dealt with in Chapter vii. We then found that, for the most part, it is the glosso-kinæsthetic centre which is at fault in these cases, though more rarely there may also be implication of the auditory word centre giving rise to amnesic defects. The visual centre is rarely or ever disturbed; while the cheiro-kinæsthetic centre seems never at fault, from mere functional defects, unless in common with the hand and arm centres generally—as would be the case in a right functional monoplegia.

The predisposing causes of "hysterical mutism" have been fully discussed, and the probability enforced that in this condition we may have to do with functional disturbances simultaneously affecting the left and the right third frontal convolutions. It only remains to suggest the possibility that some of these cases might be instances of anarthria of a functional type—that is, due to functional disturbances in the bulbar motor centres.

Of the other defects of functional type the most common exciting causes are fright or emotional shocks of various kinds (possibly associated with traumatisms); also, at times, over literary work, especially when combined with worry.

In other cases, again, temporary aphasic defects may precede or follow epileptic attacks, especially (though not only) when they are of right-sided Jacksonian type; or they may be associated with attacks of migraine. They have likewise been known not unfrequently to follow snake bites, and to occur also as sequences of poisoning by stramonium or belladonna.

These cases of speech defect of functional type are not unfrequently of an intermittent or recurrent character. They mostly exist alone and without paralysis of limbs, though at times such speech defects are associated with hemiplegia, also more or less temporary in duration.

If we look to the speech disorders due to "organic lesions" we find that the cause originally assigned by Hughlings Jackson[1] occurs by far the most frequently, namely, embolism or thrombosis of the left middle cerebral artery, or of some of its cortical branches, leading to the occurrence of foci of softening. On some of the variations thus induced I have elsewhere written as follows:[2]—

[1] *London Hosp. Reports*, vol. i., 1864, p. 388.
[2] "Paralyses: Cerebral, Bulbar, and Spinal," 1886, p. 271.

"The main portion of the Rolandic region is supplied by three branches of the Sylvian artery, each of which nourishes a special area (Fig. 22). The artery may be occluded (by embolism or thrombosis) in its main trunk, or in one or more of its separate offsets. In the latter case we may have monoplegias of different kinds according to the particular branches affected, while, where the whole of the cortical branches, or the main trunk, is involved we may have a complete hemiplegia of the opposite side (the trunk muscles and other bilaterally acting muscles only being exempt), with or without aphasia, according as the left or the right hemisphere is

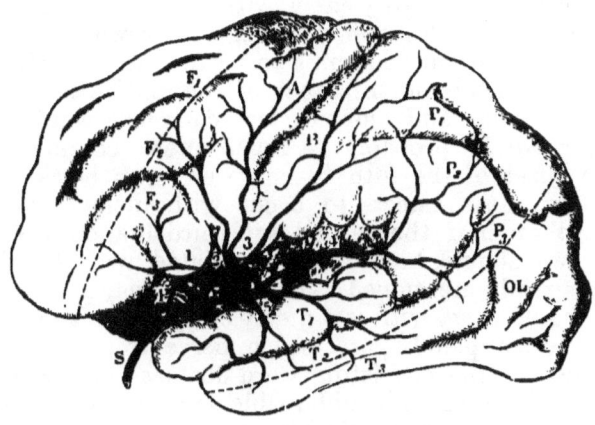

FIG. 22 (after Duret).—Diagram showing the area of Distribution of the Middle Cerebral Artery. *A.* Sylvian, or middle cerebral artery; *P.* Perforating branches of basal system; *1.* Inferior frontal branch; *2.* Ascending frontal branch; *3.* Ascending parietal branch; *4.* Parieto-sphenoidal branch; *5.* Sphenoidal branches.

affected in right-handed individuals (or *vice versâ* for left-handed persons). All this may occur from an occlusion of the middle cerebral beyond the first part of its course (two-thirds of an inch); that is, beyond the part from which the twigs of the basal system are given off to the internal capsule and the corpus striatum (and consequently in the absence of any subsequent softening in these parts). The clinical phenomena are not, however, appreciably different when this first part of the Sylvian artery becomes occluded—even though this entails softening of the motor segment of the internal capsule and of the greater portion of the striate body, as well as of the cortex in the Rolandic area."

"The first of the cortical branches of the Sylvian artery is supplied to the posterior part of the third frontal convolution, and if this vessel only becomes occluded or becomes the seat of an enduring functional spasm, on the left side, we may have as a result an aphasic condition without other paralytic complication."

Occlusion of the second and third branches of the Sylvian artery might likewise lead to the production of aphasia, owing to the cutting off of the blood supply from the anterior half of the audito-kinæsthetic commissure lying internal to the island of Reil; but then it would be associated with paralysis of the face and arm on the right side, together with a partial paralysis of the lower extremity. The paralysis of the hand and arm would necessarily carry with it an inability to write with the right hand (agraphia), though there would be nothing to prevent the patient's learning to write with the left hand.

Occlusion of the fourth branch of the middle cerebral leads to word-blindness and more or less complete word-deafness. The latter would probably be much more complete if the posterior offset of the fifth branch were also blocked. It is possible for these defects to occur with little or no motor paralysis, though if the second and third branches were also occluded we should have them associated with the kind of paralysis already mentioned. This hemiplegia, due to cortical disease, should carry with it a pretty complete loss of muscular sense in the completely paralysed upper extremity, though in consequence of the coexisting word-deafness and word-blindness no evidence would be obtainable as to its existence.

From what has been already said it must be clear that a thrombosis or embolism of the trunk of the middle cerebral artery, so situated as simultaneously to cut off the blood supply from all the five main cortical branches, would lead to right hemiplegia associated with what has been described as "total aphasia"—that is, in addition to aphasia proper there would also be agraphia as well as complete word-deafness and word-blindness, carrying with them that amount of mental degradation which is inseparable from a blotting out of all the word centres in the leading hemisphere.

Again, in cases where thrombosis simultaneously affects the left anterior, as well as the middle, cerebral artery (perhaps, owing to the process beginning in the common carotid artery), we might expect to find right hemiplegia and "total aphasia" together with an altogether unusual amount of mental degradation, in addition to blindness and loss of smell on the side of the lesion. The extra mental degradation would be due to the fact of the cutting off of the blood supply from the corpus callosum, seeing that this is mainly supplied from the anterior cerebral. As to this I have elsewhere said:[1] " The sudden cutting off, in an adult, of the functional activity of such a part as the corpus callosum, might well be supposed to produce marked symptoms—even though well-defined effects may have been absent in some of the recorded cases, in which there has been a congenital deficiency (in whole or in part) of the corpus callosum. These two sets of cases stand on a different level."

It will be seen, therefore, that very many variations in the clinical condition of patients may be associated with occlusion in whole or in part even of the cortical branches alone of the left middle cerebral artery. Some of these differences may depend upon slight variations in different individuals in the extent of the cortex supplied by the middle cerebral artery, as compared with the territories fed by the anterior and the posterior cerebral arteries.

Even more of the variations, however, would probably depend upon differences in the freedom of the anastomosis existing between the terminal ramifications of these three principal arteries, as well as between the different cortical branches of the middle cerebral itself. On this subject much difference of opinion has existed. Thus, Duret contends that anastomoses between the main trunks or the secondary branches are either absent or extremely rare. On the other hand, Heubner believes the anastomosis to be free between the different main vessels of the cortex, and also between the secondary branches, basing his opinion upon the result of his injections. This seems to be undoubtedly true in certain persons, and the existence of such a condition in a case in which there is obstruction by embolism or thrombosis, would permit the establishment of a collateral circulation, and, consequently, the staving off of softening of the cortex or the narrowing of its area in the territory corresponding with the blocked vessel. Such an exemption from softening undoubtedly occurs in certain cases of arterial obstruction, as *post-mortem* examinations have shown.

On rare occasions similar lesions occur on the two sides of the brain, for instance in the territory of the fourth, or of the fourth and fifth branches of the middle cerebral arteries, as in Cases lv., lvii., and xcix.

It is only on rare occasions that vascular lesions are pre-

[1] "Paralyses: Cerebral, Bulbar, and Spinal," 1886, p. 299. Anastomosis with the corresponding branch of the right anterior cerebral, might, of course, somewhat diminish the softening of the left half of the corpus callosum.

cisely limited to the seats of particular word centres, they are much more frequently irregular in their distribution, or multiple, and thus give rise to confused or less typical forms of speech defect.

When a necropsy is made in the early stages of these cases of embolism or thrombosis of vessels supplying the cortex, red or red and white softening mixed is found in the area affected; after longer periods, yellow softening is met with, or what French writers term "plaques jaunes;" and after much more prolonged periods, pseudo-cysts are found, owing to complete atrophy and absorption of the cerebral tissue having taken place, as in one of the cases I have recorded (Fig. 14).

Aphasic and amnesic defects of speech have been met with occasionally during or after one of the acute specific diseases, or during the puerperal state; and in all these cases a thrombosis partial or complete of the left middle cerebral artery is the most common cause.

Less frequently we have cortical hæmorrhages in one or other of these situations, either alone or associated with deeper seated bleeding. These may cause aphasia or various forms of amnesia, either alone or in association with right hemiplegia.

Less frequently still, we may have meningeal hæmorrhages pressing upon one or other of the cortical word centres; or we may have tubercular meningitis, chronic meningo-cerebritis (either simple or specific), or new growths of different kinds, involving or indirectly pressing upon one or other of these regions, and thus giving rise to corresponding speech defects.

In cases of fracture of the skull or gun-shot wounds, pressure from depressed bone may produce aphasic or amnesic defects of speech, which are speedily removed when the bone is elevated. Abscesses, also, as sequences of traumatisms may lead to similar results. Blows on the head have, in fact, often been the cause of aphemia, aphasia, or amnesia, even when no fracture or depressed bone has been present.

CHAPTER XV.

DIAGNOSIS IN SPEECH DEFECTS.

Owing to the very varied way in which the several kinds of defect in the power of intellectual expression by speech and writing are combined in different cases and owing, still further, to the importance of exactly determining the nature of the combinations presented, it will be seen to be highly desirable, if we are to frame a correct clinical and regional diagnosis, to submit all such cases to a complete examination in accordance with some uniform and definite scheme. By adopting such a course we can accurately ascertain the nature of the disabilities from which the patient is suffering; and, at the same time, those who are not accustomed to make such examinations may the more readily assure themselves that no important points which ought to receive their attention have been accidentally passed over. It is unfortunately found only too often that in reports of different cases of an aphasic or of an amnesic type, the condition of the patient in reference to one, two, or perhaps several important points receives no mention at all; and thus for lack of information positive or negative in regard to this or that capacity, the record of the case is greatly diminished in value.

Such a scheme I now append. It is based upon the physiological views expressed in this work; and it is, I believe, sufficiently comprehensive to bring out all the facts of importance in reference to such cases, so far as our present knowledge extends. It is a developed and amplified form of a scheme that has been in use in University College Hospital for many years, and which I first published in 1886.[1]

SCHEMA FOR THE EXAMINATION OF APHASIC AND AMNESIC PATIENTS.

(1) Is the person right or left handed, and if the latter does he write with the right hand?

(2) What is the degree of paralysis of limbs, and especially of the hand and arm?

---

[1] "Paralyses: Cerebral, Bulbar, and Spinal," p. 125.

(3) Is he an educated person, much accustomed to read and to write?

The Activity of the Auditory Word Centre and Glosso-Kinæsthetic Centre, with their Afferent, Commissural, and Emissive Fibres.

(4) Is he deaf, and if so to what extent, and on one or both sides?

(5) Can he recognise ordinary sounds or noises?

(6) Does he comprehend speech—is he word-deaf, and if so, to what extent? Can he recognise his own name, or simple words when they are spelt letter by letter?

(7) Is his spontaneous speech good? if not, to what extent is it impaired? Does he make use of occasional or recurring utterances? if so, give examples. Does he make use of wrong words (paraphasia), or mere gibberish?

(8) Can he name the months of the year, or the days of the week? if not, can he name the letters of the alphabet, or count from 1 to 20, either by himself or after having been started?

(9) Can he repeat short sentences, or simple words uttered before him, and if so, with what degree of readiness or distinctness?

(10) Was he musical before his illness? and if so, can he now recognise different tunes?

(11) Can he sing airs, or the actual words of songs?

The Activity of the Visual Word Centre and Cheiro-Kinæsthetic Centre, with their Afferent, Commissural, and Emissive Fibres.

(12) Is his sight good or bad? Is there homonymous hemianopsia or optic neuritis?

(13) Does he recognise printed or written words—that is, is he word-blind? or can he read to himself with comprehension?

(14) If not, can he recognise individual letters or numerals?

(15) Can he read his own writing a quarter of an hour after it has been written?

(16) If not, can he recognise short words or letters by aid of kinæsthesis; that is, by tracing them over with his finger or a pencil, his movements, if necessary, being guided by another.

(17) Does he recognise common objects, and pictures of such objects?

(18) Does he understand pantomime and gestures?

(19) Can he write spontaneously, with correctness and freedom? if not, does he spell badly, omitting or transposing letters, or does he write wrong words (paragraphia)?

(20) Can he write the days of the week, the letters of the alphabet, numerals from 1 to 20, or his own name?

(21) Can he copy written words in writing, or from print into writing (transfer copying)?

(22) Can he copy numerals easily, or perform simple arithmetical calculations?

(23) Can he merely copy laboriously, stroke by stroke, as though he were copying Hebrew or some drawing?

(24) If he was a musician, can he now read music?

(25) Can he compose and write music?

(26) Can he copy music?

(27) Can he express his wants by pantomime and gesture?

The associated Activity of three Centres, with Commissures between the Auditory and Visual Word Centres, as well as one or other set of Afferent and Emissive Fibres.

(28) Can he read aloud? Does he do it well or ill? and if the latter, in what respect? Does he mispronounce words, interpolate wrong words, or utter mere jargon?

(29) Can he name at sight words, letters, or numerals?

(30) Can he name at sight common objects?[1]

(31) Can he point to common objects whose names he hears?[1]

(32) Can he write from dictation freely, or only with many mistakes? and if the latter, with what kind of mistakes?

(33) Can he write from dictation individual letters or numerals?

(34) If a musician, can he play upon any musical instrument?

In framing these questions I have had in view the testing of the degree of integrity of speech functions, and not that of processes of perception in general. Thus no reference has been made to the patient's ability to recognise objects by touch, taste, and smell, and to respond thereto in various ways, as such abilities or disabilities cannot, in the great majority of persons, have any bearing upon the elucidation of their speech defects. It is true that if we had to do with congenitally blind persons, the importance of the integrity of the sense of touch, and of the commissural relations of its centre with other sensory centres would then be great.[2] But that is an additional complication which does not now require

---

[1] Here we may have to do with commissures between the general visual centre (rather than the visual word centre) and the auditory word centre.

[2] See also what has been said concerning "Touch Paralysis" on p. 213.

to be considered. It need not form part of the examination of any ordinary aphasic person.

These remarks are made because in a scheme for the examination of aphasic patients given by Allen Starr,[1] and also in one which is an elaboration of the same published by J. T. Eskridge,[2] the first point they mention for investigation is "the power of recognising objects seen, heard, felt, tasted, or smelt, and their uses." Again, Eskridge includes two enquiries which can, I think, lead to no appreciable result; one of them being as to "the power to call up mentally the sound of a note, figure, letter or word," presumably in a word-deaf person; while the other is as to "the power to call up mentally the appearance of an object, a figure, note, letter, or word, when word-blind." It seems to me that very few patients of the kind indicated could be made to comprehend what was wanted, and even if this could be done, the enquirer would have no means of satisfying himself whether the replies made concerning such merely subjective capacities were correct or not.

Supposing that the patient has been thoroughly examined, and that his various capacities and incapacities have been revealed, the observer then has to endeavour to interpret them, or in other words, to come to a **Clinical Diagnosis**. He has to say, that is, what more or less simple form of speech defect exists, or what combination of them is probably present in the patient whom he has just examined.

With the view of helping him to arrive at this diagnosis, it will be well to give a brief differential characterisation of the principal known conditions commonly included under the name of aphasia or amnesia, or which may be mistaken for them. In doing this I shall name no more characters under each of the forms than may suffice to differentiate it from the others; and shall at the same time indicate briefly the probable anatomical site and nature of the damage, so as to connect each clinical grouping of symptoms as far as may be possible with a *regional diagnosis*.

---

[1] *Brain*, vol. xii., 1890, p. 95.
[2] *The Medical News*, June 6, 1896.

The Clinical Characteristics, together with the Sites in which Lesions are to be looked for, of Different Forms of Speech Defect.

Name of Speech Defect.	Its Clinical Characters.	Site and Nature of Lesion.
1. ANARTHRIA.	Defective power of articulation—speech more or less unintelligible. Mind clear—understands all that is said to him. Can express himself freely by writing, when hand not paralysed. Will attempt to repeat any word when bidden to do so. There is commonly difficulty in deglutition, and there may be more or less bilateral paralysis of limbs.[1]	*Damage to Bulbar Speech Centres.*
2. APHEMIA.	(a) *Complete.* Patient is absolutely dumb. Mind clear and understands everything said to him. Writes freely and correctly when not paralysed on the right side (Cases xi., xii.). (b) *Incomplete.* Articulation more or less defective, but no tendency to use wrong words. Understands everything that is said to him. Will attempt to repeat any word when bidden to do so. Can write freely and correctly, when not paralysed on the right side.	*Damage to Efferent Fibres from Broca's Centre.*
3. APHASIA.	(a) Patient may be completely dumb. Mind clear, and understands everything said to him. Will not attempt to repeat any word. Can write freely and correctly, when not paralysed on the right side.	*Due either* (1) *to Functional Defects in both Third Frontal Convolutions (Hysterical Mutism); or* (2) *to an Organic Lesion absolutely limited to Broca's Centre (Case* xxii.*).*

[1] These cases are common and well known. References to illustrative cases will only be given where the types of speech defect are less common and less generally recognised.

Name of Speech Defect.	Its Clinical Characters.	Site and Nature of Lesion.
	(*b*) Speech represented only by utterance of some two or three mere sounds or words (recurring or occasional utterances) more or less indiscriminately. Understands everything that is said to him. Will not attempt to repeat any word when bidden.	*Destruction (1) of Broca's Centre, or (2) of Audito-Kinæsthetic Commissure (Cases* lxxxv.-lxxxvii.).
4. AGRAPHIA.[1]	(1) May be able to write fairly well (Cases xxi., xxiii.). (2) Most frequently cannot write with right hand. This may be due to paralysis of the right hand, or owing to co-existing damage to the visuo-kinæsthetic commissure or to a separate cheiro-kinæsthetic centre (if the latter exists).	
5. AMNESIA VERBALIS.	(*a*) *Slight.* Difficulty in speaking, from inability to recall nouns, with more or less use of wrong words (paraphasia). Often not paralysed in right hand, and then shows amnesic and paraphasic defects in writing. Understands all that is said to him. Can almost always repeat words when bidden.	*Lowered Excitability of Auditory Word Centre.*
	(*b*) *Severe.* Inability to recall words so great that (apart from mere counting or repeating the alphabet) not more than five or six consecutive words can be spoken. Ability to express himself in writing may be almost equally defective. Can almost always repeat words when bidden, and is often able to read aloud well and fluently (Cases xlv.-xlviii.).	
6. WORD-DEAFNESS.	(*a*) Is unable to comprehend speech. Is speechless himself or paraphasic.	*Destruction of Auditory Word Centre.*

[1] This defect may also exist alone, either from one of the causes now mentioned or from damage to the *Audito-Visual Commissure* (see No. 9a).

Name of Speech Defect.	Its Clinical Characters.	Site and Nature of Lesion.
	Unable to repeat words. Is unable to write either from dictation or spontaneously.	
	(b)[1] Is unable to comprehend speech. Is speechless himself or paraphasic. Is able to repeat words. Is unable to write from dictation, but may be able to do so spontaneously.	*Partial isolation of Auditory Word Centre in a "visual," but afferent fibres and Audito - Kinæsthetic Commissure intact. (Case lix.).*
7. PURE WORD-DEAFNESS.	Is unable to comprehend speech. Speaks himself well and freely. Cannot repeat words. Cannot write from dictation, but is able to write spontaneously.	*Isolation of left Auditory Word Centre from all afferent impressions, both direct and by way of Corpus Callosum (Case lviii.); or, Destruction of both Auditory Word Centres in a strong "visual" (Case lvii.).*
8. WORD-BLINDNESS.	(a) Is unable to comprehend written or printed words. Speech good or only slightly paraphasic. Complete agraphia — cannot write spontaneously or do transfer copying.	*Destruction of the Visual Word Centre (Cases xxxii. and xxxiii.).*
	(b)[1] Is unable to comprehend written or printed words, but may be able to read aloud. Speech good or only slightly paraphasic. May be unable or able to write spontaneously. Should be able to do transfer copying.	*Partial isolation of Visual Word Centre in an "auditive," but afferent fibres and Visuo-Kinæsthetic Commissure intact (Case lxiii.).*
9. PURE WORD-BLINDNESS.	Is unable to comprehend written or printed words. Speech good or only slightly paraphasic.	*Isolation of left Visual Word Centre from all afferent impressions, both direct*

[1] This is a very rare form, and what is said as to the site and nature of the lesion is merely conjectural. Similar remarks apply to No. 8 (b).

Name of Speech Defect.	Its Clinical Characters.	Site and Nature of Lesion.
	Is able to write from dictation and spontaneously. Can do transfer copying.[1] Cannot read own writing after interval, except by revival of kinæsthetic writing impressions.	and by way of Corpus Callosum (Cases lxvii., lxviii., lxix.). (Occipital Type); or Destruction of the left Visual Word Centre in a strong "auditive" much accustomed to write (Cases (?) lxiv., lxv., lxvi.). (Parietal Type).
9a. AGRAPHIA.	Inability to write from dictation or spontaneously. Can do transfer copying. Can read aloud well, and understands what he reads. Is able to speak fluently.	Destruction of the Audito-Visual Commissure. (Cases lxxxix., xc., xci.).
10. COMMISSURAL AMNESIA.	Inability to write from dictation or spontaneously, though ability to copy writing may be preserved. Inability to read aloud. Inability to name at sight.[2] Ability to understand all that is said. Ability to comprehend what he reads. Ability to speak more or less well spontaneously. Ability to repeat words when bidden.	Destruction of both Commissures between the Auditory and the Visual Word Centres. (Cases xcii., xciii., xciv.).

It happens not unfrequently that cases of speech defect are met with which do not admit of a very definite diagnosis. Such cases often belong to one or other of two categories.

Some of them are complicated cases owing (*a*) to the existence of two or more lesions, and perhaps with involvement of both cerebral hemispheres; others (*b*) are complicated and indefinite because though there is only one lesion, this is a large one, involving two or more centres, together with

---

[1] In a case of the "parietal type" transfer copying would probably not be possible, and the individual might not be able to write spontaneously. For other possible differences see p. 202.

[2] Naming at sight may be brought about directly from the general visual centre to the auditory word centre, by means of the occipito-temporal commissure, as indicated on p. 212.

several important commissural or associational tracts. There must be sites in which several of these tracts either cross or come into such close relations with one another that they may be damaged simultaneously, and thus give rise to considerable mental degradation, and yet perhaps of a kind not easy to specify. In some of such cases we meet with "intoxication" by words or letters, "echolalia," or other evidences of ill-defined mental impairment.

The Pathological Diagnosis.—The pathological diagnosis in speech defects, if fully considered, would occupy considerable space. This, however, seems scarcely necessary here, since it must be settled by reference to the same set of general considerations as would guide us if we were considering the pathological diagnosis in a simple case of hemiplegia.

The first problem that presents itself is to settle, as far as we are able, whether the defect is due to some so-called functional perturbation, or whether it is the result of actual organic disease. For the decision of this question, and also for the diagnosis of the nature of the organic disease when this is presumed to exist, we must obtain information upon the following points:—The age, sex, and family history of the patient; his past personal history (including diseases and accidents experienced), occupation, habits, and mode of life; his state of nutrition generally, and especially the condition of his heart and arterial system; the date of the first signs of his present illness, together with its rate and mode of increase; the precise nature of the signs and symptoms presented at the onset of the patient's malady and at subsequent stages. Information concerning such facts must be supplemented by a thorough examination of the patient's condition in regard to nervous functions at the time of his coming under observation (which will often be not during the first stages of his illness). All such data have to be taken into account; and then the net total of resulting information has to be mentally compared with what we know concerning the modes of onset of the several possible pathological or functional causes of such kinds of disease, in order that we may judge which of them has been, or has most probably been, operative in the case under consideration.

The sex and age of the patient may be of much importance in guiding us to a diagnosis that certain speech defects are of functional origin. Still, it must be remembered that it is not always young females who are the subjects of such defects; they show themselves also in males, and even in middle-aged males. A neurotic constitution with the history of some traumatism or mental shock, some overwork or worry, the taking of some vegetable poison, the occurrence of an attack of Jacksonian epilepsy or of migraine, or convalescence from some acute specific disease—one or other of these events is to be looked for as amongst the most common precursors of functional speech defects. It should also be borne in mind that they show themselves principally either in the form of "hysterical mutism," or else as an aphasic condition of such a kind as to indicate disturbance in the functions of Broca's centre.

In reference to the organic lesions causative of speech defects, by far the most common are those induced by occlusions of the left middle cerebral artery or of some of its cortical branches, giving rise to a great variety of clinical conditions, such as have been enumerated in Chapter xiv.

These occlusions are due partly to embolisms occurring in children or young people suffering from chronic valvular disease; and partly to thromboses which may occur at any period of life. Thus, in childhood and in early adult life thromboses may be favoured by certain alterations in the blood consequent upon acute specific diseases, or, later on, in association with the puerperal state—especially when at either of these periods the patient is suffering from a weak and irregular cardiac action. While in elderly persons, or even in young adults when suffering from syphilis, we have to do rather with local degenerative changes in the cerebral arteries, conjoined with weak and irregular cardiac action, as the most potent causes of thrombosis.

Where the occlusion is due to embolism of the middle cerebral itself, the onset of the attack is abrupt; but where it is caused by thrombosis we may look for a more gradual and progressive mode of commencement. In either case there may or may not be convulsions with the initiation of the symptoms. There is usually, if any loss of consciousness, stupor rather than coma; and patients may complain for some hours or even for a day or two of pain in the head on the side affected, with possibly some slight tenderness to percussion.

Where the occlusion only affects the fourth cortical branch, the general effects are comparatively slight, though the speech troubles and mental disturbance may be most marked—especially during the first few days.

Hæmorrhages occur very much less frequently as causes of speech defects, except in association with marked hemiplegia. The onset of symptoms is then more or less abrupt; the patients are mostly beyond middle age; and they may be the subjects of granular kidney, with degenerated arteries and an hypertrophied left ventricle. Though cerebral hæmorrhages are most frequently deep-seated they may also extend into and involve certain cortical areas. At other times, though more rarely, small hæmorrhages may be limited to cortical regions corresponding with one or other of the word centres; or small meningeal hæmorrhages may press thereupon.

In other cases new growths may be the pathological causes of this or that form of speech defect; but here we should probably have to do with a much more chronic history of cerebral troubles—the symptoms being of a progressive nature, and with the accompaniment possibly of optic neuritis, more or less localised headache, and vomiting.

Another cause of speech trouble of slow and progressive onset is the chronic degenerative change that takes place in bulbar motor centres in the condition known as glosso-labio-laryngeal paralysis. Here we have to do with a gradually increasing anarthria associated with dysphagia, but with an absence of the general symptoms indicative of tumour.

Traumatisms, again, are frequent causes of speech defects, and then they are mostly of rapid onset. Under this head we may have to do with gun-shot wounds, or fractures of the skull, resulting in depression of bone, localised hæmorrhages, abscesses, or inflammation with thickening and adhesion of membranes, as other causes of well-marked speech defects. In all these cases the history and nature of the traumatism goes far towards settling the question as to the pathological cause of any speech trouble that may have followed thereupon.

CHAPTER XVI.

PROGNOSIS IN SPEECH DEFECTS. CAPACITY FOR EXERCISING CIVIL RIGHTS.

THIS is a difficult subject to deal with, looking to the great variety of speech defects and to the great diversity in the brain lesions with which they may be associated.

It is sometimes a question whether a patient suffering from some speech defect will live or die, but in grave cases of this type the danger to life never depends upon the mere lesion in the speech centres—it would always be due to the lesion having a much wider area; so that in such cases the general prognosis (as to life or death) would depend upon the degree of severity of the general symptoms associated with the speech defects.

For the most part, however, the question in prognosis with which we have to deal, is not as to life or death, but as to the chances of recovery from, or of increase of, the particular speech defects or mental disabilities from which the patient may be suffering.

Speaking generally, it will be found that the prospects of recovery from speech defects, though they may be much influenced by age—that is, favoured by the fact of the early age or youth, and lessened by the advanced age of the patient—must depend upon two sets of considerations. These are: (1) the chances of the occurrence of what may be termed functional restitution; or (2) of functional compensation, taking place in the patient with whom we are concerned, either singly or in combination. A few words must, therefore be said, in the first place, concerning each of these modes of recovery.

(1) **Functional Restitution.**—By this phrase I would refer to the chances of the lesion, causing the defect in the patient with whom we have to do, undergoing some diminution, in one or other way, so as to permit of a resumption of functional activity in the parts of the brain that have been damaged.

Looking to the fact of the great frequency with which

softenings of brain tissue due to occlusions of cerebral vessels, either by embolism or thrombosis, figure among the causes of speech defects, much will depend in these cases (and therefore in a large number of cases) upon the degree of anastomosis between the cerebral vessels in the brain region affected, and in the particular individual—which, as already stated (p. 305), is subject to much variation. Apart from individual peculiarities, this will often depend upon the question whether the speech region affected occupies the centre of a rather large area whose vascular supply has been at first occluded, or more upon the confines of such an area. Thus, I think it will be found generally that the visual word centre is nearer the centre, and the auditory word centre nearer the confines of the territory of the fourth cortical branch of the middle cerebral artery—which may, in part at least, account for the fact that word-deafness often clears up earlier than word-blindness where both have at first coexisted. In cases of arterial occlusion, therefore, whilst the danger arising from the spread of thrombosis, and the consequent extension of the area of brain tissue cut off from its blood supply, has to be considered, on the other hand, in recent cases where the symptoms have been pretty stationary for a few days, we may look to the chances of their diminution by the gradual establishment of a collateral circulation, and the consequent narrowing of the area of brain tissue actually deprived of its proper blood supply.

Again, in the case of cortical or meningeal hæmorrhages, speech defects are often due to mere pressure upon, rather than to destruction of, centres or commissural fibres; and the same may hold good for many deeper seated hæmorrhages. In cases of this kind, therefore, when injurious pressure is diminished, either by gradual absorption (and consequent diminution in bulk) of some of the elements of the blood effused, or by the intervention of the surgeon, we may witness a gradual or a much more rapid improvement in the patient's condition. Moreover, where we have to do with specific new growths, a diminution in bulk under the influence of judicious treatment may also be brought about, and with it a gradual disappearance of symptoms often occurs.

In the ordinary cases of speech defect of acute onset (not due to traumatisms) the improvement which is so commonly met with during the first few days or even weeks, is commonly due in part to a restitution of functional activity, more or less rapidly brought about, owing to one or more of the defectively acting speech centres gradually recovering from the sort of inhibitory shock that may have been caused by a contiguous brain lesion; or it may be due to the vascular supply of such centres, at first cut off, being gradually restored; or again, owing to the gradual diminution of some injurious pressure that was at first exercised upon one or other of the speech centres, or upon the commissural fibres by which they are functionally associated. Such causes of improvement may be in operation singly or conjointly.

For these various reasons it behoves us to be cautious, and not to express any positive opinions of too gloomy a character until some time has elapsed after the onset of the patient's malady, in order that we may ascertain how soon, and to what extent, if any, improvement begins to set in. In some cases, indeed, the speech defect may clear up altogether in the course of a few weeks or months; but in others, after a time, the improvement ceases, or almost ceases, and such a patient may remain for an indefinite period, often for many years, in substantially the same condition—either aphemic, aphasic, or amnesic in one or other of its modes. It not unfrequently happens that what was, at first, a case of complete aphasia may, after this partial recovery, remain as one of amnesia, or of amnesia and aphemia combined.

In the so-called "functional varieties" of aphasia, favourable results are to be expected after longer or shorter intervals, whether we have to do with cases occurring as a result of worry with over application to business or literary work (and mostly without the coexistence of any hemiplegia), or with cases resulting from some of the other numerous causes that have been referred to in Chapter vii. In most of these varieties recovery may be looked for in the course of a few days, or at most, in a week or two. Cases, however, of hysterical mutism may last for months or years without change, and then may suddenly, or at times more gradually, disappear under the influence of some form of treatment, some strong emotion, or even after the occurrence of an attack of convulsions.

In cases of insanity where pseudo-mutism supervenes, this may last for a long series of years; or it may be broken for several brief periods in the course of a long series of years, as was shown in cases that have been previously referred to.

(2) **Functional Compensation.**—When improvement goes on after months, or even years, we must for the most part consider that this is brought about by an entirely different method—that is, not so much by changes taking place at the seat of lesion, as by a process of functional compensation, mostly brought about by the undamaged hemisphere gradually taking on new or greatly developed functions, of a kind similar

to those that have been annulled in the hemisphere previously taking the lead in regard to speech functions. Many of the various modes by which this may be brought about I have already referred to pretty fully in Chapter xii., as well as on pages 167 and 260.

For the full investigation of this subject we should require to know the way in which age would influence the curability of the different kinds of speech defects, but for this no sufficient materials are as yet available. Then, again, the relative degrees of curability by compensation of the different kinds of speech defect, *inter se* would have to be considered. Here, however, even if the cases were numerous enough there would be the difficulty of deciding whether any, or how much of the amelioration that occurred was due to functional "restitution" rather than to "compensation." Further, there would be the complications arising from the different natural endowments of individuals—that is, whether they are "visuals" or "auditives;" whether their nerve tissues are more or less plastic; and as to the different degrees in which centres in the right hemisphere are in a fit state and apt to take on new functions. All these inherent individual possibilities of variation may lead to different degrees of curability, even where the extent and site of the lesions are as nearly as possible similar, and where the amount of judicious treatment to which the individuals have been subjected could also be shown to be fairly equal.

The difficulties in arriving at a conclusion as to the relative degrees of curability of the different forms of speech defect by a process of functional compensation will thus be seen to be extreme; so that with the present paucity of available cases it would be quite useless to attempt to follow up this subject.

It only remains, therefore, to quote a few illustrations, or to refer to those that have been already cited, of the mode and degree of recovery that may be met with in the different forms of speech trouble, so as to give some information as to how much improvement may be looked for, or as to the forms in which recovery may possibly be met with.

As regards the influence of *age* on curability it is probably in the main true (as commonly stated) that defects of speech occurring in early life stand a much better chance of being recovered from than defects of the same kind occurring late in life. Yet even this is a rule by no means free from exceptions, as we shall presently see.

Barlow's case (Case xv.) is often referred to as a proof that the right third frontal convolution may, in a young subject, completely and rapidly, take over the functions formerly carried on by the left third frontal convolution. This, however, seems to me open to very grave doubt. This patient was ten years old when he was seized with right hemiplegia and "aphasia." But from this attack, it is said, "he had apparently recovered at the end of a month, the power of speaking having returned on the tenth day." Three months later the second attack occurred, due to the lesion on the right side of the brain. Now I very much doubt whether complete transference of function could possibly have occurred in the short period of ten days. We are not told whether the boy was left-handed or not. The right hemisphere might have been the leading hemisphere for speech, and the first lesion on the left side may have merely occasioned some functional disability in the right centre, from which in a very short time he recovered. With the occurrence of the second lesion, however, both third frontal convolutions would have been damaged, and thus the "pseudo bulbar symptoms" would have been produced.

This case stands in very remarkable contrast to one published by Ange Duval, of which an abstract is given by Bateman[1] in the following terms.

CASE C.—A little boy, aged 5 years, fell from a window upon his forehead; the result of the fall being a fracture of the frontal bone on the left side. The intelligence of this child continued unaffected, and there was no paralysis, but he never uttered another articulate sound.

This boy was accidentally drowned thirteen months after his fall, when an examination of the encephalon disclosed a cyst of the size of a walnut which was full of serum, and was evidently the result of a former contusion of the left frontal lobe. This cyst was situated principally in the third left frontal convolution.

It is very strange that no recovery of speech should have taken place in this little boy during a period of thirteen months; and there is also the difficulty as to why the boy should have remained absolutely mute during the whole of this period. It suggests the possibility that there may have been a functional complication (affecting the right third frontal convolution), caused by the shock of the injury, and persisting during the short remainder of the boy's life, as in Case xxxv., where mutism persisted, after an injury, also for a period of thirteen months, and as in a case recorded later on (Case cvii.), in which mutism persisted after a traumatism for no less than ten years. Such an explanation would account not only for the mutism, but also for the absence of recovery by compensation, when the conditions were apparently most favourable for such an occurrence.

On the other hand, there are some remarkable cases on

[1] "On Aphasia," 2nd. ed., 1890, p. 196.

record in which, in persons of adult or even of advanced age, a pretty complete recovery from aphemic or aphasic speech defects has taken place.

It may be noted here that recovery from aphemia is generally easier than recovery from aphasia, partly because no speech centres are damaged in the former case, and partly because the third right frontal convolution may apparently be more easily trained by the third left frontal than by the left or right auditory word centre—that is, the process represented after a fashion in Fig. 23, is more easily brought about than that indicated either in Fig. 24 or Fig. 25, though either of these must also be considered as a possible mode of recovery for some persons.

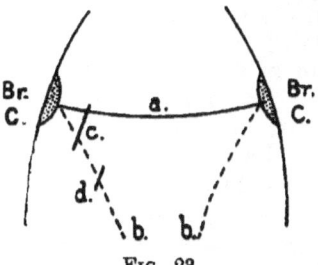

Fig. 23.

Fig. 23.—Diagram showing possible mode of recovery from Aphemia when lesion occurs in some part of the pyramidal fibres (*b*), as, for instance, in the position of line marked *d*. When lesion is in position of line *c*, recovery can only take place as in Aphasia, that is, in the same way as it must occur when Broca's Centre (Br. C.) is destroyed.

As instances of recovery from aphemia where a definite organic lesion was subsequently found at a necropsy in such a situation as to account for the original malady, I may refer to three cases quoted in an earlier chapter. One of them is Case iv., recorded by Dieulafoy, which was that of a man, aged 44 years, who was aphemic, but completely recovered his speech in about one month; though at his death, two months later, a lesion was found which was said to have cut across the left geniculate fasciculus. The rapidity of recovery here is very difficult to understand. Then, again, Case x. is that of a woman, aged 23 years, who was said to have been completely aphasic for nine months, but who then began to be able to repeat words. During the following two years she completely recovered her power of speaking, so that Déjerine says in regard to her condition three years from the date of the attack: "Now it is difficult to believe when speaking to her that one has to do with an old aphasic, since she pronounces easily and without any trouble all the words that she employs." At the necropsy the third frontal convolution was found to be healthy, but a lesion involved the left geniculate fasciculus.

Another case, just as striking, is our Case xiv., recorded by Pitres, of a woman, aged 65 years, who after the attack did not utter a single word for fifteen days, and in the following six months only certain monosyllables, though after this her power of speaking very gradually returned. About two years after the onset of her attack, it is said, "she could express her thoughts by speech, but she often hesitated before pronouncing certain words, and was obliged to make a real effort in order to do so." But at the necropsy, twelve months later, the third left frontal convolution was found to be healthy, while a focus of softening existed in the left centrum ovale which must have involved the geniculate fasciculus.

In the same Chapter vii., in which these cases are recorded, details will be found of a few other cases where, although the lesion was unilateral, death took place before there was time for recovery to occur. But where the lesion is bilateral (as in Case xv.), and we have to do with "pseudo bulbar symptoms," recovery is not to be expected by way of compensation—though it is always possible, when the lesion or lesions are not very complete, that a certain degree of improvement may take place at the damaged sites, permitting the occurrence of some amount of "functional restitution."

Much the same thing has to be said concerning the incurability by compensation of dysarthric or anarthric defects, which, apart from stammering and a few other allied troubles, are due to organic lesions of the bulbar motor centres for speech. The centres here situated, on each side of the middle line, require to be called into action simultaneously for the production of articulate speech, so that with lesions in this region no recovery by compensation is possible—unilateral lesions being capable of producing effects almost as marked and as lasting as bilateral lesions.

Although recovery seems, as I have already said, to be more readily brought about in aphemia than in aphasia, it must be borne in mind that this process is undoubtedly capable of occurring in some of these latter cases, even when Broca's convolution has been completely destroyed. The observation of cases of well marked aphasia, followed after variable periods by gradual recovery of speech, and later still by necropsies revealing destruction of the third left frontal convolution, has clearly established this.

A positive prognosis in speech defects must, as I have said, never be too hurriedly given, however gloomy the outlook may seem for the first week or two. Thus, a short time since

I had under my care in the National Hospital for the Paralysed and Epileptic, a middle-aged woman who was admitted two or three days after the onset of an apoplectic attack which left her speechless and completely paralysed on the right side. She remained in this condition for just three weeks, never having spoken a single word and the paralysis of the limbs remaining absolute, when, whilst a poultice was being applied to her back on account of an attack of bronchitis, she suddenly said, with perfect distinctness, "What is the use of it all?" and in the course of the next day or two she began to speak quite freely, and as well as if there had never been any suspension of her powers in this direction. She had previously been greatly depressed and very emotional, but this was the most unlikely sort of case in which to meet with an arrest of speech due to mere functional causes. The apoplectic attack, as the history and subsequent progress of the case showed, was clearly due to grave organic disease.

Where aphasia occurs in a young subject, and especially in one who is ambi-dextrous, the conditions may be said to be most favourable for the re-acquirement of speech by a process of compensation. Of this we have an excellent example in Wadham's case, which I have previously reported (Case xxiv.).

It is, however, not only in young subjects that recovery after the complete destruction of Broca's centre may occur. Many years ago Batty Tuke and Fraser[1] published an important case which they, at that time, regarded as adverse to Broca's views, but which may now be cited as a good instance of recovery from aphasia by a process of functional compensation. The following are a few details concerning this case.

CASE CI.—A woman, aged 54 years, was admitted to the Fife and Kinross Lunatic Asylum on December 14, 1868. Eleven years previously, whilst working in a field, she suddenly fell down and became unconscious. Her friends affirmed that she remained insensible "for some weeks," but no evidence of the existence of paralysis could be elicited from them. Complete speechlessness was present for a time. She remained weak-minded as well as aphasic from the day of the attack, but, after some years, had gradually and to a considerable extent regained her speech.

She had become actually insane only four weeks previous to her admission to the asylum, whilst after her admission her mental condition rapidly improved and she became almost sane.

"During the whole period of her residence two peculiarities in her speech were observed — a thickness of articulation resembling that of general paralysis, and a hesitancy when about to name anything, the latter increasing very much some months previous to her death." She forgot nouns in a striking way. For example "she would say, 'Give me a glass of——.' If asked if it was water, she said 'No.' Wine? 'No.' Whisky? 'Yes,

[1] *Journ. of Ment. Science*, 1872, p. 46.

whisky.' Never did she hesitate to articulate the word when she heard it." She could read, but was never observed to write. She died from caries of the vertebræ.

The necropsy showed that the old apoplectic extravasation which had occurred eleven years before her death had resulted in the complete destruction of the posterior part of the left third frontal convolution, the posterior part of the second left frontal, the foot of the ascending frontal and part of the foot of the ascending parietal. (The exact extent of the lesion is shown in a plate.)

It will be observed that after she recovered speech to a considerable extent, the defect which remained was of an amnesic rather than of an aphasic type, and precisely the same thing holds good in regard to another case of aphasia

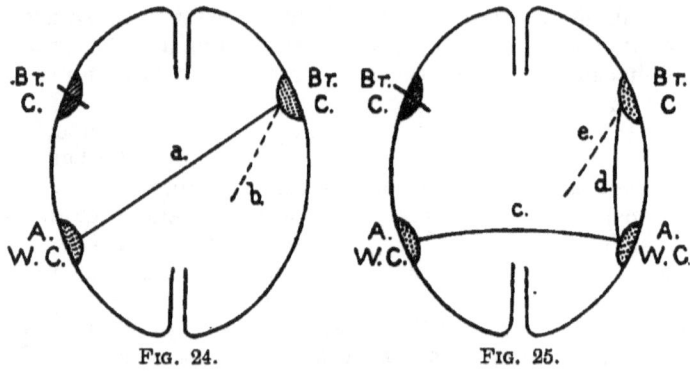

FIG. 24.—One possible mode of recovery from Aphemia or Aphasia. *a*. Commissure between the left Auditory Word Centre and Broca's Centre on the right side; *b*. Outgoing, or "Internuncial fibres" from latter Centre.

FIG. 25.—Another possible mode of recovery from Aphemia, or Aphasia. *c*. Commissure between the two Auditory Word Centres; *d*. Commissure between the right Auditory Word Centre and Broca's Centre on the right side; *e*. Internuncial fibres from the latter Centre.

in which recovery must have taken place by a process of compensation. I allude to a case recorded by Bernard of which I have given an abstract as Case xcvi., and in which the extent of the lesion is also shown by a figure. In this last case there was some slight damage in the neighbourhood of the auditory word centre, though in the other there was none. Of course the amnesic defect in this case might have been due to general degenerative changes (the woman being 54 years of age) involving the auditory word centre as well as other regions. Otherwise it might possibly be explained by the process of compensation having taken place in the manner indicated in Fig. 25, rather than by that indicated by Fig. 24, and if so the amnesia might have been due to the imperfect way in which the right auditory word centre had been able to take on its increased functional activity.

Other instances of recovery from aphasia by a process of compensation will be found recorded in Cases xxiv. and xcvi.

Another case worth mentioning here, although the details are very scanty and there was no necropsy to show the position and extent of the lesion, was long ago published by Banks.[1] This case too, after the early days of the speech trouble, seems to have been more of an amnesic than of an aphasic type.

CASE CII.—A gentleman, aged 54, had eight years previously been rendered hemiplegic on the right side and aphasic. At first the loss of speech was complete, but after twelve days he could say a few words.

Before his attack he was a ripe scholar, and had taken much pleasure in reading the best classical authors; but when his stock of words had increased, so as to enable him to converse a little on ordinary subjects, it was observed that his memory had quite failed him in regard to Greek and Latin.

For six years he continued without improving to any considerable extent, but still he was gradually acquiring new words. For the next two years his progress was more rapid; he laboured hard and almost learned over again all that he had forgotten, so as to be able once more to read his old favourite authors.

This case is interesting and encouraging from a prognostic point of view when we bear in mind the comparatively advanced age of the patient, and the very late period at which any notable improvement set in. Of course, as a rule, the longer the aphasia or other speech defect has lasted without any notable improvement, the worse is the prognosis. It is therefore well not to lose sight of exceptional cases of this sort.

A slow recovery by compensation also takes place in some cases of word-deafness and word-blindness where the left auditory and visual word centres have been destroyed. An instance of this has been recorded by Ball,[2] of which the following are a few details.

CASE CIII.—A man, aged 52, had been exposed to cold and thoroughly chilled. Some days after he was found on February 1, 1879, paralysed on the right side, and muttering unintelligibly.

After six weeks there was only slight paresis on the right side, together with a complete aphasia and agraphia. But gradually he began to talk, and little by little extended his vocabulary till, in the summer of 1879, he could succeed in expressing his ideas fairly well. It was sometimes not noticeable that he was aphasic at all. He did not substitute wrong words much in speaking.

He had partial word-deafness from the commencement; and later on he said, " The words I can't pronounce are the words I can't hear."

[1] *Dublin Quart. Jrnl.*, February, 1865.
[2] Quoted by Amidon in *New York Med. Jrnl.*, 1885, Case 4, p. 114.

At first also he could not read, but he re-acquired this power partly. Still there were always some long words in every sentence that he could only understand from the context. After several months he could copy and even write short sentences of his own composition, though with great difficulty. Writing from dictation was impossible. In talking or writing he had a great tendency to substitute numerals for other words.

In March 1880, he had another cerebral attack, beginning with a noise in the head like a pistol shot, and immediately he seemed to hear some one talking over his right shoulder, and turned to see who was addressing him.

In May 1880, he had another slight cerebral attack during which he was blind for two hours, and half an hour after the expiration of this time he died.

At the necropsy an area of softening was found in the left hemisphere, which involved the inferior parietal lobule and the posterior part of the first temporal convolution.

Probably neither the word-blindness nor the word-deafness was very complete in this case, and it will be observed that the visual word centre was only partially destroyed; still, looking to the extent of the recovery it seems almost certain that it must have been by way of functional compensation, rather than by a process of functional restitution.

Some years since Lichtheim[1] called attention to "the usually rapid disappearance of word-deafness, and the greater persistence of troubles connected with written language." There can be no doubt that it is often seen in cases where word-deafness and word-blindness have been simultaneously produced by an occlusion of the fourth cortical branch of the middle cerebral artery, that the word-deafness is recovered from more quickly and to a greater degree than the word-blindness. This I have seen on many occasions, and I am inclined to think, as I have already intimated (p. 319), that here we have to do with relative degrees of recovery by functional restitution dependent upon the varying extent to which a collateral circulation is capable of being established in the auditory and in the visual word centres respectively, when both of them have at first been deprived of their vascular supply. But whether recovery from word-deafness is more rapid than recovery from word-blindness when it takes place by a process of functional compensation must, I think, be regarded as a more doubtful matter. I allude to this question here in order to point out that these two modes of recovery must be carefully distinguished from one another, and that I am now speaking to the latter mode of recovery only—that is, concerning cases of recovery in which the auditory and the visual word centres have been actually destroyed.

[1] *Brain*, 1885, p. 467.

Another instance of this is the very remarkable case of which I have published full details (Case xciv.), where the patient came under observation three months after the original apoplectic attack, continued under observation from time to time during the next eighteen years, and who, during the whole of this period, was neither word-deaf nor word-blind. Yet on examination of this patient's brain after death, we found that the whole of the left auditory and visual word centres as well as other contiguous parts had completely disappeared, as may be seen by reference to Fig. 14. The difficulties standing in the way of any satisfactory interpretation of this case are extreme, as I have previously pointed out. Still it must always be borne in mind that the powers and aptitudes of individuals vary. Thus destruction of the left auditory word centre most frequently either annuls or very greatly impairs the patient's power of speaking. Yet this is not invariably so; there are a few cases on record in which, where the necropsy has subsequently shown that destruction of the auditory word centre has been more or less complete, such individuals though word-deaf have nevertheless preserved their voluntary speech to a remarkable extent. I have previously referred to Amidon's Cases 1 and 2, as examples of this apparent anomaly (p. 203), and also to a still more remarkable case recorded by Pick (Case lvii.), in which voluntary speech was preserved even where the auditory word centre was destroyed in each hemisphere. There is also another case, comparable with those recorded by Amidon, which came under the observation of Hitzig, and is thus quoted by Mirallié.[1]

CASE CIV.—An old lady, supposed to be suffering from softening of the brain, was at a loss in speaking for a certain number of words, whilst she was also very slightly paraphasic. Nevertheless she was able to express herself so well that at a first examination no speech trouble might be noticed.

She was completely unable to understand what was said to her. But after a time, when her condition had somewhat improved, Hitzig says, "She took notice when one pronounced certain words, though I believe she did not understand them, but that she recognised rather by analogy the sound of what was uttered, looking to her previous experiences."

She had, however, very completely preserved her comprehension of music; she appreciated airs that were sung or whistled; and she herself sang and reproduced airs, though not always very correctly.

After a time she showed symptoms which pointed to the existence of a new focus of softening—this time in the right hemisphere.

At the necropsy an area of softening was found in the left hemisphere, occupying principally the temporal lobe and more especially the posterior two-thirds of the first temporal convolution. This old softening was

[1] *Loc. cit.*, p. 78.

probably the cause of her word-deafness. In the right hemisphere there was a symmetrical focus of recent date in the temporal lobe.

Now in these exceptional cases the power of voluntary speech has not been gradually re-acquired, on the contrary, it seems never to have been lost after the patients had had a brief period in which to recover from the immediate effects of the original brain lesion. In Pick's case with the double lesion it would appear that the patient must have been a strong visual, and therefore able to incite Broca's centre from the left visual word centre. In the other cases just referred

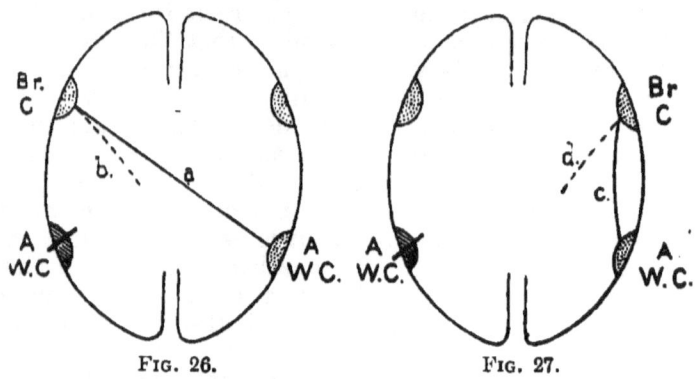

FIG. 26.—Diagram showing one possible mode of recovery from Word-Deafness, due to destruction of the left Auditory Word Centre. *a.* Commissure between the right Auditory Word Centre and Broca's Centre on left side; *b.* Internuncial fibres issuing from the latter Centre.

FIG. 27.—Diagram showing another possible mode of recovery from Word-Deafness caused in the same way. *c.* Commissure between the right Auditory Word Centre and Broca's Centre of the same side; *d.* Internuncial fibres issuing from the latter Centre.

to we need not resort to this hypothesis; it may be supposed that they were persons in whom the right auditory word centre had been naturally more active than usual, and whose incitations (after destruction of the left auditory word centre) had been sufficient of themselves to call into play Broca's centre in the left hemisphere (as in Fig. 26). It seems just possible that considerations of the same kind may be applicable in part for the explanation of my very puzzling case above referred to (Case xciv.).

In some instances, however, we have a simultaneous destruction of the left auditory word centre and of Broca's region, in which, nevertheless, not only has the word-deafness gradually passed away, but also the speechless condition that was for a long time present. Of this a very interesting example has been recorded by Bernard,[1] as occurring in one of Charcot's patients. The following are the essential details, so far as they could be gathered.

CASE CV.—A woman, aged 49 years, came under observation at the Salpêtrière, on August 26, 1883, suffering from hemiplegia with contracture of the right side, from whom the following history was obtained.
When she was six years old she was seized with this paralysis during the night—so that on awaking she found herself paralysed and speechless. For two years she could only pronounce the words "oui" and "non." Gradually, however, the power of speaking returned.
At the commencement of her illness, for a period which she cannot exactly define but which was certainly much shorter than that of her comparative speechlessness, she had been able to understand only a few of the words spoken to her, and these only after they had been several times repeated, and she looked intently at the persons addressing her. Those about her regarded her as deaf. Words were for her only indistinct sounds "like conversation in a crowd." She heard, however, all ordinary sounds that took place around her. She had previously been fond of music and liked to hear a certain military band play. But for a time after the commencement of her illness, when taken to hear this same band play she no longer appreciated it—she heard only a new noise, but one deprived of all musical characters.[2]
The results of an examination made on her admission to the Salpêtrière, revealed the following particulars:—She replied with facility and correctness to all questions put to her, both in regard to her actual condition, and her previous states of health. Her speech was, in fact, as correct and intelligible as if she had never been deprived of it. She could write a little with the left hand. She has always been able to read, and on examination she was found to read very well both print and writing. She perfectly comprehended all questions, and never answered them incorrectly.
She died in April, 1884, of an intercurrent affection, namely, obliteration by thrombosis of the abdominal aorta.
At the necropsy the right hemisphere of the brain was found to be quite normal, but in the left hemisphere Broca's region, as well as the whole of the island of Reil, the whole of the upper temporal convolution, and a part of the inferior parietal lobule were completely destroyed, as a consequence of an old focus of softening.

Here recovery was favoured by the early age of the patient, and it must have been produced by a complete transference of speech functions from the left to the right hemisphere. In the bringing of this about it was only natural—it was, in fact, inevitable—that there should have been (as this patient said there was) recovery first of all from the word-deafness. This could, of course, only have been caused by the gradual education of the right auditory word centre, and only after this had progressed to a certain extent would it be possible for this

---

[1] "De l'Aphasie," 1885, p. 159.
[2] It is interesting to note that in this case where there was tone-deafness the whole of the upper temporal convolution was found to be destroyed, while in the last case (Case civ.), where only the posterior two-thirds of this convolution were destroyed, the musical faculty was retained.

centre to aid in the development of the right third frontal convolution. Ultimately the conjoint activity of these two centres would permit the patient again to acquire the power of expressing her thoughts in articulate speech (after the manner indicated in Fig. 27).

I do not refer here to cases of recovery from word-deafness and word-blindness in which there has been no necropsy, because it is of great importance in dealing with a subject so complicated as that with which we are concerned in this work, that all important conclusions should rest only upon cases the exact nature of which, on their pathological side, has been ascertained. In the absence of such a check we are unable to discriminate between recovery by a process of functional restitution and one by functional compensation, nor can we tell to what extent mere functional disabilities (or such disabilities in the main) may have had to do with the production of many of the leading features of the morbid clinical condition that has been recorded. Such remarks apply especially to a case that has been published by Mantle,[1] in which recovery took place, after a rather prolonged period, from aphasia associated with word-deafness, word-blindness, and object-blindness. The case differs in several important respects in its clinical characteristics from those which are due to organic disease in its ordinary forms, that is, when the lesion is so extensive as to be capable of bringing about all the defects enumerated.

In regard to the important class of cases in which children are brought to us because they have never spoken at all, these will mostly be found to belong to one or other of two categories: either (a) the children have been liable to fits in infancy, or (b) the children are semi-demented or semi-idiotic.

It is important to bear in mind that some of the children belonging to the first of these classes may, as I have known, remain perfectly dumb up to the fourth, fifth, and even the sixth year, and then begin to talk, making rapid progress till all defect disappears. I have previously referred to two very remarkable cases of this kind (pp. 6 and 8), in which the children did not begin to speak till six and eight years old respectively, and then, after a very brief interval spoke quite well, though they had been untaught in the ordinary sense of the word.

In the second class of cases there is generally not much ground for hope, though occasionally marked improvement does take place when continuous efforts are made, and the attention of the child can be aroused. According to Bateman,[2]

---

[1] *Brit. Med. Jour.*, Feb. 6, 1886. [2] *Loc. cit.*, p. 280.

indeed, instances have occurred "in the Eastern Counties' Asylum for Idiots, where children, who for many years had passed for deaf and dumb, unexpectedly gave evidence of the possession of the power of speech. One boy, supposed to be a deaf-mute was heard one night to sing a chant which had been used at public worship, pronouncing the words distinctly, and giving the tune correctly. Another boy, also passing for a deaf-mute, broke into a violent passion on account of something on his slate being rubbed out, and demanded of another boy why he had done it." It is a pity that no further details are forthcoming concerning these cases, in explanation of doubts that naturally suggest themselves.

Two instances will, however, be given in the next chapter in which steady and long continued efforts of a judicious nature were successful in developing the faculty of speech, one case being that of a semi-idiotic boy over eight years of age, and the other that of a boy of weak intellect who had been completely dumb till nearly twelve years of age.

Capacity for Exercising Civil Rights.

It has been truly said by Bateman that "it is impossible to formulate any precise rules, as to how far an individual deprived of the power of speech is incapacitated for the full exercise and enjoyment of his civil rights—each case must be determined upon its own merits, and its own peculiar features."[1]

This question arises not unfrequently in regard to the capacity of an aphasic patient to make a will, or to manage his own affairs.

The same question frequently presents itself in regard to persons suffering from cerebral lesions or diseases of different kinds, when speech has not been specially interfered with. That is, it often becomes a question whether (a) the cerebral disease has left the person with a sufficiently sound memory and understanding to make a valid will, or to manage his own affairs. And this general question always remains one that has also to be considered in the cases with which we are now concerned—that is, where the brain lesion has produced aphasia or some other defect in the power of expressing thought, or of understanding the signs by which the thoughts of others are communicated. Only in these latter cases we have to

---

[1] "On Aphasia," 2nd ed., 1890, p. 311. In the chapter from which this passage is taken, Bateman cites a few instances illustrating the circumstances under which wills have been executed in this class of cases.

take into consideration an additional question, namely (*b*), the extent to which the diminished power of expressing himself, or of understanding the spoken or written language of others, of itself (and apart from (*a*) the general injury to mental power caused by brain lesions that may exist beyond the confines of the different word centres) renders him incapable of exercising these civil rights.

It may be, and often is, extremely difficult or even impossible to separate these two aspects of the question as to the effects produced on the mental powers of an individual by a particular brain lesion. In other cases, however, it is not so difficult, and that there are these two sides to the question is a point that must always be borne in mind. It is the fact, indeed, of the varying and very unequal extent to which the factor (*a*) coöperates with the factor (*b*) in different cases, coupled with the additional fact of the varying combinations of defect included under this second head, met with in different individual cases, that makes it impossible to lay down definite rules having any general application concerning the testamentary capacity of this class of patients.

It will be useless, therefore, to attempt to do anything more here than to indicate the kind of cases in which the patient would be incapable of making a valid will, and, on the other hand, the kind of cases in which the person should be capable of disposing of his property—that is, if free from disabilities under (*a*), which it would be out of place for me to attempt to deal with in this work.[1]

There are three special classes of cases in which will-making would be attended with the greatest difficulty, or would be actually impossible. These are (1) cases in which complete aphasia and agraphia exist; (2) cases in which complete word-deafness and word-blindness coexist; and (3) cases of "total aphasia" in which these two classes of defect are met with in the same individual.

(1) In the first set of cases where the patient is unable to speak or to write, it might be impossible for him, even with every help that could be given, to make a will except one of the very simplest description—though his chances of being able to do this would be very much improved if it could be

---

[1] The mere theoretical point of view as to the abilities and disabilities of different persons supposed to be suffering from each of the known individual forms of speech defect, existing separately, has been gone into with great elaboration by Byrom Bramwell (*Brit. Med. Journ.*, May 15, 1897), although in several of the cases mentioned it would be highly improbable that disposable property would be possessed, of a kind and amount likely to lead the holder to desire to make a will.

clearly shown that he was able to understand what he read, as well as what was said to him.

(2) In the second set of cases it would be almost impossible for the patient to make even the simplest will, seeing that he could neither understand what was said to him, nor read what was written; whilst he would, in all probability, be unable to write at all, and unable to speak intelligibly.

(3) In the third set of cases it would be clearly impossible for such patients to make a will, since they would be as unable to understand anything said or written for them, as to make intelligible communications to others.

There are, however, many other forms of speech defect in which it would be quite possible for the patient to make a valid will—in some cases with comparative ease, and in others with increasing degrees of difficulty. In many cases it would be needful that the terms of the will should be simplified as far as possible; and in all cases where an attempt is to be made to make a will by one of these patients, it is clear that except in the simplest cases to be presently mentioned, a lawyer should never undertake to do it alone. He should draw the will up with the assistance (and under direction as to the best method of proceeding with a view to elicit the wishes of the testator) of the ordinary medical attendant; or, in cases of difficulty, he should act under the direction of some physician specially conversant with this class of cases, and in the presence of some of the relatives of the patient. In cases of this latter type several interviews might be necessary for the completion of the will, as such patients are apt soon to become fatigued, and, as a result, confused by prolonged attention to any one subject. When this confusion once shows itself, it would be useless to attempt to proceed further at that sitting. The patient must be allowed an adequate rest before the subject is again broached.

A very full and useful sketch of the difficulties besetting such an attempt, and as to the mode of procedure to be adopted in such cases has lately been given by Byrom Bramwell.[1]

Will-making would be attended by no particular difficulties in cases of anarthria, cases of aphemia, and in those of "pure word-blindness" where the patient's power of writing is undisturbed, and where there need be, and probably would be, no diminution in mental power. Then again, in ordinary "cortical word-blindness" the patient might be able to dictate his wishes, and also to understand what had been written

[1] *Brit. Med. Jrnl.*, 1897, i., pp. 1208—1210.

when it was subsequently read aloud to him. Moreover, in the very rare cases of "pure word-deafness" the patient might be able to dictate a short will, and subsequently read it himself.

It would probably, however, be much more difficult, and perhaps impossible, for a person suffering from ordinary cortical word-deafness to make a valid will, because he would not only be unable to understand what was said to him, but would probably be unable to speak intelligibly—and might also be unable to communicate his wishes in any more intelligible fashion by writing. Further, the destruction of the auditory word centre would of itself carry with it a more marked degree of mental degradation than would be brought about by destruction of the visual word centre, or of Broca's region, either alone or in combination with that of the writing centre.

In regard to the capacity of a patient for managing his own affairs, it must be borne in mind that a will may, and probably would, only require to be made once, while the managing of an estate, or even of much more modest private affairs, would necessitate daily exertions for the purpose of giving directions, or receiving reports and other communications as to domestic details. This amount of daily recurring mental exertion would only be possible with a comparatively limited number of aphasic patients—certainly a much more limited number than those who might be permitted to exercise their testamentary capacity, and who would be likely to do so in a way that the law would recognise.

Still, there are some patients suffering from typical aphasia who would be quite capable of managing their own affairs. Take, for instance, the patient whose case was recorded by Broadbent (Case xxviii.), or the rich proprietor in the department of Landes, who was seen by Trousseau in 1863 after he had been aphasic for three years, and whose speech was reduced to the single word "oui." As evidence of his mental power, Trousseau says:[1]—"He played every day at 'all fours,' hiding his cards behind a pile of books, and using his left hand. He often won when playing with the curate, the doctor, or his son, without their allowing him to do so out of kindness. Whenever he played a trump, he laid his hand on it with an air of authority which showed that he knew its value. His son and Dr. Lafitte declared to me that he played as well as he ever used to do. Sometimes his son sits by his side to advise him, and stops him when he takes a card which is not the proper one, but he insists on playing as he likes, and by

[1] Lectures (Translation by Bazire), p. 231.

winning the game proves to his adviser that if he sacrificed a card, it was because he could thus improve his game. Although his son manages all his affairs, he insists on being consulted about the leases and contracts, etc.; and the son stated to me that his father indicates perfectly well, by gestures which are understood by those habitually around him, when certain portions of the deeds do not please him, and that he is not satisfied until alterations are made, which are, as a rule, useful and reasonable." Yet, as I have said, this gentleman's speech was restricted to the word "oui," and he was unable to write or even to read a single word.

Cases of this kind are by no means singular, although it is rather rare for aphasic patients to retain so much mental power. But if such a patient wished to manage his own affairs, and was willing to undergo the amount of recurring trouble needful for the communication of his wishes to his agents or servants, there would be no reason for vetoing it on the ground of insufficient intelligence. If, however, the daily business to be transacted were at all complicated, it would probably be much better for the sake of his health that it should not be undertaken.

The capacity for exercising other civil rights has rarely to be considered in regard to persons suffering from defects of speech, and the same may be said in reference to the responsibility of such persons for criminal acts, and their capacity to give evidence in courts of law. Nothing further, therefore, need be said here upon these subjects except to call attention to other lectures or memoirs, apart from those already cited, in which such subjects have been discussed.[1]

---

[1] The principal additional communications of this kind to which reference may be made are as follows:—Legrand du Saulle: *Gaz. des Hôpitaux*, June and July, 1868, and a course of lectures in the same Journal in 1882. Bernard: "De l'Aphasie," 1885, Chap. x. C. K. Mills: *The Review of Insanity and Nervous Disease*, Sept. and Dec., 1891. F. W. Langdon: "The Aphasias and their Medico-Legal Relations," Ohio, 1898.

CHAPTER XVII.

THE TREATMENT OF SPEECH DEFECTS.

WHAT has been said in the last chapter concerning prognosis in speech defects will have shown that from the point of view of treatment we have also to look to two distinct sets of indications—first, as to the best means for bringing about functional restitution; and, secondly, when all that is possible has been done, or is being done, in that direction to do all in our power that may be calculated to further functional compensation. We have to strive to restore the lost powers of intellectual expression by speech or writing, or by both; as well as the power of comprehending what is said and of reading with understanding. In some cases it will be found that this latter kind of treatment has proved efficacious, even when it has been had recourse to at very long periods after the initiation of the patient's malady.

Treatment to facilitate Functional Restitution.—In the great majority of cases where speech defects are caused by organic disease of the brain, and where they are associated with partial or complete hemiplegia, attention must be directed to improving as far as possible the general condition of the patient. This holds good whatever the nature of the disease—whether it has been caused by thrombosis or by embolism of one or more of the cerebral vessels, by hæmorrhage into or on the surface of the brain, by syphilitic pacchymeningitis, by an abscess, or by a new growth.

The treatment suitable for these several conditions need not be entered upon here. But it may be stated as a general rule (to which, however, many exceptions occur) that some amount of amelioration is to be expected in speech defects as in other associated symptoms, whenever the condition of the patient as a whole improves under treatment.

The degree of amelioration that can be brought about will naturally vary widely in different instances of these multiform brain affections. Thus, in cerebral hæmorrhage the ultimate improvement will be greatest where the mechanisms concerned with speech (whether cortical centres or conducting channels) have been merely pressed upon rather than more severely damaged by the effused blood. Then, again, in embolism and also in thrombosis there is always the chance that the effects first produced may after a time be diminished or narrowed in their range, owing to the partial establishment of a collateral circulation. This occurs more in some individuals than in others, but in all it is especially apt to occur, as I pointed out near the commencement of the last chapter, where the speech mechanisms interfered with are situated on the confines rather than near the centre of the brain area supplied by the occluded vessel. In regard to both embolism and thrombosis, however, it must likewise be borne in mind that aggravation of symptoms may also be caused in the early stages of the disease by extension of the area of vascular occlusion, owing to the setting up of thrombosis on the proximal side of an embolus, or the extension in the same direction of a primary thrombus.

Again, it is worthy of special mention that where the speech defects have been caused by syphilitic pacchymeningitis, much good may be expected to result from large doses of iodide of potassium together with mercury, taken during several weeks or even months; and, further, that in cases where fracture of the skull with depressed bone exists, where there is an abscess in the brain, as well as in some cases of cerebral tumour, much good may be done by judicious surgical interference.

Of the latter class of cases an interesting example is given by Bateman[1] of a case related at a meeting of the Royal Academy of Medicine in Ireland, in which a traumatic aphasia was cured by operative measures that were had recourse to several weeks after the original injury.

CASE CVI.—A boy had been struck with a small knife over the squamous portion of the left temporal bone ten days before he came under observation at Sir Patrick Dun's Hospital. The wound was healed, but he had some aphasia. Pain in the head and ear supervened, and the aphasia increased.

It was determined to explore, and he was trephined some weeks after the original injury, when a wound was found in the dura mater corresponding to the puncture of the bone. A sinus forceps was passed in, the wound opened up, and some blood-clot escaped; the patient was decidedly better, but the next morning he was again aphasic. The wound was washed out, and more blood-clot escaped; the aphasia almost immediately disappeared, but two days later it returned, and the wound was again washed out. After this the patient progressed favourably and entirely recovered.

[1] *Loc. cit.*, p. 296.

Finally, it should be stated that some of the most typical cases of functional restitution are to be met with among the very varied class of so-called "functional aphasias," which are many of them speedily curable by appropriate remedies, by rest and attention to the general health, by moral treatment and various measures that need not now be particularised as they have been pretty fully mentioned in Chapter vii., when citing numerous instances of this important class of speech defects. No method is more striking or more worthy of being remembered than that of the administration of ether in refractory cases of the comparatively common condition known as "hysterical mutism," examples of which have been quoted (p. 128). Sometimes speech may be suddenly restored in cases of this type after it has been lost for several years. We have seen too that insane persons will sometimes speak again, on comparatively slight provocation, after intervals of dumbness lasting for ten, twenty, or more years (see p. 122).

It is very surprising, however, to learn that dumbness may supervene after a traumatism in a young child, that it may last for a period of ten years, and then may be suddenly recovered from after a few applications of the electric current. A remarkable case of this kind has, according to Bateman,[1] been recorded by Charpentier.

CASE CVII.—The patient was an inmate of the Bicêtre, without any cerebral, neurotic or diathetic antecedents. The only abnormal symptom in early childhood was that he only began to talk at the age of three years; but from that period to the age of five years and a half, he spoke like other children. He had never been deaf, never had convulsions, and had never stammered; moreover, his language, intelligence, and general behaviour were those of other children of his age.

When he was five years and a half old, he was one day brought home from school with his head covered with blood-stained bandages; his parents had never been able to ascertain what had happened. The child got well without any loss of consciousness, delirium, convulsions or paralysis; but from that moment he became dull, speechless, and to use the words of his mother, he had become "like a mummy, silent and concentrated upon himself." This state of things persisting, at the age of 14 he was placed at the Bicêtre.

After he had been at the hospital about a year, M. Charpentier wished to try the effect of electricity. On the third application of the electric current, he said several times "Pardon, Monsieur." On the fourth day on entering the ward, M. Charpentier was surprised to see the patient sitting up in bed with a smiling countenance; he welcomed his physician with a volley of words spoken very rapidly—too rapidly even ("A la manière de nos persécutés loquaces"); the style of his conversation was childish and silly, but what astonished those around him was to hear him employ expressions which he could not have known at the commencement of his disorder, when he was only five years old.

[1] *Loc. cit.*, p. 289.

It looks as if the mental shock associated with the traumatism (whatever may have been its precise nature) must have set up a functional disturbance of the brain similar to that which occurs in hysterical mutism, and if this be the proper interpretation to put upon the case it is a most remarkable one from two points of view more especially—that is, on account of the early age at which the malady occurred, and from its very long duration. I know of no other case of hysterical mutism, certainly, comparable with it in either respect, though, as regards early age, Cases xxxv. and c. may also be examples. The case is worthy of being borne in mind from the point of view of prognosis as well as in regard to treatment, bearing out as it does the important curative agency of the electric current in this as well as in other functional disturbances of the nervous system.

The various improvements that may be brought about by the process of functional restitution, either through mere lapse of time (during which ordinary reparative changes occur), under the influence of drugs or other measures, or through the skill of the surgeon, mostly, though as we have seen not invariably, take place at a period not very remote from the onset of the patient's malady.

Treatment to facilitate Functional Compensation.—On the other hand, almost all the ameliorations in the patient's state that are brought about slowly and after long periods from the onset of the illness—sometimes even after years have elapsed—are due to a process of functional compensation. In this process existing centres are knitted together in previously unaccustomed ways (for new modes of functional activity) coincidently with the development of new sets of commissural fibres; or else centres hitherto of comparatively low organisation and slight activity, gradually take on a higher organisation and a greatly developed functional vigour; or, again, these two processes may occur in association. The development of commissures may be between word centres in the damaged hemisphere, or else between some of these and other centres in the uninjured hemisphere.

In all probability the facility with which such changes can be brought about in different individuals, other things being equal, would be subject to considerable variation. Similarly, individuals vary in the facility with which they acquire new knowledge or new aptitudes—vary, that is, in their educability. And this educability must be regarded as a function of the plasticity of the nervous system, of the greater or less facility with which new nervous paths and nervous units are capable of developing in response to oft-repeated stimuli.

We have every reason for believing that this plasticity is greatest in childhood, though it continues to a diminished extent well on into adult life, and is present still, for varying periods, even long beyond middle age. This receptivity of the brain for new knowledge varies much, however, as we all know, in different individuals. The mental activity of some persons comparatively early moves in set grooves; while in others well on into a ripe old age a marked receptivity is shown, with interest in and power of acquiring new knowledge.

It would seem quite justifiable to conclude that an individual of the latter type might, and probably would, stand a much better chance of recovering to a greater degree from any speech defects from which he may be suffering, than would have been the case had he belonged to the former type—that is, had he been an individual with a less plastic nervous system, and one which tended at an earlier age to lapse into a condition of developmental quiescence.

Of course it must be fully recognised that the cure of speech defects by functional compensation would be a process of great relative difficulty, as compared with the mere acquirement of new knowledge or new motorial aptitudes by a person with a healthy and undamaged brain. It is only contended that the processes must be similar in kind, however extensively they may differ in degree.

The attempt to treat patients with a view to promote functional compensation must necessarily vary much in mode of procedure in accordance with the particular nature of the defect we are endeavouring to remedy.

In all cases where we have to do with *Anarthria* or *Dysarthria* the methods required are essentially similar—and such cases have been found, not unfrequently, to prove very amenable to steady and persistent treatment. The patient has to be assiduously drilled in the pronunciation of simple vowel sounds and consonants, at the same time that he is shown how to arrange the organs of articulation for the production of such sounds, whenever such instruction is needed. He is then schooled in the same way in the articulation of monosyllables, and subsequently in the combination of monosyllables into short words. A laborious process of tuition has, in fact, to be carried out from day to day—the patient has slowly to be taught anew how to speak.

In some cases, however, where we have to do with bulbar defects, the patients may never have spoken at all, by reason of some imperfect development of these centres, in consequence of which speech has been greatly delayed. These cases stand in a class apart—they cannot be said, after tuition, to be instances of functional restitution, since the function had previously never been developed; nor are they cases of functional compensation, for, as I have already pointed out (p. 324), this process cannot occur in the case of bulbar lesions. What happens is nevertheless most akin to that which occurs in functional compensation.

Some of these patients are to be found in deaf and dumb institutions, and others in asylums for idiots and imbeciles. Some particulars concerning one such case have been given by Wyllie.[1] The patient was seen by him when his speech was almost perfect, he being then a lad aged 18 years, in employment as a tailor, though formerly he had been an inmate of an institution for the deaf and dumb in Edinburgh. The superintendent, Mr. Illingworth, furnished Wyllie with the following notes of the case.

CASE CVIII.—The boy H. R. entered this institution at the age of seven years. His hearing was almost normal, but he had no power of speech and did not attempt to articulate a single word. His general intelligence was below the average. After he had been at school for four years an attempt was made to teach him to speak by the oral method (previously to this he had always communicated by signs and the manual alphabet).

Exactly the same means were adopted as with a totally deaf child, and the difficulties in obtaining the correct sounds were almost as great.

After four years' instruction by the oral method, H. R. was able to speak very distinctly but with slight hesitation, which however was daily wearing off; and when he left school, a year ago, he gave promise of becoming (with encouragement and practice) a very good, free speaker.

Hadden also recorded[2] three cases of boys (aged respectively seven, eleven, and four), suffering from what he termed "idioglossia," in whom, though the intelligence was fairly good, speech was very defective, the characteristic features of "lalling" being very strongly pronounced. These children were taught, while in St. Thomas's Hospital, by the oral method, with excellent results. A very similar case has likewise been recorded by Frederic Taylor,[3] in which great improvement also took place under "careful oral instruction."

Then, again, Bateman[4] speaks of a case of "tardy development of the faculty of speech" which came under his own observation, though he gives no details as to the mode in which speech was acquired. He says: "The child never spoke at all till he was six years old, and it was thought he

---

[1] Loc. cit., p. 133.
[2] Jrnl. of Mental Science, January, 1891.
[3] Med. Chir. Trans., vol. lxxiv., 1891, p. 191.
[4] Loc. cit., p. 252.

would remain dumb. At six years of age he began to talk, and was able to receive an education suitable to his condition in life, but he grew up to manhood a person of feeble intellectual and also of feeble physical power."

He likewise says (p. 288) that Benedikt, of Vienna, had mentioned to him " an instance of tardy development of speech in a case of congenital aphasia " in which this faculty began to manifest itself as late as the age of twelve years.

The same author quotes[1] from Wildbur, the Superintendent of the State Asylum of New York, another case which, though different in nature and also in its mode of recovery, may well be mentioned here.

CASE CIX.—A boy, aged 8 years, good-looking and well-formed, came under observation, in whom idiotcy supervened in infancy. He looked intelligent, and was very gentle and obedient. He understood any simple language addressed to him, or spoken in his hearing; he could repeat the sentences heard or the question spoken, but could originate nothing in the way of speech under any circumstances. The control over his vocal organs was complete; he spoke quite distinctly and with appropriate emphasis.

He soon began to learn rapidly the exercises given to him, but the power or disposition to originate speech, even within the range of his wants or his affections, was wanting.

In this case the defect was overcome at last, through reading exercises, and eventually the boy could speak spontaneously.

Except for the fact that the power of voluntary speech was wanting up to some period beyond the eighth year, this case is one of an altogether different type. Here it would seem that we had rather to do with a defectively developed auditory word centre, which had its nutrition improved and was gradually further developed under the influence of physiological promptings emanating from the visual word centre—and therefore by an entirely different method from that employed in the other cases previously referred to.

Cases of *Aphemia* and *Aphasia* must be treated on principles very similar to those adopted in dealing with cases like those just referred to of undeveloped speech in children. The process is, however, generally much less laborious in these cases where speech has already been well acquired, but has been lost on account of a lesion in the course of the fasciculus geniculatus, or on account of damage to Broca's region— apparently because it is easier to bring about functional compensation in these cases through further development of the right third frontal convolution, than to educate previously undeveloped bulbar speech centres. This principle of treat-

[1] *Loc. cit.*, p. 93.

ment was long ago recognised by Osborn[1] when dealing with the remarkable case of jargon speech which I have previously quoted in part (Case lxxxi.). Its precise nature is, as I have said, very doubtful. It might be a case of incoördinate action either of the bulbar speech centres, or else of Broca's centre. Osborn says: "Having explained to the patient my view of the peculiar nature of his case, and having produced a complete conviction in his mind that the defect lay in his having lost, not the power, but the art of using the vocal organs, I advised him to commence learning to speak like a child, repeating first the letters of the alphabet, and subsequently words, after another person. The result has been most satisfactory, and affords the highest encouragement to those who may labour under this peculiar kind of deprivation; there being now very little doubt, if his health is spared, and his perseverance continues, that he will obtain a perfect recovery of speech."

A case of aphemia probably due to a lesion affecting the left fasciculus geniculatus pretty close to the bulbar speech centres, if we may judge from the other associated symptoms, was many years ago recorded by Bristowe[2] in which the patient was taught to speak again, although he had been nine months in an entirely speechless condition.

CASE CX.—A steward of a steam packet, aged 36, was after a few hours of malaise, seized on March 7, 1869 with a series of epileptiform attacks following one another quickly. On recovering consciousness four hours after their commencement he found himself lying on the floor of the cabin unable to move a limb, speechless, and stone deaf, though he could see and understand everything that was going on. He remained in this condition as nearly as possible up to the time of his arrival at Singapore on March 20.

At that time his right leg and arm were still weak, and his left leg and arm were numb and quite powerless. He had considerable difficulty in masticating his food, and he was still perfectly deaf and speechless. He gradually improved in the Singapore Hospital. "In the first week he regained the complete use of his right side, and audition so far returned that he could hear when spoken to loudly. His hearing was completely restored by April 22. He also regained to a great extent the use of his left arm, and improved remarkably in general health." He left the hospital in the middle of June, and was put on board a sailing vessel homeward bound. On November 1, he was admitted into St. Thomas's Hospital, still speechless and dragging his left leg much in walking.

In reference to his condition Dr. Bristowe gave the following particulars: "I found that he was perfectly intelligent, that he understood everything that was said to him, that he could read well and comprehend everything that he read, and that he could maintain a conversation of any length, he writing on a slate and his interlocutor speaking. He wrote indeed with remarkably facility a very excellent and legible hand, expressing himself

[1] *Dublin Journl. of Med. Science*, Nov., 1833, p. 169.
[2] *Trans. of the Clin. Soc.*, 1870, p. 92.

with perfect point and accuracy, except for an occasional error of spelling and construction, due evidently to defective education. But he could not speak, he could not utter a single articulate sound. I ascertained, however, that he could perform with his lips, tongue and cheeks all possible forms of voluntary movement, and also that he was capable of vocal intonation, in other words, that he could produce musical laryngeal sounds."

Bristowe resolved to teach him to speak anew, and this he proceeded to do, after explaining his intentions and the nature of the factors concerned with articulation to the patient. He first got him to sound a laryngeal note, subsequently explaining to him and showing him how to modify the shape and size of the oral passage and aperture; and getting him at the same time to expire, either with or without laryngeal intonation, he made him utter, both in a whisper and in a loud voice, certain of the more simple and obvious vowel sounds.

At his next lesson he set to work to teach him the labials, and he subsequently taught him by the same process the lingual and guttural consonantal sounds; and thus in the course of four or five lessons, principally by making the patient watch the movements of his (Dr. Bristowe's) lips, he regained the power of articulating all the simple vowel and consonantal sounds. He then began to teach him to combine letters, and eventually succeeded in making him talk well, although he spoke somewhat slowly and evidently had to give more thought to the pronunciation of his words than healthy people ordinarily do.

Dr. Bristowe adds: "The lessons which I gave him were, as I have shown, few and short. But he himself, as soon as ever he had appreciated the fact that he had organs capable of evolving articulate sounds, supplemented my instruction with the most zealous practice. Thus the vowels and consonantal sounds which he uttered somewhat imperfectly during a lesson were learnt accurately by my next visit; and as soon as he had begun to combine sounds, he practised them in various combinations with great industry; the sister of the ward and nurses, and more especially three or four intelligent patients who were friendly with him and interested in his progress, giving him constant assistance."

I have quoted this case very fully not only on account of its many points of interest, but because it is perhaps the first in which the full oral method was ever practised in a general hospital (that is, outside the walls of an institution for the deaf and dumb) for the restoration of speech in a case of aphemia or aphasia, and because of the full details that are given as to the method pursued with so much success. It is, as I have said, the mode of treatment that should be adopted in all cases of aphemia as well as of aphasia, with a view to bring about as much improvement as may be possible in this class of cases. It is, however, a method which requires some skill in the teacher, and not a little enthusiasm and perseverance, to carry it to a successful issue. The fact that the patient himself was very intelligent and persevering, and that several willing helpers also lent their aid, goes far to account for the rapid and complete success in this particular case.

Even without any special tuition of this kind cases of

aphemia are after a more or less lengthened period apt to recover their speech. Three instances of this were quoted in the last chapter (p. 323). It will also be found to hold true generally that recovery is much more easily to be brought about in aphemia than in aphasia.

The combination of conditions most favourable for recovery in aphasia is probably where we have to do with a young patient who is ambi-dextrous, or, what comes to much the same thing, is left-handed but has been taught to write with the right hand. Under these circumstances the inequality between the functional activity of the left and the right hemisphere is distinctly diminished, and as a consequence the right third frontal convolution all the more easily takes on the performance of speech functions. The speedy cure, with a minimum amount of trouble, that was brought about in the interesting case recorded by Wadham (Case xxiv.), may be taken as an illustration of the truth of these propositions.

We have seen, however, from the case recorded by Batty Tuke and Fraser (Case ci.), and also from the two cases recorded by Bernard (Cases xcvi. and cv.), that aphasia occurring in adults and indubitably caused by destruction of Broca's centre may, nevertheless, in the course of a few years be completely recovered from, even without any special tuition or drilling.

Some observations of much interest have recently been published by A. Thomas[1] which were carried out in Déjerine's wards at the Salpêtrière, with the object of ascertaining to what extent improvement could be brought about in cases of aphasia of considerable duration—employing the oral method as originally adopted by Bristowe.

He records two cases, one of them being a typical case of aphasia which, he says, had "supervened in circumstances that left no doubt as to the seat of the lesion," and which had already been in existence for five years. The following are a few details.

CASE CXI.—A woman, aged 34 years, suffering from acute otitis, about a fortnight after parturition was seized with violent pains in the head, and other symptoms which gave rise to the suspicion that she was suffering from an abscess of the brain. And, although there was neither hemiplegia nor aphasia, she was trephined over the region of the third frontal convolution. Several punctures of the brain were made, but no abscess was found. As a sequence a large cerebral hernia developed, and with it appeared right hemiplegia and aphasia. From the commencement the aphasia was complete, the patient being unable to utter a word.

Five years from the date of the commencement of this aphasia she

[1] *Compt. Rendus de la Société de Biologie*, November 6, 1897.

entered the Salpêtrière, in the month of April, 1896, and then her condition was found to be as follows.

The only words she uttered were "oui" and "non." She could neither repeat words nor read aloud. She could only write her name and her age with the left hand; not being able to write at all from dictation, though she could do transfer copying. As regards reading to herself, she understood common words, and the value of numerals; but she could not grasp the meaning of sentences of less familiar words or even of common words when written with separated letters or syllables, so as to present an unusual appearance.

Dr. Thomas says : "We commenced the re-education on April 22, 1896. We taught her to repeat successively by sight of the movements of the tongue and lips. vowels and simple syllables (association of a consonant with a vowel) ; by May 4, the patient could articulate all syllables. She then re-learned by the same method the spelling and the reading of certain syllables, and at the same time she made attempts at writing. When the constituents of words were re-acquired, she learned to repeat words of one syllable, then of two and of several syllables ; and to read words in a syllabic fashion. At the end of six weeks the re-education was terminated, and the patient left to herself.

"Now, one year from the commencement of the re-education, the patient can reply without hesitation, and in words correctly pronounced to all questions put to her; she does not yet frame actual sentences, but she makes herself understood. She can repeat short sentences well, but long sentences are only partially repeated. She can read aloud correctly, pronouncing each syllable distinctly. Her power of understanding what she reads has much improved. The power of writing spontaneously and to dictation has not sensibly improved, but during the year the patient has made few attempts in this direction."

The second case recorded by Thomas is one in which he endeavoured to re-educate in the same manner a patient who had been suffering from "cortical motor aphasia for fifteen years." The following brief details are given.

CASE CXII.—A woman, who at the commencement of the treatment was only able to articulate three or four words, could, after a month's treatment methodically carried out, repeat nearly all the words that she heard without being aided by the sight of the movements of the tongue and lips.

Since this period she has been examined on several occasions. She can now pronounce a good number of words spontaneously, but the emission of each word only takes place after several hesitations. She is incapable of pronouncing a sentence, or of reading aloud, but her power of understanding what she reads has improved. Her power of writing has remained stationary; she copies well, but is incapable of writing spontaneously or from dictation.

As Thomas says, the progress has not been very considerable " but this very interesting fact remains that a patient who had been aphasic for fifteen years, has been able, after a relatively short period of re-education, to repeat all words and to pronounce spontaneously a certain number of them."

These results with such old cases of aphasia are certainly very remarkable and important, and, as Thomas truly says, there is reason to think that if similar treatment were carried out within a few months from the commencement of an attack of aphasia, the amelioration in the patient's condition might be very considerable. The fact that in the first of these two cases the patient's power of reading aloud was so much greater than her power of speaking spontaneously is a very interesting fact. This may have been due to the circumstance that a direct route had been opened up from the visual word centre to Broca's region (Fig. 12, *e, e*); or, on the other hand, the auditory word centre may have been sluggish, as it is in many cases of marked amnesia verbalis, though capable of being roused by strong incitations coming to it from the visual word centre, as would be the case in reading.

It is clear, therefore, that in all cases of this latter type (*amnesia verbalis*) great good might result from frequently stimulating the auditory word centre by incitations coming to it from the visual word centre. Of improvement brought about in this way we have an excellent example in a well-known case recorded many years ago by Hun of Albany[1] before anything definite was known on the subject of aphasia. The following are a few details concerning this case.

CASE CXIII.—A blacksmith, aged 35 years, who before his illness could read and write with facility, had an apoplectic attack, attributed to "cerebral congestion," and remained in a state of stupor for several days.

On recovering from this condition he understood what was said, but it was observed that he had great difficulty in expressing himself in words, and for the most part could only make his wants known by signs.

He knew the meaning of words spoken before him, but could not recall those needed to express himself, nor could he repeat words when he heard them pronounced; he was conscious of the difficulty under which he was labouring, and seemed surprised and distressed at it. If Dr. Hun pronounced the word he needed, he seemed pleased, and would say, "Yes, that is it," but was unable to repeat the words after him.

After fruitless attempts to repeat a word Dr. Hun wrote it for him, and then he would begin to spell it letter by letter, and after a few trials was able to pronounce it; if the writing were now taken from him he could no longer pronounce the word; but after a long study of the written word, and frequent repetition, he would learn it so as to retain it, and afterwards use it. He kept a slate on which the words he required most were written, and to this he referred when he wished to express himself. He gradually learned these words and extended his vocabulary, so that after a time he was able to dispense with his slate. He could read tolerably well from a printed book, but hesitated about some words. When he was unable to pronounce a word, he was also unable to write it till he had seen it written; and then he could learn to write as he learned to pronounce, by repeated trials.

At the end of six months, by continually learning new words, he could make himself understood pretty well, often however, employing circumlocution when he could not recall the proper word, somewhat as if he were speaking a foreign language imperfectly learned.

[1] *American Journal of Insanity*, April, 1851.

This case seems very closely to resemble those of amnesia verbalis which I have recorded in a previous chapter (Cases xlv. and xlviii.), the only irregularity being the inability to repeat words after hearing them pronounced. This irregularity which is met with occasionally, is not very easy to explain. It was present also in my Case xcvii.

It seems to me clear that in all cases of amnesia verbalis, where ability to read aloud is preserved, the best chance of improving the patient's condition would be found to be daily practice of this kind, in conjunction with suitable dietetic and medicinal treatment. In this way we should be doing our best, not only to improve the nutrition of the brain as a whole, but also to stimulate the functional activity of the weak auditory word centre by means of incitations coming to it from its closely related visual word centre.[1]

The treatment of ordinary *Cortical Word-Deafness* by functional compensation must obviously be directed to exercising, and thus favouring the development of, the auditory word centre pertaining to the right or undamaged cerebral hemisphere, and the bringing of it into more complete relation with Broca's centre in the left hemisphere (Fig. 28). This is a process which will, in most cases, require the expenditure of much time and patience. Efforts must be made to concentrate the patient's attention upon simple vowel sounds, and afterwards upon monosyllables spelt letter by letter; whilst, later, the patient must be encouraged to try and repeat them by the oral method, that is, after he has been shown the position of the lips and tongue needful for their utterance. The readiness with which this can be done will depend much upon the amount of preservation of general intelligence, and upon the degree to which the patient's attention can be concentrated upon the efforts being made by his teacher. When once even a little progress has been achieved in this direction, we may expect that improvement will, after a time, become more rapid; and then, having slowly learned to recognise and pronounce monosyllables, the exercises must go on to deal with longer words—words of two syllables and of three syllables—the separate syllables being pronounced first and then the word as a whole. Where these patients have preserved some ability to read, the task of teaching them will probably be lightened, because in addition to being able to make the patient in this

---

[1] Treatment of this kind would, of course, be directed to the bringing about of functional restitution rather than of functional compensation. It is more convenient, however, to refer to it here in this connection.

way understand what we wish him to do, he may also, by such efforts, help to open up a direct functional relationship between the left visual word centre and Broca's region.

An excellent example of improvement brought about in a case of word-deafness by having recourse to such methods (employed in a more or less empirical manner), was recorded many years ago by Schmidt,[1] of which the following are the most important details.

CASE CXIV.—A young woman, aged 25 years, whilst straining, ten days after her first confinement, suddenly lost consciousness. She soon recovered and with assistance she moved into an adjoining room, and was able to ask what had happened to her, though she had much trouble in pronouncing these few words. She also only understood with difficulty and after several repetitions the words that were addressed to her. There was no paralysis of the tongue or of either of the limbs.

The next day the word-deafness and speech had become much worse. The patient did not understand a single word, even when spoken to loudly. The difficulty in speaking was also very pronounced. She had to make many efforts to find the word she wished and to pronounce it; sometimes she altered it or pronounced another word in its place, or pronounced it as though it were spelt quite differently. Her other senses were unaltered; she looked well and her mental condition seemed not to have suffered. She took notice of all that went on around her, but in spite of every effort she could not express herself as she wished; and in order to make her understand what others wished to say to her, recourse was had to writing. If a question was written upon the slate, she looked attentively at one word after the other, and tried to pronounce them separately, then together, and afterwards she replied.

Although she seemed not to hear words (or, at least, not to understand them), it was soon ascertained that she heard and could discriminate other common sounds such as a knock at the door, the difference between two bells, and the ticking of a watch.

She heard and gave attention when vowel sounds were pronounced separately, and she repeated them. If one pronounced a monosyllabic word she did not seem to understand it, but if one separated the different letters and accentuated them distinctly, she could then repeat it. For words of several syllables, it was necessary at first to pronounce the first syllable, then the next, and then the two together in order that she should understand the word.

Little by little she learned to apprehend words more quickly; but six months passed away before she could comprehend an entire sentence, however short (even when it was pronounced slowly and with accentuation of each word), without it being necessary to repeat it to her.

This patient seems to have been word-deaf but not completely letter-deaf. The case was observed at a time before word-deafness had been generally recognised, and it is obvious from some expressions used in the detailed account of the patient's condition, that it was for a long time regarded as a peculiar

---

[1] *Allg. Zeitschr. f. Psychiatrie*, 1871, Bd. xxvii., p. 304, and quoted by Mdlle. Skwortzoff in her thesis "De la Cécité et de la Surdité des Mots dans l'Aphasie," Paris, 1881, p. 77.

case of defective hearing, but without any adequate comprehension of the exact nature of the defect. It was for this reason that I previously intimated that the right methods of treatment had been stumbled upon empirically. Those around this patient simply availed themselves of the two abilities that were preserved, namely, (*a*) her ability to discriminate and repeat simple vowel sounds, and (*b*) her limited ability to read with understanding and, though with great difficulty, to pronounce the words aloud. Still the fact remains that this treatment, empirically carried out, ultimately led to the cure of the patient, and is, moreover, the very treatment which a more complete knowledge of the defect leads us to recommend as that best calculated to bring about a cure.[1]

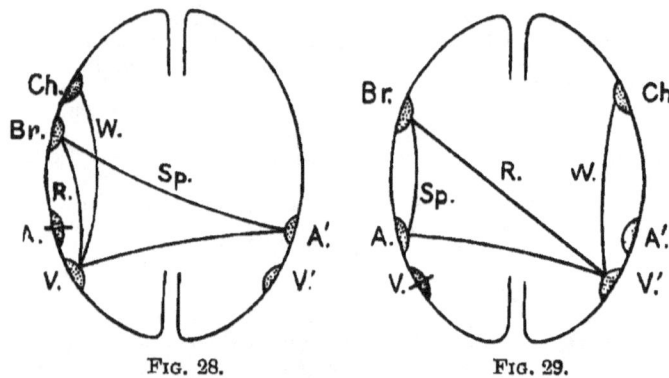

Fig. 28.   Fig. 29.

Fig. 28.—Diagram indicating the mode in which Speaking (Sp.), Reading (R.), and Writing (W.), may be brought about in a case where there has been destruction of the left Auditory Word Centre.

Fig. 29.—Diagram indicating the mode in which Speaking (Sp.), Reading (R.), and Writing (W.), may be ultimately brought about where there has been destruction of the left Visual Word Centre.

The ability exhibited by this patient of (in a laboured way) making out the import of written words, and then in a slow, hesitating way pronouncing them, is only likely to occur in an exceptionally strong "visual." Such a subject, however, by constant attempts to read aloud might gradually open up a direct channel of communication between the left visual word centre and Broca's region, or perfect channels that were already in existence between these two word centres, by the further development of commissural fibres. The process of

---

[1] She explained to Dr. Schmidt after her cure that "she heard perfectly when she was spoken to (during her illness), but that she only perceived the words as a confused noise."

reading aloud might be brought about in this way; though for ultimate correct speaking, the right auditory word centre and the commissures between it and Broca's centre would require to be further developed (Fig. 28, Sp.)

In the cases in which Broca's centre is destroyed as well as the left auditory word centre, and where a cure is ultimately brought about by compensation, as in Bernard's case previously quoted (Case cv.), the first stage of this process must be accomplished by the gradual development of the right auditory word centre. When this development has advanced to a certain extent, the patient must subsequently be treated as though one had to do with a simple case of aphasia—that is, by the oral method. In such a case, however, the treatment would probably be more tedious, because commissural relations would have to be established, not as in ordinary aphasia between the perfectly developed left auditory word centre and the third right frontal convolution (as in Fig. 24), but between this latter convolution and the, as yet, imperfectly developed auditory word centre of the right side (Fig. 27). Still it must be remembered, that in the case recorded by Bernard, a cure took place after some years, although no special training had been carried out.

Cases of *Pure Word-Deafness* being so exceedingly rare, no case is yet known in which recovery has taken place, and it would, in all probability, be exceedingly difficult to bring it about. Owing to the patient's ability to speak, to read, and to write by means, in the main, of the nervous mechanisms contained in the left hemisphere, this side of the brain would still continue very fully to perform its functions. Moreover, the situation of the lesion might be such as to prevent functional relations becoming established between the left visual and the right auditory word centres, and this being so the right auditory word centre would remain in an isolated condition. If, as in such a case, this centre should be thus cut off from all functional relations either with its fellow or with the visual word centre of the left hemisphere, it seems evident that comprehension of speech could never be restored—that is, no process of functional compensation would be possible.

In ordinary *Cortical Word-Blindness* the object to be aimed at, in order to bring about a cure by compensation, would be gradually to develop the right visual word centre by bringing it into functional relation with the left auditory word centre. This must be done by treating the patient as though he were a young child learning to read for the first time, and I need scarcely point out that the process would be more easily brought about in patients who were only word-blind, and not letter-blind.[1] Otherwise, the patient's attention must be concentrated upon individual letters one by one, at the same time that the teacher pronounces their names. After a certain amount of practice of this kind, the patient must be encouraged to try and name the letters himself (making use here of the oral method), in order to endeavour gradually to build up a commissural and functional relationship between the right visual word centre and Broca's region —as in the deaf-mute who is taught to speak (Fig. 29). Then, from letters the teacher should pass to monosyllables, naming each individual letter, and following it by the pronunciation of the sound as a whole. Here again efforts to read such monosyllables should be made by the patient (the teacher assisting by the oral method) till some facility has been acquired. Then an advance must be made with words of one and two syllables. Exercises of this kind carried out methodically and over long periods will be necessary for this education of the right visual word centre, and for the establishment of functional relations between it and the two undamaged word centres in the left hemisphere—namely, the auditory word centre and Broca's centre. At this period, or earlier, efforts should also be made to call the right visual word centre into further activity, by teaching the patient to write with the left hand. The three new functional relations that would thus have been established are roughly and diagrammatically indicated in Fig. 29.

The opinion has been expressed by Lichtheim and others that word-deafness is recovered from much more quickly than word-blindness, and certainly there are not many good cases on record in which it would seem that a cure of word-blindness has been brought about by functional compensation.

A case of partial word-blindness recorded by Bertholle is quoted by Bernard,[2] in which even after three years of attempted re-education the recovery of the power of reading, although considerable, was far from complete. Further on (p. 104) he records another case in which there was only partial improvement for a time (under not very zealous efforts however), with a subsequent relapse. Madlle. Skwortzoff, again, records[3] a case in which there was only a very partial

---

[1] As in Case lxix., recorded by Hughes Bennett, and also in the case of a boy referred to after Case lx., where the patient was able to read short words by spelling them letter by letter, and thus rousing the auditory word centre as well as Broca's centre.

[2] *Loc. cit.*, pp. 96-99.

[3] "De la Cécité et de la Surdité des Mots," Paris, 1881, p. 86.

recovery after attempts had been made for several months to teach the patient to read afresh; and on the following page she alludes to another case, occurring in an old gentleman, with whom all the efforts that were made to enable him to recognise letters again were unsuccessful. Her own attempts to teach another word-blind patient to recognise letters anew by aid of the sense of touch were only successful to a very limited extent (*loc. cit.*, p. 47), although the subject was a much more favourable one, being a young woman 33 years of age.

In cases of *Pure Word-Blindness* the difficulty of bringing about a cure by compensation would probably be even greater than in ordinary cortical word-blindness. Here a new track and mode of producing writing would not require to be established. All that is necessary would be, if possible, to develop the visual word centre in the right hemisphere, and to establish functional relations between it and the auditory word centre, as well as with Broca's centre, in the left hemisphere. Attempts must be made to do this in the manner previously indicated, when speaking of ordinary cortical word-blindness; that is, the patient must be taught anew as a child would be taught. The power of reading letters and words through the initiation of kinæsthetic impressions, possessed by these patients, is not likely to aid them very much, except that it would afford a means by which an intelligent patient could teach himself much, and have to rely less upon aid given by others for the gradual re-acquirement of the art of reading letters, words and, ultimately, sentences.

In some of these cases, however, no amount of teaching may be of any avail for the restoration of the lost power of reading. There may be the same kind of reason for this as for that which might sometimes cause the incurability of cases of pure word-deafness, that is, the complete isolation of the right visual word centre from the left auditory word centre (with which, for a cure, it would require to be brought into functional relation), owing to one of the lesions productive of the original malady lying directly in the course that any new commissural fibres connecting these two centres would have to take.

A case apparently of this type has been recorded by Pringle Morgan,[1] as one of so-called "congenital word-blindness." Yet even here, where one might think the conditions would be most favourable for a cure, up to the age of fourteen years the boy had only succeeded in learning his letters, and reading short words by spelling them letter by letter, and thus getting

See p. 181.

at their meaning through the auditory word centre. The boy was said to be bright and of average intelligence, and the schoolmaster who had taught him for some years went so far as to say "that he would be the smartest lad in the school if the instruction were entirely oral." Probably the right method had not been adopted in the attempts to teach him to read, but even then, that he should not have learned more during so many years must be considered remarkable.

A case apparently of pure word-blindness is recorded by Bernard[1] at great length, in which Charcot directed the treatment with a partial amount of success, making use of the patient's ability to read letters and words by re-awaking kinæsthetic impressions. Although the patient acquired a good deal of facility in reading words in this fashion, he could not read in the ordinary way by sight — he read only by writing movements, and that slowly and with difficulty.

A similar, though greater, ability to read in this manner was attained by one of Charles L. Dana's patients, referred to on p. 236. It seems possible that another case to which Bernard alludes in these terms may also have been a case of pure word-blindness:—"A theatrical manager, according to Piorry, still capable of writing, but incapable of reading, even what he himself had written, set himself to learn his *a b c* like a child. After great trouble and a long time, and by proceeding as though he had never been able to read before, he became able to read again."

Cases of combined *Word-Deafness and Word-Blindness*, when complete, are probably incurable by functional compensation. These patients are practically unteachable, and in the rare cases in which a more or less complete cure has taken place, it must always remain doubtful whether an alleviation in their condition by a process of functional restitution, more or less continuous, has not been more potent than a process of functional compensation, in bringing about their recovery.

In the last chapter I have recorded one case observed by Ball (Case cii.) where there was almost complete word-deafness and word-blindness, and in which the power of understanding speech and of reading was pretty fully regained within a period of fifteen months. Here the improvement set in so early as to make it possible that it was due in part to functional restitution. Then, again, there is the very remarkable case recorded by Mantle, also referred to in the last chapter, where recovery ultimately took place from very complete word-blindness and word-deafness. Here I am strongly inclined to think that the recovery was due to functional restitution rather than to functional compensation.

[1] *Loc. cit.*, pp. 77-89.

TREATMENT OF SPEECH DEFECTS

In cases of *Agraphia* the readiness with which a cure can be brought about will vary much in accordance with the seat of the lesion by which it is occasioned. On page 112 I have tabulated the various causes of agraphia, and a reference to this table will show that the cases are divisible into two categories, according as the agraphia is "complete" or "partial." As compensation in agraphia is brought about (if at all) by the training of a new writing centre in the right hemisphere, and since this acts, as I think, under the guidance of the visual centre of the same side, the agraphic person simply has to learn to write with the left hand, in the same laborious fashion that he originally mastered the art with the left side of his brain and his right hand. Some different views, however, in regard to details are held as to the exact mode in which writing with the left hand is brought about. Thus, Déjerine

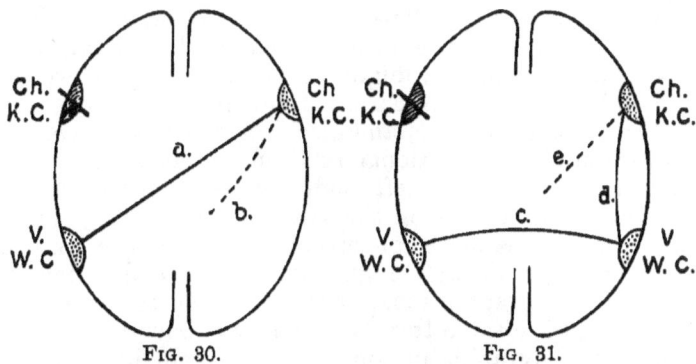

Fig. 30.—The mode of recovery from Agraphia suggested by Déjerine, when the left visual word centre is not destroyed.

Fig. 31.—The mode of recovery from Agraphia, under similar conditions, which I deem more probable.

thinks that in all cases in which the left visual word centre is intact, a direct relation is established between this visual word centre and the writing centre (that is, the part of the hand and arm centre that is concerned with writing movements) which is being developed in the right hemisphere. He believes this to be the way in which writing with the left hand would be brought about if a man had, for instance, his arm amputated, or was compelled merely in consequence of the right arm being paralysed to learn to write with the left hand. I have previously expressed the reasons why I do not accept this view.[1]

[1] See p. 274 and also what is said (p. 92) concerning left-handed persons suffering from left hemiplegia and aphasia, and their ability to write with the right hand.

With uncomplicated destruction of the left cheiro-kinæsthetic centre, or of the visuo-kinæsthetic commissure, there ought to be no difficulty in the patient learning to write with the left hand. Where we have to do, however, with the third cause of "complete" agraphia, namely, destruction of the left visual word centre, it is commonly believed that the patient is unable to learn to write with the left hand. But this is a statement that requires some qualification. Writing is an art that presupposes the power of reading. It seems perfectly clear, therefore, that so long as the patient remained word-blind as a consequence of destruction of the left visual word centre he would be unable to learn to write with his left hand. But whenever his word-blindness became cured by compensation, there would, in accordance with my views, no longer be any obstacle to his learning to write with his left hand as indicated in Fig. 29.

If we look now to the "partial" forms of agraphia (see p. 112), I may say that the form caused by destruction of the audito-visual commissure, might be cured by compensation in quite a different way—especially if the patient were a person much accustomed to write, and also a strong "auditive." In such a person direct functional relations might after a time be established between the left auditory word centre and the cheiro-kinæsthetic centre on the same side (Fig. 12, f, f). On the other hand, it is quite conceivable that such a lesion occurring in a strong "visual" might not after the first few days give rise to any agraphia except as regards ability to write from dictation; and the same thing may be said concerning the next cause of partial agraphia in some persons, namely, destruction of the left auditory word centre.

But where the person in whom the auditory word centre is destroyed is not a strong "visual," agraphia will be produced in association with word-deafness. The cure by compensation of the latter trouble would be brought about by gradual development of the right auditory word centre, and in that case it might, perhaps, be easier for the patient gradually to learn to write with the left hand than to attempt to re-acquire the power with the right hand.

SUBJECT INDEX.

Abductors of larynx, 135
Achromatopsia, 197
Adductors of larynx, 135
Agraphia absent, 87-93
,,   and aphasia, 80
,,   ,,   ,,   cases of, 84-93
,,   ,,   ,,   views concerning, 81
,,   ,,   destruction of Broca's region, 100
,,   ,,   left-handed aphasics, 102
,,   causes of classified, 112
,,   partial, 249, 251
,,   produced by hypnotism, 111
,,   recovery from, 273-275
,,   treatment of, 357
,,   uncomplicated, no cases of, 107
,,   with aphasia, cases of, 108-111
Alexia, 235
,,   absent with word-deafness, 96
,,   aphasia, and hemianæsthesia, 98-100
,,   sub-cortical, 193
,,   with aphasia, 95-100
Alternation of aphonia and mutism, 127
Amimia, 298
Amnesia, 140
,,   nature of, 12
,,   of nouns, 147, 151
,,   of touch, 113
,,   various forms of, 145
,,   verbalis, 94, 146, 152, 179
,,   ,,   treatment of, 349
Amnesics, reading aloud by, 149-152
Amusia, 284-298
Analysis of voluntary movements, 15
Anarthria, 55, 61
,,   treatment of, 342
Anastomosis of cerebral arteries, 305
Annexes of sensory centres, 47
Aphasia, alexia and hemianæsthesia, 98-100
,,   and agraphia, 80

Aphasia and agraphia, views concerning, 81
,,   ,,   general paralysis of the insane, 121
,,   ,,   hysteria, 125-137
,,   ,,   intestinal worms, 124
,,   ,,   neuralgia, 124
,,   ,,   lesions in island of Reil, 241-245
,,   ,,   word-blindness, 95-100
,,   after and before convulsions, 120
,,   a paralytic affection, 275-277
,,   associated with fevers, 118
,,   ,,   ,,   catalepsy, 122
,,   ,,   ,,   insanity, 122
,,   cases of, 84-93
,,   caused by emotion, 123
,,   ,,   ,,   narcotic poisons, 117
,,   ,,   ,,   spasm of vessels, 116
,,   ,,   ,,   snake poison, 118
,,   diagnosis difficult, 89
,,   during puerperal period, 118
,,   etiology of, 301-306
,,   from reflex irritation, 124
,,   gibberish, 152
,,   mode of recovery from, 264-267
,,   motor, 140
,,   optic, 212, 261
,,   of recollection, 150
,,   recovery from, 78, 324-327, 330
,,   sensory, 140
,,   ,,   symptoms of, 143-145
,,   ,,   various forms of, 145
,,   treatment of, 344-350
,,   with alexia, 95-100
,,   ,,   hemianæsthesia, 240
,,   without agraphia, cases of, 87-93
Aphasics, able to sing, 286-293
,,   and will making, 334-336

Aphasics, civil rights of, 333
,, left-handed, 90-93
,, ,, ,, and agraphia, 102
,, mental condition of, 88, 109
,, schema for examination of, 307
Aphemia, 140
,, and mutism, diagnosis of, 134
,, cases of, 64-66, 73, 76
,, characters of, 62, 72
,, complete, cases of, 68-70
,, described as aphasia, 63
,, duration of, 67
,, from disease of right hemisphere, 63
,, incomplete, 68
,, mode of recovery from, 67, 263, 267
,, recovery from, 73, 76, 323
,, treatment of, 344
Aphonia and mutism, alternation of, 127
,, hysterical, 126
,, mode of production of, 133
,, pathogenesis of, 136
Aphthongia, 59
Arteries of cortex, anastomosis of, 305
Articulation and heredity, 4
,, the learning of, 5
Auditives, 23, 145, 180, 186, 203
Audito-kinæsthetic commissure, damage to, 239-245
,, visual commissure, damage to, 247-252
Auditory and visual word centres, damage to commissures between, 247-260
Auditory and visual word centres, left, destruction of, 202-208
Auditory and visual word centres, destroyed on both sides, 162-164
Auditory and visual word centres, unequally damaged, 208
Auditory centre, left, isolation of, 174-178
,, memory, 12
,, nerves, semi-decussation of, 175
,, verbal images, recall of, 21
Auditory word centre, cases of destruction of, 154
Auditory word centre, defects from disease of, 145-162
Auditory word centre, destruction of, without word-deafness, 252-260
Auditory word centre, effects of destruction of, 153-159, 239
Auditory word centre, right, action of, 167-170, 201
Auditory word centre, site of, 15
,, ,, centres, both destroyed, 165, 166

Auditory word centres, commissure between, 169
,, ,, ,, destruction of both, 159-162
Autonomy limited of kinæsthetic centres, 224

Bilaterally acting muscles, 71
Bilateral representation in cortex, 135
,, ,, of movements in cortex, 171
Blind, congenitally, writing of, 29
Blindness to musical notes, 294
Brain, acts as a whole, 30
Broca's centre, destruction of, and agraphia, 100
,, ,, effects of destruction of, 91-101
,, ,, lesions of, 84-93
Bulbar speech centres, disease in, 55

Cases of agraphia with aphasia, 108-111
,, ,, aphasia, 84-91
,, ,, ,, and agraphia, 84-93
,, ,, ,, without agraphia, 87-93
,, ,, aphemia, 64-66
,, ,, complete aphemia, 69, 70
,, ,, functional speech defects, 115-139
,, ,, uncomplicated agraphia absent, 107
Catalepsy and aphasia, 122
Centre for concepts, separate, 42, 46, 278, 279
,, ,, naming, 43
,, ,, propositionising, 43
,, ,, writing, 82
Centres, cerebral, nature of, 14
,, cortical, excitability of, 31
,, ,, motor, 15
,, ,, nature of, 104
,, bulbar speech, disease in, 55
,, sensory, annexes of, 47
Cerebral cortex, constitution of, 17
Characters of aphemia, 62, 72
Civil rights, and aphasics, 333
Cheiro-kinæsthetic centre, 82
Cheiro-kinæsthetic centre, existence of, 101-108
Cheiro-kinæsthetic centre, is it topographically separate? 103-105
Cheiro-kinæsthetic centre, supposed localisation of, 18
Cheiro-kinæsthetic impressions, 13
Chorea, 55
Classification of causes of agraphia, 112
,, ,, speech defects, 51, 53
Clinical diagnosis of speech defects, 310-314

SUBJECT INDEX

Commissure, audito-kinæsthetic, damage to, 239-245
,, audito-visual, damage to, 247-252
,, between auditory word centres, 169
,, occipito-temporal, 261
,, temporo-occipital, 261
,, visuo-kinæsthetic, damage to, 245-246
Commissures between word centres, 19, 28, 234-238
,, between auditory and visual word centres, damage to, 252-260
,, defects due to lesions of, 234-261
,, imaginary, 238
,, occasionally used, 235
Compensation, functional, 320, 341-358
Complex thoughts and words, 40
Comprehension of writing, 96
Concept centre, separate, 42, 46, 54, 278, 279
Conception of voluntary movements, 173
Concepts and percepts, 39
,, thinking by, 49
Congenital disease of visual word centre, 180
,, word-blindness, 181
,, word-blindness, treatment of, 355
Congenitally blind persons, writing of, 29
Convulsions, preceded or followed by aphasia, 120
Corpus callosum, great development of in man, 173
,, ,, oblique fibres in, 173
,, ,, softening of, 305
Cortex, cerebral, constitution of, 17
,, bilateral representation of movements in, 135, 171
Cortical centres, nature of, 14, 104
,, executive motor mechanisms, 137
,, motor centres, 15
,, ,, ,, non-existent, 16, 18
,, representation of laryngeal movements, 135
,, word-blindness, 182
,, ,, ,, cases of, 183-185
,, ,, ,, treatment of, 353
,, word-deafness, 153
,, ,, ,, treatment of, 350
Counters for thought, 23

Cramp, writers', 60

Deaf-mutes, thought images of, 29
Deafness to musical tones, 295
Defects due to lesions of commissures, 234-261
,, from disease of auditory word centre, 145-162
,, from disease of visual word centre, 179-202
,, in musical faculty, 284-298
Destruction of auditory word centre, effects of, 153-159, 239
,, ,, both auditory word centres, 159-162
,, ,, Broca's centre, effects of, 91-101
,, ,, Broca's centre and agraphia, 100
,, ,, left auditory and visual word centres, 202-208
Delayed speech, 6-8, 332
Diagnosis in speech defects, 307-317
,, of aphasia sometimes difficult, 89
,, ,, aphemia and mutism, 134
,, ,, pure word-blindness, parietal type, 202
,, ,, pure word-blindness, occipital type, 202
Diagrams relating to speech, 45
Dictation, writing from, 29
Disease in bulbar speech centres, 55
,, pons Varolii, 57
,, of auditory word centre, defects from, 145
,, ,, visual word centre, defects from, 179-202
,, structural in kinæsthetic centres, 80-113
,, functional, in kinæsthetic centres, 114-139
Dumbness caused by traumatisms, 340
,, sudden cessation of, 6-8
Duration of aphemia, 67
Dysarthria, 55
,, with lesions of right hemisphere, 263

Echolalia, 152
Echo-speech and word-deafness, 177, 178
Effects of destruction of auditory word centre, 153-159
Embolisms, minute, 116
Emotion, aphasia caused by, 123
Ether, hysterical mutism cured by, 128
Etiology of aphasia, 301-306
Excitability of word centres, 31
Excitation of word centres, modes of, 31
Explanation of word-deafness, 142
,, ,, word-blindness, 142

Expression of thought by speech, 27
,, ,, ,, ,, writing, 28

Fasciculus geniculatus, 62
Fevers, aphasia associated with, 118
Fibres, pyramidal, lesions of, 61
Forceps, major, 196, 201
Fright, aphasia caused by, 123
Functional compensation, 320-332, 341-358
,, defect of visual word centre, 233
,, restitution, 318-320, 338-341
,, speech defects, cases of, 115

Geniculate fasciculus, 62
Gesture language, 298
Gibberish aphasia, 153, 220, 223
,, speech, 204
,, writing, 152
Glosso-kinæsthetic impressions, 13
,, ,, centre, localisation of, 18
Glosso-laryngeal paralysis, 56
Graphic centre, 82, 101-108
,, ,, localisation of, 105
,, ,, whether isolated or not, 103-105

Hemiachromatopsia, 197
Hemianæsthesia, aphasia, and alexia, 98-100
Hemianæsthesia with aphasia, 240
Hemianopsia and word-blindness, 182
Hemisphere left, prepotency of, 31, 33-35, 262
,, right, organisation of, 35
Heredity and articulation, 4
Hypnotism, as cause of agraphia, 111
,, ,, ,, ,, hysterical mutism, 124
Hysteria and aphasia, 125-137
Hysterical aphonia, 126
,, mutism, 68, 125-137, 202
,, ,, caused by hypnotism, 124
,, ,, characters of, 126
,, ,, cure of by ether, 128
,, ,, pathogenesis of, 138-139

Ideal recall of speech movements, 10
,, ,, ,, words, 20
,, ,, ,, writing movements, 10
Idioglossia, 52
Images, auditory verbal, recall of, 21
,, psycho-motor, 137
Incomplete word-deafness, 158
Initial letters, memory of, 148
Insular sclerosis, 55
Intellect and sense, 38

Intelligence of aphasics, 109, 336
Insane, general paralysis of, and aphasia, 121
Insanity associated with aphasia, 122
Intestinal worms and aphasia, 124
Intoxication with a letter, 228
Isolated speech-deafness, 174
Isolation of left auditory word-centre, 174-178
,, ,, visual word centre, 193-202

Jargon aphasia, 220, 221
,, speech, 204

Kinæsthetic centres, limited autonomy of, 224
,, ,, structural disease in, 80-113
,, ,, functional disease in, 114-139
,, impressions, reading by, 187, 191, 197, 198, 235, 236, 249
,, impressions, recoverability of, 10
,, memory, 12
,, verbal memories, 24

Lalling, 57
Language, varieties of, 2
Laryngeal movements, representation in cortex, 135
Laryngeal paralysis, unilateral, 70
Larynx, abductors of, 135
,, adductors of, 135
Left auditory word centre, isolation of, 174-178
Left-handed aphasics, 90-93
Left-hand writing, cerebral processes in, 201
Left hand, writing with, 92, 299
Left hemisphere, prepotency of, 31, 33-35, 262
Lesions of island of Reil and aphasia, 241-245
Lesions of pyramidal fibres, 61
Letter blindness, 183
Lichtheim's types of speech defects, 277-281
Localisation of cheiro-kinæsthetic centres, 18, 104-107
,, of glosso-kinæsthetic centres, 18
,, of word centres, 13
Loss of memory, 11
Loss of recollection, 11

Mechanism, cortical executive motor, 137
,, oral articulative, 137
Memories, kinæsthetic verbal, 24

Memories of words, 10
Memory, auditory, 12
   „ kinæsthetic, 12
   „ loss of, 11
   „ of initial letters, 148
   „ visual, 12
Mental condition of aphasics, 88, 109, 336
Middle cerebral artery, distribution of, 303
Mind-blindness, 210
Minute embolisms, 116
Mirror writing, 298-300
Motor aphasia, 140
   „ centres, cortical, non-existent, 16, 18
   „   „ no ideal recall in, 20
Movements, bilateral representation of, 171
   „ of larynx, cortical representation of, 135
   „ of speech and writing, ideal recall of, 10
   „ voluntary, conception of, 172
Movement, voluntary, analysis of, 15
Muscles, bilaterally acting, 71
Musical faculty, defects in, 284-298
Mutism and sub-cortical lesions, 71
   „   „ aphonia, alternation of, 127
   „   „ aphemia, diagnosis of, 134
   „ cure of by electricity, 341
   „ from fright, 123
   „ hysterical, 68, 125-137, 202
   „ hysterical, caused by hypnotism, 124
   „   „ characters of, 126
   „   „ cure of by ether, 128
   „   „ cure of by electricity, 340
   „   „ from traumatisms, 322, 340
   „   „ long duration of, 341
   „   „ pathogenesis of, 133-139
   „   „ recurrent, cases of, 130-133
   „ pseudo, 113
   „ recurrent utterances absent in, 136

Names, amnesia of, 147
Naming at sight, 248, 261
Naming centre, 43
Narcotic poisons, causing aphasia, 117
Nature of cortical centres, 14, 104
   „   „ perceptic processes, 26, 44

Nature of word centres, 14, 104
Neuralgia and aphasia, 124
Neural processes underlying voluntary movements, 15
Note-blindness, 294
Nouns, amnesia of, 147, 151
Numerals, reading and writing of, 282-284

Object blindness, 210
Oblique fibres in corpus callosum, 173
Occasional utterances, production of, 267-273
Occipital type of pure word-blindness, 193-202
Occipital type of pure word-blindness, diagnosis of, 202
Occipito-temporal commissure, 261
Optic aphasia, 212, 261
Optic radiations, 182, 197
Oral articulative mechanism, 137
Organisation of right hemisphere, 35

Pantomime, 298
Paragraphia and paraphasia, 216-233
   „ modes of production of, 223
   „ simple, 226-232
   „   „ pathogenesis of, 230, 232
Paralysis general of the insane and aphasia, 121
   „ glosso-laryngeal, 56
   „ laryngeal, unilateral, 70
   „ pseudo-bulbar, 78
Paraphasia and paragraphia, 216-233
   „ example of, 205
   „ grades of, 216
   „ modes of production of, 221-224
   „ simple, 224
   „   „ relations of, 225
   „ with word-blindness, 181
   „   „ word-deafness, 164, 165-167
Paraphasic speech, 204
Parietal type of pure word-blindness, 186-192
   „ type of pure word-blindness, diagnosis of, 202
Partial agraphia, 249, 251
   „ word-deafness, 158
Pathogenesis of aphonia, 136
   „   „ hysterical mutism, 133-139
   „   „ paraphasia, 221-224
   „   „ simple paragraphia, 230, 232
   „..   „ transitory aphasia, 119
Pathological diagnosis of speech defects, 314

Perception, process of, 27, 44
Perceptive elements, words as, 30
„ process, nature of, 26, 44
Percepts and concepts, 39
„ thinking by, 49
Phonation, centre for, 137
Poisons, narcotic, causing aphasia, 117
Pons Varolii, disease in, 57
Predominant use of right hand, 32
Prepotency of left hemisphere, 31
Primary revival of words, 22, 24
Process of perception, 27, 44
Production of aphonia, 133
„ „ whispering, 133
Prognosis in speech defects, 318-333
Propositionising centre, 43
Pseudo-bulbar paralysis, 78
Pseudo-mutism, 130
Psycho-motor images, 137
Puerperal period, aphasia during, 118
Pure word-blindness, 182
„ „ „ lesions causing, 199-201
„ „ „ of occipital type, 193-202
„ „ „ of occipital type, diagnosis of, 202
„ „ „ of parietal type, 186-192
„ „ „ of parietal type, diagnosis of, 202
„ „ „ treatment of, 355
Pure word-deafness, 174-178, 281
„ „ „ explanation of, 175-177
„ „ „ mode of production of, 201
„ „ „ treatment of, 353
Pyramidal fibres, lesions of, 61

Reading aloud, 28
„ „ by amnesics, 149-152
„ „ „ kinæsthetic impressions, 187, 191, 197, 198, 235, 236, 249
„ „ „ spelling, 196, 197
„ „ without comprehension, 185
Reading and writing of numerals, 282-284
Reading, processes involved in, 8
Recall, ideal, of words, 20
„ no ideal, in motor centres, 20
„ of auditory verbal images, 21
Recollection, aphasia of, 150
„ loss of, 11
„ of words, 21
„ process of, 11

Recoverability of kinæsthetic impressions, 10
Recovery from agraphia, 273-275
„ „ aphasia, 78, 324-327, 330
„ „ aphemia, 67, 73, 76, 323
„ „ word-blindness, 327-332
„ „ word-deafness, 327-332
„ mode of, from aphemia and aphasia, 263-267
Recurrent mutism, cases of, 130-133
Recurring utterances, 82, 83
„ „ absent in mutism, 136
Recurring utterances, production of, 267-273
Reflex irritation, causing aphasia, 124
Reil, island of, lesions in, 241-245
Representation, bilateral, in cortex, 135
„ in cortex of laryngeal movements, 135
Restitution, functional, 318-320, 338-341
Results of right-handedness, 33
Revival, primary, of words, 22, 24
Right auditory word centre, 201
„ „ „ „ action of, 167-170
Right-handedness, cause of, 32
„ „ results of, 33
„ hand, predominant use of, 32
Right hemisphere and dysarthria, 263
„ „ aphemia, from disease of, 68
„ „ organisation of, 35
Right visual word centre, 170, 201

Scanning utterance, 56
Schema for examination of aphasics, 307
Semi-decussation of auditory nerves, 175
Sensations, simple and complex, 39
Sense and intellect, 38
Sensory aphasia, 140
„ „ cases of, 154
„ „ symptomatology of, 143-145
Sensory aphasia, various forms of, 145
„ centres, annexes of, 47, 48
Separability of thought and words, 42
Simple and complex sensations, 39
Singing by aphasics, 286-293
Site of auditory word centre, 15
„ „ visual word centre, 15
Snake poison, causing aphasia, 118
Sounds, as thought counters, 3
Spasm of vessels, causing aphasia, 116
Speech, acquisition of, 3
„ and writing, 79
„ centres, bulbar, disease in, 55

SUBJECT INDEX

Speech, condition of, with word-deafness, 153-159
Speech-deafness, isolated, 174
Speech delayed, 332
" defects, auditory, study of, 19
" " classification of, 51, 53,
" " diagnosis of, 307-317
" " etiology of, 301-306
" " functional, 114
" " Lichtheim's types of, 272-281
" " prognosis in, 318-333
" " study of, 19
" " treatment of, 338-358
" diagrams relating to, 45
" emotional, 270
" expression of thought by, 27
" movements, ideal recall of, 10
" spontaneous with word-deafness, 155
" untaught, 6-8
Spelling, processes involved in, 227, 232, 233
Stammering, 58
Structural disease in kinæsthetic centres, 80-113
Stuttering, 58
Sub-cortical alexia, 193
" " lesions and mutism, 71
" " " signs of, 72
" " word-deafness, 174
Syllable stumbling, 217
Sylvian artery, distribution of, 303

Tapetum, 196, 201
Temporo-occipital commissure, 261
Thinking by concepts, 49
" " percepts, 49
" different grades of, 37
Third frontal convolution, destruction of and agraphia, 100
Third frontal convolution, lesions of, 84-93
Tone-deafness, 295-297
Thought and language, 37
" " words, 41
" " " separability of, 42
" complex and words, 40
" counters, 3, 23
" expression of by speech, 27
" " " writing, 28
" images of deaf-mutes, 29
" symbols, words as, 30
Touch amnesia, 213
" paralysis, 213
Transfer copying, 202
Transitory aphasia, pathogenesis of, 119
Traumatisms, causing mutism, 322, 340
Treatment of agraphia, 357
" " amnesia verbalis, 349
" " anarthria, 342

Treatment of aphasia, 344-350
" " aphemia, 344
" " cortical word-blindness, 353
" " cortical word-deafness, 350
" " dysarthria, 342
" " mutism, 340
" " pure word-blindness, 355
" " " word-deafness, 353

Unilateral laryngeal paralysis, 70
Untaught speech, 6-8
Utterances, recurring, 82, 83
" " absent in mutism, 136
" " and occasional, 267-273

Various forms of amnesia, 145
" " sensory aphasia, 145
Verbal amnesia, 146-152
" images, auditory, recall of, 21
" memories, kinæsthetic, 24
Vessels, spasm of, causing aphasia, 116
Visuals, 23, 25, 145, 180, 186, 203
Visual and auditory word centres destroyed on both sides, 162-164
Visual and auditory word centres, left, destruction of, 202-208
Visual and auditory word centres unequally damaged, 208
Visual memory, 12
" verbal images, 25
" word centre, congenital disease of, 180, 181
Visual word centre, defects from disease of, 179-202
Visual word centre, destruction of, without word-blindness, 259
Visual word centre, functional defect of, 233
Visual word centre, isolation of, 193-202
" " " right, 170, 201
" " " site of, 14
Visual word centres, damage to commissures between, 252-260
Visuo-kinæsthetic commissure, damage to, 245, 246
Volition, neural processes underlying, 15
Voluntary movement, analysis of, 15
" movements, conception of, 173

Whispering, production of, 133
Will-making by aphasics, 334-336
Word-blindness, 14, 141
" " absent with destruction of visual word centre, 259
" " and aphasia, 95-100

Word-blindness, and agraphia, 182
,, ,, ,, hemianopsia, 182
,, ,, congenital, 181, 355
,, ,, cortical, 182, 353
,, ,, explanation of, 142
,, ,, locality of lesion in, 142
,, ,, pure, 182, 355
,, ,, ,, occipital type of, 193-202
,, ,, parietal type of, 186-192
,, ,, recovery from, 327-332
Word centre, auditory, defects from disease of, 145-162
,, ,, auditory, site of, 15
,, ,, visual, functional defect of, 233
,, ,, ,, site of, 14
,, centres, auditory, destruction of both, 159-162
,, ,, commissures between, 19, 22, 234-238
,, ,, localisation of, 13
,, ,, modes of excitation of, 31
Word-deafness, 14
,, ,, absent with destruction of auditory word centre, 259
,, ,, cases of, 154
,, ,, condition of speech with, 153-159
,, ,, cortical, treatment of, 350
,, ,, first explanation of, 141
,, ,, locality of lesion in, 141
,, ,, mode of production of, 201
,, ,, paraphasia with, 164, 165-167

Word-deafness, partial, 158
,, ,, pure, 174-178, 281, 353
,, ,, recovery from, 327-332, 350
,, ,, spontaneous speech with, 155
,, ,, sub-cortical, 174-178
,, ,, without alexia, 96
,, ,, ,, tone-deafness, 329
Words and complex thought, 40
,, ,, thought, 41
,, as elements of perception, 30
,, ,, thought symbols, 30
,, ideal recall of, 20
,, memories of, 10
,, primary revival of, 22, 24
,, recollection of, 21
Worms, intestinal, and aphasia, 124
Writers' cramp, 60
Writing, ability to copy, 107
,, and reading of numerals, 282-284
,, ,, speech, 79
,, centre, 82
,, ,, existence of, 101-108
,, ,, is it topographically separate? 103-105
,, ,, localisation of, 105
,, comprehension of, 96
,, expression of thought by, 28
,, from dictation, 29
,, movements, ideal recall of, 20
,, of gibberish, 152
,, processes involved in, 9
,, typographic, 81, 107
,, unusual guidance in, 192
,, with left hand, 92
,, with left hand, cerebral processes involved in, 201, 357

October, 1894.

CATALOGUE OF WORKS
PUBLISHED BY
H. K. LEWIS
136 GOWER STREET, LONDON, W.C.

Established 1844.

A. C. ABBOTT, M.D.
First Assistant, Laboratory of Hygiene, University of Pennsylvania.
THE PRINCIPLES OF BACTERIOLOGY: A Practical Manual for Students and Physicians. With Illustrations, post 8vo, 7s. 6d.

H. ALDER-SMITH, M.B. LOND., F.R.C.S.
Resident Medical Officer, Christ's Hospital, London.
RINGWORM: Its Diagnosis and Treatment.
Third Edition, enlarged, with Illustrations, fcap. 8vo, 5s. 6d.

HARRISON ALLEN, M.D.
Consulting Physician to Rush Hospital for Consumption.
A HANDBOOK OF LOCAL THERAPEUTICS. General Surgery by R. H. HARTE, M.D., Surgeon to the Episcopal and St. Mary's Hospitals; Diseases of the Skin by A. VAN HARLINGEN, M.D., Professor of Diseases of the Skin in the Philadelphia Polyclinic; Diseases of the Ear and Air Passages by H. ALLEN, M.D.; Diseases of the Eye by G. C. HARLAN, M.D., Surgeon to Wills Eye Hospital. Edited by H. ALLEN, M.D. Large 8vo, 14s. *nett.*

JAMES ANDERSON, M.D., F.R.C.P.
Late Assistant Physician to the London Hospital, &c.
MEDICAL NURSING. Notes of Lectures given to the Probationers at the London Hospital. Edited by E. F. LAMPORT, Associate of the Sanitary Institute. With an Introductory Biographical Notice by the late SIR ANDREW CLARK, BART. Crown 8vo.
[*Nearly ready.*

E. CRESSWELL BABER, M.B. LOND.
Surgeon to the Brighton and Sussex Throat and Ear Hospital.
A GUIDE TO THE EXAMINATION OF THE NOSE, WITH REMARKS ON THE DIAGNOSIS OF DISEASES OF THE NASAL CAVITIES. With Illustrations, small 8vo, 5s. 6d.

JAMES B. BALL, M.D. LOND., M.R.C.P.
Physician to the Department for Diseases of the Throat and Nose, and Senior Assistant Physician, West London Hospital.
INTUBATION OF THE LARYNX. With Illustrations, demy 8vo, 2s. 6d.

G. GRANVILLE BANTOCK, M.D., F.R.C.S. EDIN.
Surgeon to the Samaritan Free Hospital for Women and Children.

I.
ON THE USE AND ABUSE OF PESSARIES. Second Edition, with Illustrations, 8vo, 5s.

II.
ON THE TREATMENT OF RUPTURE OF THE FEMALE PERINEUM IMMEDIATE AND REMOTE. Second Edition, with Illustrations, 8vo, 3s. 6d.

III.
A PLEA FOR EARLY OVARIOTOMY. Demy 8vo, 2s.

FANCOURT BARNES, M.D., M.R.C.P.
Physician to the Chelsea Hospital for Women; Obstetric Physician to the Great Northern Hospital, &c.

A GERMAN-ENGLISH DICTIONARY OF WORDS AND TERMS USED IN MEDICINE AND ITS COGNATE SCIENCES. Square 12mo, Roxburgh binding, 9s.

JAMES BARR, M.D.
Physician to the Northern Hospital, Liverpool; Medical Officer of Her Majesty's Prison, Kirkdale, &c.

THE TREATMENT OF TYPHOID FEVER, and reports of fifty-five consecutive cases with only one death. With Introduction by W. T. GAIRDNER, M.D., LL.D., Professor of Medicine in the University of Glasgow. With Illustrations, demy 8vo, 6s.

ASHLEY W. BARRETT, M.B. LOND., M.R.C.S., L.D.S.E.
Dental Surgeon to, and Lecturer on Dental Surgery in the Medical School of, the London Hospital.

DENTAL SURGERY FOR MEDICAL PRACTITIONERS AND STUDENTS OF MEDICINE. Second Edition, with Illustrations, cr. 8vo, 3s. 6d. [LEWIS'S PRACTICAL SERIES.]

ROBERTS BARTHOLOW, M.A., M.D., LL.D.
Professor Emeritus of Materia Medica, General Therapeutics, and Hygiene in the Jefferson Medical College of Philadelphia, &c., &c.

I.
A PRACTICAL TREATISE ON MATERIA MEDICA AND THERAPEUTICS. Eighth Edition, revised and enlarged, large 8vo, 21s.

II.
A TREATISE ON THE PRACTICE OF MEDICINE, FOR THE USE OF STUDENTS AND PRACTITIONERS. Fifth Edition, with Illustrations, large 8vo, 21s.

H. CHARLTON BASTIAN, M.A., M.D., F.R.S., F.R.C.P.
Professor of the Principles and Practice of Medicine in University College, London; Physician to University College Hospital, &c.

I.
PARALYSES: CEREBRAL, BULBAR, AND SPINAL. A MANUAL OF DIAGNOSIS FOR STUDENTS AND PRACTITIONERS. With numerous Illustrations, 8vo, 12s. 6d.

II.
VARIOUS FORMS OF HYSTERICAL OR FUNCTIONAL PARALYSIS. Demy 8vo, 7s. 6d.

GEO. M. BEARD, A.M., M.D.
AND
A. D. ROCKWELL, A.M., M.D.
Formerly Professor of Electro-Therapeutics in the New York Post Graduate Medical School; Fellow of the NewYork Academy of Medicine, &c.

I.
ON THE MEDICAL AND SURGICAL USES OF ELEC-
TRICITY. Eighth Edition. With over 200 Illustrations, roy. 8vo, 28s.

II.
NERVOUS EXHAUSTION (NEURASTHENIA) ITS
HYGIENE, CAUSES, SYMPTOMS AND TREATMENT. Second Edition, 8vo, 7s. 6d.

W. M. BEAUMONT.
Surgeon to the Bath Eye Infirmary.

THE SHADOW-TEST IN THE DIAGNOSIS AND
ESTIMATION OF AMETROPIA. Post 8vo, 2s. 6d.

E. H. BENNETT, M.D., F.R.C.S.I.
Professor of Surgery, University of Dublin,
AND
D. J. CUNNINGHAM, M.D., F.R.C.S.I.
Professor of Anatomy and Chirurgery, University of Dublin.

THE SECTIONAL ANATOMY OF CONGENITAL
CŒCAL HERNIA. With coloured plates, sm. folio, 5s. 6d.

A. HUGHES BENNETT, M.D., M.R.C.P.
Physician to the Hospital for Epilepsy and Paralysis, Regent's Park, and Assistant Physician to the Westminster Hospital.

I.
A PRACTICAL TREATISE ON ELECTRO-DIAGNOSIS
IN DISEASES OF THE NERVOUS-SYSTEM. With Illustrations, 8vo, 8s. 6d.

II.
ILLUSTRATIONS OF THE SUPERFICIAL NERVES
AND MUSCLES, WITH THEIR MOTOR POINTS; a knowledge of which is essential in the Art of Electro-Diagnosis. 8vo, cloth, 2s.

HORATIO R. BIGELOW, M.D.
Permanent Member of the American Medical Association; Fellow of the British Gynæcological Society, &c.

I.
GYNÆCOLOGICAL ELECTRO-THERAPEUTICS. With an Introduction by DR. GEORGES APOSTOLI. With Illustrations, demy 8vo, 8s. 6d.

II.
PLAIN TALKS ON ELECTRICITY AND BATTERIES,
WITH THERAPEUTIC INDEX, FOR GENERAL PRACTITIONERS AND STUDENTS OF MEDICINE. With Illustrations, crown 8vo, 4s. 6d.

DR. THEODOR BILLROTH.
Professor of Surgery in Vienna.

GENERAL SURGICAL PATHOLOGY AND THERA-
PEUTICS. With additions by Dr. ALEXANDER VON WINIWARTER, Professor of Surgery in Luttich. Translated from the Fourth German edition, and revised from the Tenth Edition, by C. E. HACKLEY, A.M., M.D. 8vo, 18s.

DRS. BOURNEVILLE AND BRICON.
MANUAL OF HYPODERMIC MEDICATION.
Translated from the Second Edition, and Edited, with Therapeutic Index of Diseases, by ANDREW S. CURRIE, M.D. Edin., &c. With Illustrations, crown 8vo, 3s. 6d.

ROBERT BOXALL, M.D. CANTAB., M.R.C.P. LOND.
Assistant Obstetric Physician at the Middlesex Hospital, &c.
ANTISEPTICS IN MIDWIFERY. 8vo. 1s.

RUBERT BOYCE, M.B., M.R.C.S.
Assistant Professor of Pathology in University College, London.
A TEXT-BOOK OF MORBID HISTOLOGY for Students and Practitioners. With 130 coloured Illustrations, royal 8vo, 31s. 6d.
[*Now ready.*

GURDON BUCK, M.D.
CONTRIBUTIONS TO REPARATIVE SURGERY: Showing its Application to the Treatment of Deformities, produced by Destructive Disease or Injury; Congenital Defects from Arrest or Excess of Development; and Cicatricial Contractions from Burns. Large 8vo, 9s.

MARY BULLAR & J. F. BULLAR, M.B. CANTAB., F.R.C.S.
RECEIPTS FOR FLUID FOODS. 16mo, 1s.

CHARLES H. BURNETT, A.M., M.D.
Emeritus Professor of Otology in the Philadelphia Polyclinic ; Clinical Professor of Otology in the Woman's Medical College of Pennsylvania ; Aural Surgeon to the Presbyterian Hospital, Philadelphia.
SYSTEM OF DISEASES OF THE EAR, NOSE AND THROAT. By 45 Eminent American, British, Canadian and Spanish Authors. Edited by CHARLES H. BURNETT. With Illustrations, in two Imperial 8vo vols., half morocco, 48s. nett. [*Just published.*

STEPHEN SMITH BURT, M.D.
Professor of Clinical Medicine and Physical Diagnosis in the New York Post-graduate School and Hospital.
EXPLORATION OF THE CHEST IN HEALTH AND DISEASE. With Illustrations, crown 8vo, 6s.

DUDLEY W. BUXTON, M.D., B.S., M.R.C.P.
Administrator of Anæsthetics and Lecturer in University College Hospital, the National Hospital for Paralysis and Epilepsy, Queen's Square, and the Dental Hospital of London.
ANÆSTHETICS THEIR USES AND ADMINISTRATION. Second Edition, with Illustrations, crown 8vo, 5s.
[LEWIS'S PRACTICAL SERIES.]

HARRY CAMPBELL, M.D., B.S. LOND., M.R.C.P.
Physician to the North-West London Hospital.

I.
THE CAUSATION OF DISEASE: An exposition of the ultimate factors which induce it. Demy 8vo, 12s. 6d.

II.
FLUSHING AND MORBID BLUSHING: THEIR PATHOLOGY AND TREATMENT. With plates and wood engravings, royal 8vo, 10s. 6d.

III.
DIFFERENCES IN THE NERVOUS ORGANISA- TION OF MAN AND WOMAN, PHYSIOLOGICAL AND PATHOLOGICAL. Royal 8vo, 15s.

IV.
HEADACHE AND OTHER MORBID CEPHALIC SEN- SATIONS. Royal 8vo, 12s. 6d. [*Just published.*]

R. E. CARRINGTON, M.D., F.R.C.P.
Late Assistant Physician and Senior Demonstrator of Morbid Anatomy at Guy's Hospital.

NOTES ON PATHOLOGY. With an Introductory Chapter by J. F. GOODHART, M.D. (ABERD.), F.R.C.P., Physician to Guy's Hospital, Edited by H. EVELYN CROOK, M.D. LOND., F.R.C.S. ENG., and GUY MACKESON, L.R.C.P., M.R.C.S. Crown 8vo, 3s. 6d.

ALFRED H. CARTER, M.D. LOND.
Fellow of the Royal College of Physicians; Physician to the Queen's Hospital, Birmingham; Professor of Therapeutics in Queen's College, Birmingham.

ELEMENTS OF PRACTICAL MEDICINE. Sixth Edition, crown 8vo, 9s.

F. H. CHAMPNEYS, M.A., M.D. OXON., F.R.C.P.
Physician-Accoucheur and Lecturer on Obstetric Medicine at St. Bartholomew's Hospital: Examiner in Obstetric Medicine in the University of Oxford, &c.

I.
LECTURES ON PAINFUL MENSTRUATION. THE HARVEIAN LECTURES, 1890. Roy. 8vo, 7s. 6d.

II.
EXPERIMENTAL RESEARCHES IN ARTIFICIAL RESPIRATION IN STILLBORN CHILDREN, AND ALLIED SUBJECTS. Crown 8vo, 3s. 6d.

W. BRUCE CLARKE, M.A., M.B. OXON., F.R.C.S.
Assistant Surgeon to, and Senior Demonstrator of Anatomy and Operative Surgery at St. Bartholomew's Hospital; Surgeon to the West London Hospital, &c.

THE DIAGNOSIS AND TREATMENT OF DISEASES OF THE KIDNEY AMENABLE TO DIRECT SURGICAL INTERFERENCE. With Illustrations, demy 8vo, 7s. 6d.

JOHN COCKLE, M.A., M.D.
Physician to the Royal Free Hospital.

ON INTRA-THORACIC CANCER. 8vo, 4s. 6d.

ALEXANDER COLLIE, M.D. ABERD., M.R.C.P. LOND.
Secretary of the Epidemiological Society for Germany and Russia, &c.

ON FEVERS: THEIR HISTORY, ETIOLOGY, DIAG- NOSIS, PROGNOSIS, AND TREATMENT. Illustrated with Coloured Plates, crown 8vo, 8s. 6d. [LEWIS'S PRACTICAL SERIES.]

M. P. MAYO COLLIER, M.B., M.S. LOND., F.R.C.S. ENG.
Professor of Comparative Anatomy and Physiology at the Royal College of Surgeons, England, &c.

THE PHYSIOLOGY OF THE VASCULAR SYSTEM.
Illustrations, 8vo, 3s. 6d.

WALTER S. COLMAN, M.B., M.R.C.P. LOND.
Pathologist and Registrar to the National Hospital for the Paralysed and Epileptic.

SECTION CUTTING AND STAINING: A Practical Guide to the Preparation of Normal and Morbid Histological Specimens. Crown 8vo, 3s.

W. H. CORFIELD, M.A., M.D. OXON., F.R.C.P. LOND.
Professor of Hygiene and Public Health in University College, London; Medical Officer of Health for St. George's, Hanover Square, &c.

DWELLING HOUSES: their Sanitary Construction and Arrangements. Third Edition, with Illustrations, crown 8vo, 3s. 6d.

J. LEONARD CORNING, M.A., M.D.
Consultant in Nervous Diseases to St. Francis Hospital.

A PRACTICAL TREATISE ON HEADACHE, NEU- RALGIA, SLEEP AND ITS DERANGEMENTS, AND SPINAL IRRITATION. With an Appendix—Eye Strain, a Cause of Headache By DAVID WEBSTER, M.D. Second Edition. Demy 8vo, 7s. 6d.

EDWARD COTTERELL, F.R.C.S. ENG., L.R.C.P. LOND.
Late House Surgeon, University College Hospital.

ON SOME COMMON INJURIES TO LIMBS; their Treatment and After-treatment, including Bone-setting (so-called). With Illustrations, small 8vo, 3s. 6d.

SIDNEY COUPLAND, M.D., F.R.C.P.
Physician to the Middlesex Hospital, and Lecturer on Practical Medicine in the Medical School; late Examiner in Medicine at the Examining Board for England.

NOTES ON THE CLINICAL EXAMINATION OF THE BLOOD AND EXCRETA. Third Edition, 12mo, 1s. 6d.

CHARLES CREIGHTON, M.D.
Demonstrator of Anatomy in the University of Cambridge.

I.
A HISTORY OF EPIDEMICS IN BRITAIN FROM A.D. 664 TO THE EXTINCTION OF THE PLAGUE. Demy 8vo, 18s.

II.
ILLUSTRATIONS OF UNCONSCIOUS MEMORY IN DISEASE, including a Theory of Alteratives. Post 8vo, 6s.

III.
CONTRIBUTIONS TO THE PHYSIOLOGY AND PATHOLOGY OF THE BREAST AND LYMPHATIC GLANDS. New Edition with additional chapter, with wood-cuts and plate, 8vo, 9s.

IV.
BOVINE TUBERCULOSIS IN MAN: An Account of the Pathology of Suspected Cases. With Chromo-lithographs and other Illustrations, 8vo, 8s. 6d.

H. RADCLIFFE CROCKER, M.D. LOND., B.S., F.R.C.P.
Physician for Diseases of the Skin in University College Hospital, &c.

DISEASES OF THE SKIN; THEIR DESCRIPTION, PATHOLOGY, DIAGNOSIS, AND TREATMENT. Second Edition, with 92 Illustrations, 8vo, 24s. *[Just published.*

EDGAR M. CROOKSHANK, M.B. LOND., F.R.M.S.
Professor of Comparative Pathology and Bacteriology in, and Fellow of King's College London.

I.

MANUAL OF BACTERIOLOGY: Illustrated with Coloured Plates from original drawings, and with other Illustrations in the text. Third Edition, 8vo, 21s.

II.

HISTORY AND PATHOLOGY OF VACCINATION. Vol. I., A Critical Inquiry. Vol. II., Selected Essays, (Edited) including works by Jenner, Pearson, Woodville, Henry Jenner, Loy, Rogers, Birch, Bousquet, Estlin, Ceely, Badcock, Auzias-Turenne, Dubreuilh and Layet. Two volumes, illustrated with 22 coloured plates, including reproductions of the plates illustrating Jenner's Inquiry, of selected plates from the work of Ceely and others, and with a reduced facsimile of an engraving of Mr. Jesty, a facsimile of the first folio of the manuscript of Jenner's original paper, a facsimile of an unpublished letter from Jenner to Mr. Head. Royal 8vo, 20s. *nett.*

J. BRENDON CURGENVEN, M.R.C.S., L.S.A.
Formerly House Surgeon to the Royal Free Hospital; Honorary Secretary of the Harveian Society, the Infant Life Protection Society, &c.

THE DISINFECTION OF SCARLET FEVER AND OTHER DISEASES BY ANTISEPTIC INUNCTION. 8vo, 1s. 6d.

HERBERT DAVIES, M.D., F.R.C.P.
Late Consulting Physician to the London Hospital.

THE MECHANISM OF THE CIRCULATION OF THE BLOOD THROUGH ORGANICALLY DISEASED HEARTS. Edited by ARTHUR TEMPLER DAVIES, B.A. (Nat. Science Honours), M.D. Cantab., M.R.C.P.; Physician to the Royal Hospital for Diseases of the Chest. Crown 8vo, 3s. 6d.

HENRY DAVIS, M.R.C.S.
Teacher and Administrator of Anæsthetics at St. Mary's Hospital, and Assistant Anæsthetist to the Dental Hospital of London.

GUIDE TO THE ADMINISTRATION OF ANÆSTHETICS. Second Edition, fcap. 8vo, 2s. 6d.

J. THOMPSON DICKSON, M.A., M.B. CANTAB.
Late Lecturer on Mental Diseases at Guy's Hospital.

THE SCIENCE AND PRACTICE OF MEDICINE IN RELATION TO MIND, the Pathology of the Nerve Centres, and the Jurisprudence of Insanity, being a course of Lectures delivered at Guy's Hospital. Illustrated by Chromo-lithographic Drawings and Physiological Portraits. 8vo, 14s.

F. A. DIXEY, M.A., D.M.
Fellow of Wadham College, Oxford.

EPIDEMIC INFLUENZA: A Study in Comparative Statistics. With Diagrams and Tables. 8vo, 7s. 6d.

HORACE DOBELL, M.D.
Consulting Physician to the Royal Hospital for Diseases of the Chest, &c.
ON DIET AND REGIMEN IN SICKNESS AND Health, and on the Interdependence and Prevention of Diseases and the Diminution of their Fatality. Seventh Edition, 8vo, 5s. *nett.*

CHARLES W. DULLES, M.D.
Fellow of the College of Physicians of Philadelphia and of the Academy of Surgery; Physician to the Rush Hospital.
ACCIDENTS AND EMERGENCIES: A Manual of the Treatment of Surgical and Medical Emergencies in the absence of a Physician. Fourth Edition, with Illustrations, post 8vo, 3s. *nett.*

JOHN EAGLE.
Member of the Pharmaceutical Society.
A NOTE-BOOK OF SOLUBILITIES. Arranged chiefly for the use of Prescribers and Dispensers. 12mo, 2s. 6d.

ARTHUR W. EDIS, M.D. LOND., F.R.C.P.
Senior Physician to the Chelsea Hospital for Women; Late Obstetric Physician to the Middlesex Hospital.
STERILITY IN WOMEN: including its Causation and Treatment. With 33 Illustrations, demy 8vo, 6s.

ALEXANDER S. FAULKNER.
Surgeon-Major, Indian Medical Service.
A GUIDE TO THE PUBLIC MEDICAL SERVICES. Compiled from Official Sources. 8vo, 2s.

W. SOLTAU FENWICK, M.B., B.S. LOND.
Member of the Royal College of Physicians; Assistant Physician to the Evelina Hospital for Sick Children, &c.
THE DYSPEPSIA OF PHTHISIS: Its Varieties and Treatment, including a Description of Certain Forms of Dyspepsia associated with the Tubercular Diathesis. Demy 8vo. [*Just ready.*

DR. FERBER.
MODEL DIAGRAM OF THE ORGANS IN THE THORAX AND UPPER PART OF THE ABDOMEN. With Letter-press Description. In 4to, coloured, 5s.

J. MAGEE FINNY, M.D. DUBL.
King's Professor of Practice of Medicine in School of Physic, Ireland, &c.
NOTES ON THE PHYSICAL DIAGNOSIS OF LUNG DISEASES. 32mo, 1s. 6d.

AUSTIN FLINT, M.D., LL.D.
Professor of Physiology and Physiological Anatomy in the Bellevue Hospital Medical College, New York, Visiting Physician to the Bellevue Hospital, &c.
A TEXT-BOOK OF HUMAN PHYSIOLOGY. Fourth Edition, Illustrated by plates, and 316 wood engravings, large 8vo, 25s.

W. H. RUSSELL FORSBROOK, M.D. LOND., M.R.C.S.
Consulting Medical Officer to the Government of the Cape of Good Hope; formerly Surgical Registrar to Westminster Hospital.
A DISSERTATION ON OSTEO-ARTHRITIS. Demy 8vo, 5s.

J. MILNER FOTHERGILL, M.D., M.R.C.P.
Late Physician to the City of London Hospital for Diseases of the Chest, Victoria Park, &c.

I.
INDIGESTION AND BILIOUSNESS. Second Edition, post 8vo, 7s. 6d.

II.
GOUT IN ITS PROTEAN ASPECTS. Post 8vo, 7s. 6d.

III.
THE TOWN DWELLER: His Needs and His Wants. With an Introduction by SIR B. W. RICHARDSON, M.D., LL.D., F.R.S. Post 8vo, 3s. 6d.

PROFESSOR E. FUCHS.
Professor of Ophthalmology in the University of Vienna.

TEXTBOOK OF OPHTHALMOLOGY. Translated from the German by A. DUANE, M.D., Assistant Surgeon, Ophthalmic and Aural Institute, New York. Second Edition, with 190 Illustrations, large octavo, 21s.

SIR DOUGLAS GALTON.
Late Royal Engineers, K.C.B., Hon. D.C.L., LL.D., F.R.S., Assoc. Inst. C.E., M.I.Mech.E., F.S.A., F.G.S., F.L.S., F.C.S., F.R.G.S., &c., Formerly Secretary Railway Department Board of Trade, &c., &c.

HEALTHY HOSPITALS. Observations on some points connected with Hospital Construction. 8vo, with Illustrations, 10s. 6d.

JOHN HENRY GARRETT, M.D.
Licentiate in Sanitary Science and Diplomate in Public Health, Universities of Durham and Cambridge, &c.

THE ACTION OF WATER ON LEAD; being an inquiry into the Cause and Mode of the Action and its Prevention. Crown 8vo, 4s. 6d.

ALFRED W. GERRARD, F.C.S.
Examiner to the Pharmaceutical Society; Teacher of Materia Medica and Pharmacy at University College Hospital.

I.
ELEMENTS OF MATERIA MEDICA AND PHARMACY. Crown 8vo, 8s. 6d.

II.
NEW OFFICIAL REMEDIES, B.P., 1890. Supplement to the above. Crown 8vo, 1s.

HENEAGE GIBBES, M.D.
Lecturer on Physiology and on Normal and Morbid Histology in the Medical School of Westminster Hospital; etc.

PRACTICAL HISTOLOGY AND PATHOLOGY. Third Edition, revised and enlarged, crown 8vo, 6s.

R. T. GLAZEBROOK, M.A., F.R.S.
Assistant Director of the Cavendish Laboratory; Fellow of Trinity College, Cambridge.

HEAT AND LIGHT. Crown 8vo, 5s. The two parts are also published separately. HEAT, 3s.; LIGHT, 3s.
[CAMBRIDGE NATURAL SCIENCE MANUALS.]

JAMES F. GOODHART, M.D. ABERD., F.R.C.P.
Physician to Guy's Hospital, and Consulting Physician to the Evelina Hospital for Sick Children.
ON COMMON NEUROSES: or the Neurotic Element in Disease and its Rational Treatment. Second Edition, crown 8vo, 3s. 6d.
[*Just published.*

C. A. GORDON, M.D., C.B.
Deputy Inspector General of Hospitals, Army Medical Department.
REMARKS ON ARMY SURGEONS AND THEIR WORKS. Demy 8vo, 5s.

JOHN GORHAM, M.R.C.S.
TOOTH EXTRACTION: a Manual on the proper mode of extracting Teeth. Fourth Edition, fcap. 8vo, 1s. 6d.

GEORGE M. GOULD, A.M., M.D.
Ophthalmic Surgeon to the Philadelphia Hospital, &c.
I.
A NEW MEDICAL DICTIONARY: including all the words and phrases used in Medicine, with their proper pronunciation and definitions. 8vo, 10s. 6d.
II.
A POCKET MEDICAL DICTIONARY, Giving the Pronunciation and Definition of about 12000 of the Principal Words used in Medicine and the Collateral Sciences. 32mo, 4s. *nett.*

W. R. GOWERS, M.D., F.R.C.P., M.R.C.S.
Physician to University College Hospital, &c.
DIAGRAMS FOR THE RECORD OF PHYSICAL SIGNS. In books of 12 sets of figures, 1s.

SURGEON-CAPTAIN A. E. GRANT, M.B.
Indian Medical Service; Professor of Hygiene, Madras Medical College.
THE INDIAN MANUAL OF HYGIENE.
Vol. I., 8vo, 10s. 6d. *nett.*

LANDON C. GRAY, M.D.
Professor of Nervous and Mental Diseases in the New York Polyclinic; Visiting Physician to St. Mary's Hospital, &c.
A TREATISE ON NERVOUS AND MENTAL DISEASES FOR STUDENTS AND PRACTITIONERS OF MEDICINE. With 168 illustrations, 8vo, 21s.

J. B. GRESSWELL, M.R.C.V.S.
Provincial Veterinary Surgeon to the Royal Agricultural Society.
VETERINARY PHARMACOLOGY AND THERAPEUTICS. With an Index of Diseases and Remedies. Fcap. 8vo, 5s.

A. HILL GRIFFITH, M.D.
Surgeon, Manchester Royal Eye Hospital.
THE DIAGNOSIS OF INTRA-OCULAR GROWTHS. With 8 woodcuts, 8vo, 1s. 6d.

SAMUEL D. GROSS, M.D., LL.D., D.C.L. OXON.
Professor of Surgery in the Jefferson Medical College of Philadelphia.

A PRACTICAL TREATISE ON THE DISEASES, INJURIES, AND MALFORMATIONS OF THE URINARY BLADDER, THE PROSTATE GLAND, AND THE URETHRA. Third Edition, revised and edited by S. W. GROSS, A.M., M.D., Surgeon to the Philadelphia Hospital. Illustrated by 170 engravings, 8vo, 18s.

SAMUEL W. GROSS, A.M., M.D.
Surgeon to, and Lecturer on Clinical Surgery in, the Jefferson Medical College Hospital and the Philadelphia Hospital, &c.

A PRACTICAL TREATISE ON TUMOURS OF THE MAMMARY GLAND: embracing their Histology, Pathology, Diagnosis, and Treatment. With Illustrations, 8vo, 10s. 6d.

DR. JOSEF GRUBER.
Professor of Otology in the University of Vienna, etc.

A TEXT-BOOK OF THE DISEASES OF THE EAR. Translated from the second German edition, and Edited, with additions, by EDWARD LAW, M.D., C.M. EDIN., M.R.C.S. ENG., Surgeon to the London Throat Hospital for Diseases of the Throat, Nose and Ear; and COLEMAN JEWELL, M.B. LOND., M.R.C.S. ENG., late Surgeon and Pathologist to the London Throat Hospital. Second English Edition, with 165 Illustrations, and 70 coloured figures on 2 lithographic plates, royal 8vo, 28s.

F. DE HAVILLAND HALL, M.D., F.R.C.P. LOND.
Physician to Out-patients, and in charge of the Throat Department at the Westminster Hospital, &c.

DISEASES OF THE NOSE AND THROAT. Crown 8vo, with 2 coloured plates and 59 illustrations, 10s. 6d. [*Now ready.*]
[LEWIS'S PRACTICAL SERIES.]

WILLIAM A. HAMMOND, M.D.
Professor of Mental and Nervous Diseases in the Medical Department of the University of the City of New York, &c.

SPIRITUALISM AND ALLIED CAUSES AND CONDITIONS OF NERVOUS DERANGEMENT. With Illustrations post 8vo, 8s. 6d.

ALEXANDER HARVEY, M.D.
Late Emeritus Professor of Materia Medica in the University of Aberdeen, &c.,
AND
ALEXANDER DYCE DAVIDSON, M.D., F.R.S. EDIN.
Late Regius Professor of Materia Medica in the University of Aberdeen.

SYLLABUS OF MATERIA MEDICA FOR THE USE OF STUDENTS, TEACHERS AND PRACTITIONERS. Based on the relative values of articles and preparations in the British Pharmacopœia. Ninth Edition, 32mo, 1s. 6d.

H. HELBING, F.C.S.
MODERN MATERIA MEDICA: FOR MEDICAL MEN, PHARMACISTS, AND STUDENTS. Fourth Edition, 8vo, 8s. *nett.*
[*Just published.*]

C. HIGGENS, F.R.C.S.
Ophthalmic Surgeon to Guy's Hospital; Lecturer on Ophthalmology at Guy's Hospital Medical School.

MANUAL OF OPHTHALMIC PRACTICE.
Illustrations, crown 8vo, 6s. [LEWIS'S PRACTICAL SERIES.]

BERKELEY HILL, M.B. LOND., F.R.C.S.
Professor of Clinical Surgery in University College; Surgeon to University College Hospital and to the Lock Hospital.

THE ESSENTIALS OF BANDAGING. With directions for Managing Fractures and Dislocations; for administering Ether and Chloroform; and for using other Surgical Apparatus; with a Chapter on Surgical Landmarks. Sixth Edition, revised and enlarged, Illustrated by 144 Wood Engravings, crown 8vo, 5s.

BERKELEY HILL, M.B. LOND., F.R.C.S.
Professor of Clinical Surgery in University College; Surgeon to University College Hospital and to the Lock Hospital;

AND

ARTHUR COOPER, L.R.C.P., M.R.C.S.
Surgeon to the Westminster General Dispensary.

I.
SYPHILIS AND LOCAL CONTAGIOUS DISORDERS.
Second Edition, entirely re-written, royal 8vo, 18s.

II.
THE STUDENT'S MANUAL OF VENEREAL DISEASES. Being a Concise Description of those Affections and of their Treatment. Fourth Edition, post 8vo, 2s. 6d.

FROM HOSPITAL WARD TO CONSULTING ROOM, with Notes by the Way; a Medical Autobiography. By a Graduate of the London University. Post 8vo, 3s. 6d.

PROCTER S. HUTCHINSON, M.R.C.S.
Assistant Surgeon to the Hospital for Diseases of the Throat.

A MANUAL OF DISEASES OF THE NOSE AND THROAT; including the Nose, Naso-pharynx, Pharynx, and Larynx. With Illustrations, crown 8vo, 3s. 6d.

C. R. ILLINGWORTH, M.D. ED., M.R.C.S.
THE ABORTIVE TREATMENT OF SPECIFIC FEBRILE DISORDERS BY THE BINIODIDE OF MERCURY. Crown 8vo, 3s. 6d.

SIR W. JENNER, Bart., M.D.
Physician in Ordinary to H.M. the Queen, and to H.R.H. the Prince of Wales.

THE PRACTICAL MEDICINE OF TO-DAY: Two Addresses delivered before the British Medical Association, and the Epidemiological Society, (1869). Small 8vo, 1s. 6d.

GEORGE LINDSAY JOHNSON, M.A., M.B., B.C. CANTAB.
Clinical Assistant, late House Surgeon and Chloroformist, Royal Westminster Ophthalmic Hospital, &c.

A NEW METHOD OF TREATING CHRONIC GLAUCOMA, based on Recent Researches into its Pathology. With Illustrations and coloured frontispiece, demy 8vo, 3s. 6d.

JOHN M. KEATING,
Fellow of the College of Physicians, Philadelphia, &c.
AND
HENRY HAMILTON.

POCKET MEDICAL LEXICON.
32mo, 3s. *nett.*

T. N. KELYNACK, M.D.
Pathologist to the Manchester Royal Infirmary; Demonstrator and Lecturer on Pathology in the Owen's College.

A CONTRIBUTION TO THE PATHOLOGY OF THE VERMIFORM APPENDIX. With Illustrations, large 8vo, 10s. 6d.

HENRY R. KENWOOD, M.B., D.P.H., F.C.S.
Instructor in the Hygienic Laboratory, University College, and Assistant to Professor Corfield in the Public Health Department, University College.

PUBLIC HEALTH LABORATORY WORK, including Methods employed in Bacteriological Research, with special reference to the Examination of Air, Water and Food, contributed by RUBERT BOYCE, M.B., Assistant Professor of Pathology, University College. With Illustrations, cr. 8vo, 10s. 6d.

[LEWIS'S PRACTICAL SERIES.]

NORMAN KERR, M.D., F.L.S.
President of the Society for the Study of Inebriety; Consulting Physician, Dalrymple Home for Inebriates, etc.

INEBRIETY OR NARCOMANIA: its Etiology, Pathology, Treatment, and Jurisprudence. Third Edition, 8vo, 21s.
[*Now ready.*

NORMAN W. KINGSLEY, M.D.S., D.D.S.
President of the Board of Censors of the State of New York; Member of the American Academy of Dental Science, &c.

A TREATISE ON ORAL DEFORMITIES AS A BRANCH OF MECHANICAL SURGERY. With over 350 Illustrations, 8vo, 16s.

F. CHARLES LARKIN, F.R.C.S. ENG.
Surgeon to the Stanley Hospital; late Assistant Lecturer in Physiology in University College, Liverpool,
AND
RANDLE LEIGH, M.B., B.SC. LOND.
Senior Demonstrator of Physiology in University College, Liverpool.

OUTLINES OF PRACTICAL PHYSIOLOGICAL CHEMISTRY. Second Edition, with Illustrations, crown 8vo, paper 2s. 6d. *nett*, or cloth 3s. *nett.*

J. WICKHAM LEGG, F.R.C.P.
Formerly Assistant Physician to Saint Bartholomew's Hospital, and Lecturer on Pathological Anatomy in the Medical School.

I.
ON THE BILE, JAUNDICE, AND BILIOUS DISEASES.
With Illustrations in chromo-lithography, 719 pages, roy. 8vo, 25s.

II.
A GUIDE TO THE EXAMINATION OF THE URINE.
Seventh Edition, edited and revised, by H. LEWIS JONES, M.D., M.A., M.R.C.P., Medical Officer in charge of the Electrical Department in St. Bartholomew's Hospital. With Illustrations, fcap. 8vo, 3s. 6d.
[*Now ready*

ARTHUR H. N. LEWERS, M.D. LOND., M.R.C.P. LOND.
Obstetric Physician to the London Hospital; Examiner in Midwifery and Diseases of Women to the Society of Apothecaries of London, &c.

A PRACTICAL TEXTBOOK OF THE DISEASES OF WOMEN. Fourth Edition, Illustrations, crown 8vo, 10s. 6d.
[LEWIS'S PRACTICAL SERIES.]

LEWIS'S POCKET CASE BOOK FOR PRACTITIONERS AND STUDENTS. Roan, with pencil, 3s. 6d. *nett.*

LEWIS'S POCKET MEDICAL VOCABULARY.
Second Edition, thoroughly revised, 32mo, roan, 3s. 6d.

T. R. LEWIS, M.B., F.R.S. ELECT, ETC.
Late Fellow of the Calcutta University, Surgeon-Major Army Medical Staff, &c.

PHYSIOLOGICAL & PATHOLOGICAL RESEARCHES.
Arranged and edited by SIR WM. AITKEN, M.D., F.R.S., G. E. DOBSON, M.B., F.R.S., and A. E. BROWN, B.Sc. Crown 4to, portrait, 5 maps, 43 plates including 15 chromo-lithographs, and 67 wood engravings, 30s. *nett.*

*** A few copies only of this work remain for sale.

C. B. LOCKWOOD, F.R.C.S.
Hunterian Professor, Royal College of Surgeons of England; Surgeon to the Great Northern Hospital; Senior Demonstrator of Anatomy and Operative Surgery in St. Bartholomew's Hospital.

HUNTERIAN LECTURES ON THE MORBID ANATOMY, PATHOLOGY AND TREATMENT OF HERNIA.
36 Illustrations, Demy 8vo, 5s.

WILLIAM THOMPSON LUSK, A.M., M.D.
Professor of Obstetrics and Diseases of Women in the Bellevue Hospital Medical College, &c.

THE SCIENCE AND ART OF MIDWIFERY.
Fourth Edition, with numerous Illustrations, 8vo, 18s.

A. W. MACFARLANE, M.D., F.R.C.P. EDIN.
Examiner in Medical Jurisprudence in the University of Glasgow; Honorary Consulting Physician (late Physician) Kilmarnock Infirmary.

INSOMNIA AND ITS THERAPEUTICS.
Medium 8vo, 12s. 6d.

RAWDON MACNAMARA.
Professor of Materia Medica, Royal College of Surgeons, Ireland; Senior Surgeon to the Westmoreland (Lock) Government Hospital; Surgeon to the Meath Hospital, &c.

AN INTRODUCTION TO THE STUDY OF THE BRITISH PHARMACOPŒIA. Demy 32mo, 1s. 6d. [*Just published.*

JOHN MACPHERSON, M.D.
Inspector-General of Hospitals H.M. Bengal Army (Retired). Author of "Cholera in its Home," &c.

ANNALS OF CHOLERA FROM THE EARLIEST PERIODS TO THE YEAR 1817. With a map. Demy 8vo, 7s. 6d.

A. COWLEY MALLEY, B.A., M.B., B.CH., T.C.D.

PHOTO-MICROGRAPHY; including a description of the Wet Collodion and Gelatino-Bromide Processes, together with the best methods of Mounting and Preparing Microscopic Objects for Photo-Micrography. Second Edition, with Photographs and Illustrations, crown 8vo, 7s. 6d.

PATRICK MANSON, M.D., C.M.

THE FILARIA SANGUINIS HOMINIS; AND CERTAIN NEW FORMS OF PARASITIC DISEASE IN INDIA, CHINA, AND WARM COUNTRIES. Illustrated with Plates and Charts. 8vo, 10s. 6d.

JEFFERY A. MARSTON, M.D., C.B., F.R.C.S., M.R.C.P. LOND.
Surgeon General Medical Staff (Retired).

NOTES ON TYPHOID FEVER: Tropical Life and its Sequelæ. Crown 8vo, 3s. 6d.

EDWARD MARTIN, A.M., M.D.

MINOR SURGERY AND BANDAGING WITH AN APPENDIX ON VENEREAL DISEASES. 82 Illustrations, crown 8vo, 4s.

WILLIAM MARTINDALE, F.C.S.
Late Examiner of the Pharmaceutical Society, and late Teacher of Pharmacy and Demonstrator of Materia Medica at University College,

AND

W. WYNN WESTCOTT, M.B. LOND.
Deputy Coroner for Central Middlesex.

THE EXTRA PHARMACOPŒIA, with Medical References, and a Therapeutic Index of Diseases and Symptoms. Seventh Edition, limp roan, med. 24mo, 7s. 6d.

WILLIAM MARTINDALE, F.C.S.
Late Examiner of the Pharmaceutical Society, &c.

I.
COCA, AND COCAINE. Their History, Medical and Economic Uses, and Medicinal Preparations. Third Edition, fcap. 8vo.
[*In preparation.*

II.
ANALYSES OF TWELVE THOUSAND PRESCRIPTIONS: being Statistics of the Frequency of Use therein of all Official and Unofficial Preparations. Fcap. 4to, 2s. 6d. *nett.* [*Now ready.*

MATERIA MEDICA LABELS.

Adapted for Public and Private Collections. Compiled from the British Pharmacopœia of 1885, with the additions of 1890. The Labels are arranged in Two Divisions:—

Division I.—Comprises, with few exceptions, Substances of Organized Structure, obtained from the Vegetable and Animal Kingdoms.

Division II.—Comprises Chemical Materia Medica, including Alcohols, Alkaloids, Sugars, and Neutral Bodies.

On plain paper, 10s. 6d. *nett.* On gummed paper, 12s. 6d. *nett.*

The 24 additional Labels of 1890 only, 1s. *nett.*

*** Specimens of the Labels, of which there are over 470, will be sent on application.

S. E. MAUNSELL, L.R.C.S.I.
Surgeon-Major, Medical Staff.

NOTES OF MEDICAL EXPERIENCES IN INDIA PRINCIPALLY WITH REFERENCE TO DISEASES OF THE EYE. With Map, post 8vo, 3s. 6d.

ANGEL MONEY, M.D. LOND., F.R.C.P.
Late Assistant Physician to University College Hospital, and to the Hospital for Sick Children Great Ormond Street

I.
TREATMENT OF DISEASE IN CHILDREN: EMBODYING THE OUTLINES OF DIAGNOSIS AND THE CHIEF PATHOLOGICAL DIFFERENCES BETWEEN CHILDREN AND ADULTS. Second Edition, with plates, crown 8vo, 10s. 6d.
[Lewis's Practical Series.]

II.
THE STUDENT'S TEXTBOOK OF THE PRACTICE OF MEDICINE. Fcap. 8vo, 6s. 6d.

A. STANFORD MORTON, M.B., F.R.C.S. ENG.
Assistant Surgeon to the Moorfields Ophthalmic Hospital; Ophthalmic Surgeon to the Great Northern Central Hospital.

REFRACTION OF THE EYE: Its Diagnosis, and the Correction of its Errors. Fifth Edition, with Illustrations, small 8vo, 3s. 6d.
[*Now ready.*

C. W. MANSELL MOULLIN, M.A., M.D. OXON., F.R.C.S.
Surgeon and Lecturer on Physiology at the London Hospital; Formerly Radcliffe's Travelling Fellow and Fellow of Pembroke College Oxford, etc.

I.
ENLARGEMENT OF THE PROSTATE: its Treatment and Radical Cure. With plates, roy. 8vo. [*Nearly ready.*

II.
SPRAINS; THEIR CONSEQUENCES AND TREATMENT. Second Edition, crown 8vo, 4s. 6d.

PAUL F. MUNDE, M.D.
Professor of Gynecology at the New York Polyclinic; President of the New York Obstetrical Society and Vice-President of the British Gynecological Society, &c.

THE MANAGEMENT OF PREGNANCY, PARTURITION, AND THE PUERPERAL STATE. Second Edition, square 8vo, 3s. 6d.

WILLIAM MURRAY, M.D., F.R.C.P. LOND.
Consulting Physician to the Children's Hospital, Newcastle-on-Tyne, &c.

ILLUSTRATIONS OF THE INDUCTIVE METHOD IN MEDICINE. Crown 8vo, 3s. 6d.

WILLIAM MURRELL, M.D., F.R.C.P.
Physician to, and Lecturer on Pharmacology and Therapeutics at Westminster Hospital; late Examiner in Materia Medica to the Royal College of Physicians of London, &c.

I.
WHAT TO DO IN CASES OF POISONING. Seventh Edition, royal 32mo, 3s. 6d.

II.
MASSOTHERAPEUTICS, OR MASSAGE AS A MODE OF TREATMENT. Fifth Edition, with Illustrations, crown 8vo, 4s. 6d.

III.
CHRONIC BRONCHITIS AND ITS TREATMENT. Crown 8vo, 3s. 6d.

HENRY D. NOYES, A.M., M.D.
Professor of Ophthalmology and Otology in Bellevue Hospital Medical College; Executive Surgeon to the New York Eye and Ear Infirmary, &c.

A TEXTBOOK ON DISEASES OF THE EYE. Second and revised Edition, Illustrated by 5 chromo-lithographic plates, 10 plates in black and colours and 269 wood engravings, 28s. *nett*

GEORGE OLIVER, M.D., F.R.C.P.

I.
PULSE-GAUGING: A Clinical Study of Radial Measurement and Pulse Pressure. With Illustrations, fcap. 8vo. [*Nearly ready.*

II.
THE HARROGATE WATERS: Data Chemical and Therapeutical, with notes on the Climate of Harrogate. Addressed to the Medical Profession. With Map of the Wells, crown 8vo, 3s. 6d.

III.
ON BEDSIDE URINE TESTING: a Clinical Guide to the Observation of Urine in the course of Work. Fourth Edition, fcap. 8vo, 3s. 6d.

DR. A. ONODI.
Docent of Rhinology and Laryngology in the University of Budapest.

THE ANATOMY OF THE NASAL CAVITY, AND ITS ACCESSORY SINUSES. An Atlas for Practitioners and Students, translated by St. Clair Thomson, M.D. Lond., F.R.C.S. Eng., M.R.C.P. Lond. With plates, royal 8vo. [*In preparation.*

SAMUEL OSBORN, F.R.C.S.
Surgeon to the Hospital for Women, Soho Square; Surgeon to the Royal Naval Artillery Volunteers.

I.
AMBULANCE LECTURES: FIRST AID. Third Edition, with Illustrations, fcap. 8vo, 2s. [*Just published.*

II.
AMBULANCE LECTURES: HOME NURSING AND HYGIENE. Second Edition, with Illustrations, fcap. 8vo, 2s.

WILLIAM OSLER, M.D., F.R.C.P. LOND.
President of the Association of American Physicians; Professor of Medicine, Johns Hopkins University, and Physician-in-Chief Johns Hopkins Hospital, Baltimore.

I.

ON CHOREA AND CHOREIFORM AFFECTIONS.
Demy 8vo, 5s. [*Just ready.*

II.

THE CEREBRAL PALSIES OF CHILDREN. A Clinical Study from the Infirmary for Nervous Diseases, Philadelphia. Demy 8vo, 5s.

KURRE W. OSTROM.
Instructor in Massage and Swedish Movements in the Philadelphia Polyclinic and College for Graduates in Medicine.

MASSAGE AND THE ORIGINAL SWEDISH MOVEMENTS; their application to various diseases of the body. Second Edition, with Illustrations, 12mo, 3s. 6d. *nett.*

CHARLES A. PARKER.
Assistant Surgeon to the Hospital for Diseases of the Throat, Golden Square, London.

POST-NASAL GROWTHS. Demy 8vo, 4s. 6d. [*Just published.*

ROBERT W. PARKER.
Senior Surgeon to the East London Hospital for Children; Surgeon to the German Hospital.

I.

DIPHTHERIA: ITS NATURE AND TREATMENT, WITH SPECIAL REFERENCE TO THE OPERATION, AFTER-TREATMENT AND COMPLICATIONS OF TRACHEOTOMY. Third Edition, with Illustrations, 8vo, 6s.

II.

CONGENITAL CLUB-FOOT; ITS NATURE AND TREATMENT. With special reference to the subcutaneous division of Tarsal Ligaments. 8vo, 7s. 6d.

LOUIS C. PARKES, M.D. LOND., D.P.H.
Lecturer on Public Health at St. George's Hospital; Medical Officer of Health for Chelsea.

I.

HYGIENE AND PUBLIC HEALTH. Third Edition, with numerous Illustrations, crown 8vo, 10s. 6d.
[LEWIS'S PRACTICAL SERIES.]

II.

INFECTIOUS DISEASES, NOTIFICATION AND PREVENTION. Fcap. 8vo, roan, 4s. 6d.

JOHN S. PARRY, M.D.
Obstetrician to the Philadelphia Hospital, Vice-President of the Obstetrical and Pathological Societies of Philadelphia, &c.

EXTRA-UTERINE PREGNANCY; Its Causes, Species, Pathological Anatomy, Clinical History, Diagnosis, Prognosis and Treatment. 8vo, 8s.

THEOPHILUS PARVIN, M.D.
Professor of Obstetrics and Diseases of Women and Children at the Jefferson Medical School.
LECTURES ON OBSTETRIC NURSING, Delivered at the Training School for Nurses of the Philadelphia Hospital. Post 8vo, 2s. 6d.

E. RANDOLPH PEASLEE, M.D., LL.D.
Late Professor of Gynæcology in the Medical Department of Dartmouth College; President of New York Academy of Medicine, &c., &c.
OVARIAN TUMOURS: Their Pathology, Diagnosis, and Treatment, especially by Ovariotomy. Illustrations, roy. 8vo, 16s.

LESLIE PHILLIPS, M.D.
Surgeon to the Birmingham and Midland Skin and Lock Hospital.
MEDICATED BATHS IN THE TREATMENT OF SKIN DISEASES. Crown 8vo, 4s. 6d.

HENRY G. PIFFARD, A.M., M.D.
Clinical Professor of Dermatology, University of the City of New York; Surgeon in Charge of the New York Dispensary for Diseases of the Skin, &c.
A PRACTICAL TREATISE ON DISEASES OF THE SKIN. With 50 full page Original Plates and 33 Illustrations in the Text, 4to, £2 12s. 6d. *nett.*

G. V. POORE, M.D., F.R.C.P.
Professor of Medical Jurisprudence, University College; Assistant Physician to, and Physician in charge of the Throat Department of, University College Hospital.
LECTURES ON THE PHYSICAL EXAMINATION OF THE MOUTH AND THROAT. With an Appendix of Cases. 8vo, 3s. 6d.

R. DOUGLAS POWELL, M.D., F.R.C.P.,
Physician Extra-ordinary to H.M. the Queen; Physician to the Middlesex Hospital; Consulting Physician to the Hospital for Consumption and Diseases of the Chest at Brompton.
DISEASES OF THE LUNGS AND PLEURÆ, INCLUDING CONSUMPTION. Fourth Edition, with coloured plates and wood engravings, 8vo, 18s. [*Now ready*

TABLE OF PHYSICAL EXAMINATION OF THE LUNGS: with Note on International Nomenclature of Physical Signs (reprinted from DR. POWELL's "Diseases of the Lungs,"). On one sheet, 6d.

HERBERT A. POWELL, M.A., M.D., M.CH. OXON., F.R.C.S.E.
THE SURGICAL ASPECT OF TRAUMATIC INSANITY. Medium 8vo, 2s. 6d.

URBAN PRITCHARD, M.D. EDIN., F.R.C.S. ENG.
Professor of Aural Surgery at King's College, London; Aural Surgeon to King's College Hospital; Senior Surgeon to the Royal Ear Hospital.
HANDBOOK OF DISEASES OF THE EAR FOR THE USE OF STUDENTS AND PRACTITIONERS. Second Edition, With Illustrations, crown 8vo, 5s.   [LEWIS'S PRACTICAL SERIES.]

CHARLES W. PURDY, M.D. (QUEEN'S UNIV.)
Professor of Genito-Urinary and Renal Diseases in the Chicago Polyclinic, &c., &c.
BRIGHT'S DISEASE AND THE ALLIED AFFECTIONS OF THE KIDNEYS. With Illustrations, large 8vo, 8s. 6d.

DR. THEODOR PUSCHMANN.
Public Professor in Ordinary at the University of Vienna.
A HISTORY OF MEDICAL EDUCATION FROM THE MOST REMOTE TO THE MOST RECENT TIMES. Translated and edited by EVAN H. HARE, M.A. OXON., F.R.C.S. ENG., L.S.A. Demy 8vo, 21s.

C. H. RALFE, M.A., M.D. CANTAB., F.R.C.P. LOND.
Assistant Physician to the London Hospital; Examiner in Medicine to the University of Durham, &c., &c.
A PRACTICAL TREATISE ON DISEASES OF THE KIDNEYS AND URINARY DERANGEMENTS. With Illustrations, crown 8vo, 10s. 6d.   [LEWIS'S PRACTICAL SERIES.]

FRANCIS H. RANKIN, M.D.
President of the New York Medical Society.
HYGIENE OF CHILDHOOD. Suggestions for the care of Children after the Period of Infancy to the completion of Puberty. Crown 8vo, 3s.

AMBROSE L. RANNEY, A.M., M.D.
Professor of the Anatomy and Physiology of the Nervous System in the New York Post-Graduate Medical School and Hospital, &c.
THE APPLIED ANATOMY OF THE NERVOUS SYS- TEM. Second Edition, 238 Illustrations, large 8vo, 21s.

H. A. REEVES, F.R.C.S. EDIN.
Senior Assistant Surgeon and Teacher of Practical Surgery at the London Hospital; Surgeon to the Royal Orthopædic Hospital.
BODILY DEFORMITIES AND THEIR TREATMENT: A HANDBOOK OF PRACTICAL ORTHOPÆDICS. Illustrations, crown 8vo, 8s. 6d.   [LEWIS'S PRACTICAL SERIES.]

RALPH RICHARDSON, M.A., M.D.
Fellow of the College of Physicians, Edinburgh.
ON THE NATURE OF LIFE: An Introductory Chap- ter to Pathology. · Second edition, revised and enlarged. Fcap. 4to, 10s. 6d.

W. RICHARDSON, M.A., M.D., M.R.C.P.
REMARKS ON DIABETES, ESPECIALLY IN REFERENCE TO TREATMENT. Demy 8vo, 4s. 6d.

SAMUEL RIDEAL, D.SC. (LOND.), F.I.C., F.C.S., F.G.S.
Fellow of University College, London.

I.
PRACTICAL ORGANIC CHEMISTRY; The Detection and Properties of some of the more important Organic Compounds. 12mo, 2s. 6d.

II.
PRACTICAL CHEMISTRY FOR MEDICAL STUDENTS, required at the First Examination of the Conjoint Examining Board in England. Foolscap 8vo, 2s.

J. JAMES RIDGE, M.D.
Medical Officer of Health, Enfield.
ALCOHOL AND PUBLIC HEALTH. Second Edition, crown 8vo, 2s.

E. A. RIDSDALE.
Associate of the Royal School of Mines.
COSMIC EVOLUTION; being Speculations on the Origin of our Environment. Fcap. 8vo, 3s.

SYDNEY RINGER, M.D., F.R.S.
Professor of the Principles and Practice of Medicine in University College; Physician to, and Professor of Clinical Medicine in, University College Hospital.

I.
A HANDBOOK OF THERAPEUTICS.
Twelfth Edition thoroughly revised, 8vo, 15s.

II.
ON THE TEMPERATURE OF THE BODY AS A MEANS OF DIAGNOSIS AND PROGNOSIS IN PHTHISIS. Second Edition, small 8vo, 2s. 6d.

R. LAWTON ROBERTS, M.D. LOND., D.P.H. CAMB., M.R.C.S. ENG.
Honorary Life Member of, and Lecturer and Examiner to, the St. John Ambulance Association.

I.
ILLUSTRATED LECTURES ON AMBULANCE WORK. Fourth Edition, copiously Illustrated, crown 8vo, 2s. 6d.

II.
ILLUSTRATED LECTURES ON NURSING AND HYGIENE. Second Edition, with Illustrations, crown 8vo, 2s. 6d.

FREDERICK T. ROBERTS, M.D., B.SC., F.R.C.P.
Professor of Materia Medica and Therapeutics and of Clinical Medicine in University College; Physician to University College Hospital; Consulting Physician to Brompton Consumption Hospital, &c.

I.
THE THEORY AND PRACTICE OF MEDICINE.
Ninth Edition, with Illustrations, large 8vo, 21s.   [*Just Published.*

II.
THE OFFICINAL MATERIA MEDICA.
Second Edition, entirely rewritten in accordance with the latest British Pharmacopœia, fcap. 8vo, 7s. 6d.

III.
NOTES ON THE ADDITIONS MADE TO THE BRITISH PHARMACOPŒIA, 1890.   Fcap. 8vo, 1s.

D. B. ST. JOHN ROOSA, M.D.
Professor of Diseases of the Eye and Ear in the New York Post-Graduate Medical School; Consulting Surgeon to the Brooklyn Eye and Ear Hospital, &c.

A PRACTICAL TREATISE ON THE DISEASES OF THE EAR: Including a Sketch of Aural Anatomy and Physiology. Seventh Edition, Illustrated, large 8vo, 25s.

ROBSON ROOSE, M.D., LL.D., F.C.S.
Fellow of the Royal College of Physicians in Edinburgh, &c.

I.
GOUT, AND ITS RELATIONS TO DISEASES OF THE LIVER AND KIDNEYS.   Seventh Edition, crown 8vo, 4s. 6d.
[*Now ready.*

II.
NERVE PROSTRATION AND OTHER FUNCTIONAL DISORDERS OF DAILY LIFE.   Second Edition, demy 8vo, 18s.

III.
LEPROSY AND ITS PREVENTION: as Illustrated by Norwegian Experience.   Crown 8vo, 3s. 6d.

WILLIAM ROSE, M.B., B.S. LOND., F.R.C.S.
Professor of Surgery in King's College, London, and Surgeon to King's College Hospital.

HARELIP AND CLEFT PALATE.   With Illustrations, demy 8vo, 6s.

BERNARD ROTH, F.R.C.S.
Fellow of the Medical Society of London; Member of the Clinical and Pathological Societies and of the Medical Officers of Schools' Association.

THE TREATMENT OF LATERAL CURVATURE OF THE SPINE.   With Photographic and other Illustrations, demy 8vo, 5s.

J. BURDON SANDERSON, M.D., LL.D., F.R.S.
Jodrell Professor of Physiology in University College, London.

UNIVERSITY COLLEGE COURSE OF PRACTICAL EXERCISES IN PHYSIOLOGY. With the co-operation of F. J. M. PAGE, B.Sc., F.C.S.; W. NORTH, B.A., F.C.S., and AUG. WALLER, M.D. Demy 8vo, 3s. 6d.

W. H. O. SANKEY, M.D. LOND., F.R.C.P.
Late Lecturer on Mental Diseases, University College, London, etc.

LECTURES ON MENTAL DISEASE. Second Edition, with coloured Plates, 8vo, 12s. 6d.

JOHN SAVORY.
Member of the Society of Apothecaries, London.

A COMPENDIUM OF DOMESTIC MEDICINE AND COMPANION TO THE MEDICINE CHEST: Intended as a source of easy reference for Clergymen, Master Mariners, and Travellers; and for Families resident at a distance from professional assistance. Tenth Edition, sm. 8vo, 5s.

DR. C. SCHIMMELBUSCH.
Privat-docent and Assistant Surgeon in Prof. v. Bergmann's University Clinic at Berlin.

THE ASEPTIC TREATMENT OF WOUNDS. With a Preface by Professor DR. E. VON BERGMANN. Translated from the Second German Edition by A. T. RAKE, M.B., B.S. LOND., F.R.C.S. ENG., Registrar and Pathologist to the East London Hospital for Children. With Illustrations, crown 8vo. [*Nearly ready.*

E. SCHMIEGELOW, M.D.
Consulting Physician in Laryngology to the Municipal Hospital and Director of the Oto-Laryngological Department in the Polyclinic at Copenhagen.

ASTHMA: Especially in its Relation to Nasal Disease. Demy 8vo, 4s. 6d.

DR. B. S. SCHULTZE.
Professor of Gynecology; Director of the Lying-in Hospital, and of the Gynecological Clinic at Jena.

THE PATHOLOGY AND TREATMENT OF DISPLACEMENTS OF THE UTERUS. Translated by J. J. MACAN, M.A., M.R.C.S. and edited by A. V. MACAN, M.B., M.Ch., Master of the Rotunda Lying-in Hospital, Dublin. With 120 Illustrations, medium 8vo, 12s. 6d.

JOHN SHAW, M.D. LOND., M.R.C.P.
Obstetric Physician to the North-West London Hospital.
ANTISEPTICS IN OBSTETRIC NURSING. A Text-
book for Nurses on the Application of Antiseptics to Gynæcology and Midwifery. Coloured plate and woodcuts, 8vo, 3s. 6d.

A. J. C. SKENE, M.D.
Professor of Gynecology in the Long Island College Hospital, Brooklyn, New York.
TREATISE ON THE DISEASES OF WOMEN, FOR
THE USE OF STUDENTS AND PRACTITIONERS. Second Edition, with coloured plates and 251 engravings, large 8vo, 28s.

J. LEWIS SMITH, M.D.
Physician to the New York Foundling Asylum; Clinical Professor of Diseases of Children in Bellevue Hospital Medical College.
A TREATISE ON THE DISEASES OF INFANCY
AND CHILDHOOD. Seventh Edition, with Illustrations, large 8vo, 21s.

FRANCIS W. SMITH, M.B., B.S.
THE SALINE WATERS OF LEAMINGTON. Second Edit., with Illustrations, crown 8vo, 1s. *nett.*

JOHN KENT SPENDER, M.D. LOND.
Physician to the Royal Mineral Water Hospital, Bath.
THE EARLY SYMPTOMS AND THE EARLY TREAT-
MENT OF OSTEO-ARTHRITIS, commonly called Rheumatoid Arthritis, with special reference to the Bath Thermal Waters. Sm. 8vo, 2s. 6d.

LOUIS STARR, M.D.
Physician to the Children's Hospital, Philadelphia; late Clinical Professor of Diseases of Children in the Hospital of the University of Pennsylvania.
HYGIENE OF THE NURSERY. Including the General
Regimen and Feeding of Infants and Children; Massage, and the Domestic Management of the Ordinary Emergencies of Early Life. Third Edition, with Illustrations, crown 8vo, 3s. 6d.

JAMES STARTIN, M.B., M.R.C.S.
Senior Surgeon to the London Skin Hospital.
I.
A PHARMACOPŒIA FOR DISEASES OF THE SKIN.
Third Edition, 32mo, 2s. 6d.
II.
LECTURES ON THE PARASITIC DISEASES OF
THE SKIN. VEGETOID AND ANIMAL. With Illustrations, crown 8vo, 2s. 6d.

W. E. STEAVENSON, M.D.
Late in charge of the Electrical Department in St. Bartholomew's Hospital,
AND
H. LEWIS JONES, M.A., M.D., M.R.C.P.
Medical Officer in charge of the Electrical Department in St. Bartholomew's Hospital.
MEDICAL ELECTRICITY. A Practical Handbook for Students and Practitioners. With Illustrations, crown 8vo, 9s.
[LEWIS'S PRACTICAL SERIES.]

JOHN LINDSAY STEVEN, M.D.
Assistant Physician and Pathologist, Glasgow Royal Infirmary; Physician for Out-patients, Royal Hospital for Sick Children, Glasgow; Lecturer on Pathology, St. Mungo's and Queen Margaret Colleges, Glasgow, &c.
THE PATHOLOGY OF MEDIASTINAL TUMOURS. With special reference to Diagnosis. With Plates, 8vo, 4s. 6d.

LEWIS A. STIMSON, B.A., M.D.
Surgeon to the Presbyterian and Bellevue Hospitals; Professor of Clinical Surgery in the Medical Faculty of the University of the City of New York, &c.
A MANUAL OF OPERATIVE SURGERY. Second Edition, with three hundred and forty-two Illustrations, post 8vo, 10s. 6d.

ADOLF STRÜMPELL.
Professor and Director of the Medical Clinique at Erlangen.
A TEXT-BOOK OF MEDICINE FOR STUDENTS AND PRACTITIONERS. Second Edition translated from the German by Dr. H. F. VICKERY and Dr. P. C. KNAPP, with Editorial Notes by Dr. F. C. SHATTUCK, Visiting Physician to the Massachusetts General Hospital, etc. Complete in one large vol., with 119 Illustrations, imp. 8vo, 28s.

JUKES DE STYRAP, M.K.Q.C.P., ETC.
Physician-Extraordinary, late Physician in Ordinary, to the Salop Infirmary; Consulting Physician to the South Salop and Montgomeryshire Infirmaries, etc.

I.
THE YOUNG PRACTITIONER: WITH PRACTICAL HINTS AND INSTRUCTIVE SUGGESTIONS, AS SUBSIDIARY AIDS, FOR HIS GUIDANCE ON ENTERING INTO PRIVATE PRACTICE. Demy 8vo, 7s. 6d. *nett.*

II.
A CODE OF MEDICAL ETHICS: WITH GENERAL AND SPECIAL RULES FOR THE GUIDANCE OF THE FACULTY AND THE PUBLIC IN THE COMPLEX RELATIONS OF PROFESSIONAL LIFE. Third Edition, demy 8vo, 3s. *nett.*

III.
MEDICO-CHIRURGICAL TARIFFS. Fourth Edition, fcap. 4to, revised and enlarged, 2s. *nett.*

IV.
THE YOUNG PRACTITIONER: HIS CODE AND TARIFF. Being the above three works in one volume. Demy 8vo 10s. 6d. *nett.*

C. W. SUCKLING, M.D. LOND., M.R.C.P.
Professor of Materia Medica and Therapeutics at the Queen's College, Physician to the Queen's Hospital, Birmingham, etc.

I.

ON THE DIAGNOSIS OF DISEASES OF THE BRAIN, SPINAL CORD, AND NERVES. With Illustrations, crown 8vo, 8s. 6d.

II.

ON THE TREATMENT OF DISEASES OF THE NERVOUS SYSTEM. Crown 8vo, 7s. 6d.

JOHN BLAND SUTTON, F.R.C.S.
Lecturer on Comparative Anatomy, Senior Demonstrator of Anatomy, and Assistant Surgeon to the Middlesex Hospital; Erasmus Wilson Lecturer, Royal College of Surgeons, England.

LIGAMENTS: THEIR NATURE AND MORPHOLOGY. With numerous Illustrations, post 8vo, 4s. 6d.

HENRY R. SWANZY, A.M., M.B., F.R.C.S.I.
Examiner in Ophthalmic Surgery in the University of Dublin, and in the Royal University of Ireland; Surgeon to the National Eye and Ear Infirmary, and Ophthalmic Surgeon to the Adelaide Hospital, Dublin.

A HANDBOOK OF THE DISEASES OF THE EYE AND THEIR TREATMENT. Fourth Edition, Illustrated with wood-engravings, coloured plates, colour tests, etc., small 8vo, 10s. 6d.

EUGENE S. TALBOT, M.D., D.D.S.
Professor of Dental Surgery in the Woman's Medical College; Lecturer on Dental Pathology and Surgery in Rush Medical College, Chicago.

IRREGULARITIES OF THE TEETH AND THEIR TREATMENT. With 152 Illustrations, royal 8vo, 10s. 6d.

ALBERT TAYLOR.
Associate Sanitary Institute; Chief Sanitary Inspector to the Vestry of St. George, Hanover Square, etc.

THE SANITARY INSPECTOR'S HANDBOOK. With Illustrations, cr. 8vo., 5s.

H. COUPLAND TAYLOR, M.D.
Fellow of the Royal Meteorological Society.

WANDERINGS IN SEARCH OF HEALTH, OR MEDICAL AND METEOROLOGICAL NOTES ON VARIOUS FOREIGN HEALTH RESORTS. With Illustrations, crown 8vo, 6s.

J. C. THOROWGOOD, M.D.
Assistant Physician to the City of London Hospital for Diseases of the Chest.
THE CLIMATIC TREATMENT OF CONSUMPTION AND CHRONIC LUNG DISEASES. Third Edition, post 8vo, 3s. 6d.

D. HACK TUKE, M.D., LL.D.
Fellow of the Royal College of Physicians, London.
THE INSANE IN THE UNITED STATES AND CANADA. Demy 8vo, 7s. 6d.

DR. R. ULTZMANN.

ON STERILITY AND IMPOTENCE IN MAN. Translated from the German with notes and additions by ARTHUR COOPER, L.R.C.P., M.R.C.S., Surgeon to the Westminster General Dispensary. With Illustrations, fcap. 8vo, 2s. 6d.

W. H. VAN BUREN, M.D., LL.D.
Professor of Surgery in the Bellevue Hospital Medical College.
DISEASES OF THE RECTUM: And the Surgery of the Lower Bowel. Second Edition, with Illustrations, 8vo, 14s.

ALFRED VOGEL, M.D.
Professor of Clinical Medicine in the University of Dorpat, Russia.
A PRACTICAL TREATISE ON THE DISEASES OF CHILDREN. Third Edition, translated and edited by H. RAPHAEL, M.D., from the Eighth German Edition, illustrated by six lithographic plates, part coloured, royal 8vo, 18s.

A. DUNBAR WALKER, M.D., C.M.
THE PARENT'S MEDICAL NOTE BOOK.
Oblong post 8vo, cloth, 1s. 6d.

A. J. WALL, M.D. LOND.
Fellow of the Royal College of Surgeons of England; of the Medical Staff of H. M. Indian Army (Retired List).
ASIATIC CHOLERA: ITS HISTORY, PATHOLOGY, AND MODERN TREATMENT. Demy 8vo, 6s.

JOHN RICHARD WARDELL, M.D. EDIN., F.R.C.P. LOND.
Late Consulting Physician to the General Hospital Tunbridge Wells.
CONTRIBUTIONS TO PATHOLOGY AND THE PRAC- TICE OF MEDICINE. Medium 8vo, 21s.

W. SPENCER WATSON, B.M. LOND., F.R.C.S. ENG.
Surgeon to the Throat Department of the Great Northern Hospital; Senior Surgeon to the Royal South London Ophthalmic Hospital.

I.
DISEASES OF THE NOSE AND ITS ACCESSORY CAVITIES. Second Edition, with Illustrations, demy 8vo, 12s. 6d.

II.
THE ANATOMY AND DISEASES OF THE LACHRYMAL PASSAGES. With Illustrations, demy 8vo, 2s. 6d.

III.
EYEBALL-TENSION: Its Effects on the Sight and its Treatment. With woodcuts, p. 8vo, 2s. 6d.

IV.
ON ABSCESS AND TUMOURS OF THE ORBIT. Post 8vo, 2s. 6d.

FRANCIS H. WELCH, F.R.C.S.
Surgeon Major, A.M.D.

ENTERIC FEVER: as Illustrated by Army Data at Home and Abroad, its Prevalence and Modifications, Ætiology, Pathology and Treatment. 8vo, 5s. 6d.

W. WYNN WESTCOTT, M.B.
Deputy Coroner for Central Middlesex.

SUICIDE; its History, Literature, Jurisprudence, and Prevention. Crown 8vo, 6s.

FRANK J. WETHERED, M.D.
Medical Registrar to the Middlesex Hospital, and Demonstrator of Practical Medicine in the Middlesex Hospital Medical School; late Assistant Physician to the City of London Chest Hospital, Victoria Park.

MEDICAL MICROSCOPY. A Guide to the Use of the Microscope in Medical Practice. With Illustrations, crown 8vo, 9s.
[LEWIS'S PRACTICAL SERIES.]

E. G. WHITTLE, M.D. LOND., F.R.C.S. ENG.
Senior Surgeon to the Royal Alexandra Hospital for Sick Children, Brighton.

CONGESTIVE NEURASTHENIA, OR INSOMNIA AND NERVE DEPRESSION. Crown 8vo, 3s. 6d.

SIR JOHN WILLIAMS, BART., M.D., F.R.C.P.
Consulting Physician to University College Hospital; Physician Accoucheur to H.R.H. Princess Beatrice, &c.

CANCER OF THE UTERUS: Being the Harveian Lectures for 1886. Illustrated with Lithographic Plates, royal 8vo, 10s. 6d.

E. F. WILLOUGHBY, M.D. LOND.

THE NATURAL HISTORY OF SPECIFIC DISEASES OR STUDIES IN ŒTIOLOGY, IMMUNITY, AND PROPHYLAXIS. 8vo, 2s. 6d.

E. T. WILSON, M.B. OXON., F.R.C.P. LOND.
Physician to the Cheltenham General Hospital; Associate Metropolitan Association of Medical Officers of Health.

DISINFECTANTS AND ANTISEPTICS: HOW TO USE THEM. In Packets of one doz. price 1s., by post 1s. 1d.
[*Just thoroughly revised.*

DR. F. WINCKEL.
Formerly Professor and Director of the Gynæcological Clinic at the University of Rostock.

THE PATHOLOGY AND TREATMENT OF CHILDBED: A Treatise for Physicians and Students. Translated from the Second German Edition, with many additional notes by the Author, by J. R. CHADWICK, M.D. 8vo, 14s.

BERTRAM C. A. WINDLE, M.A., M.D. DUBL.
Professor of Anatomy in the Queen's College, Birmingham; Examiner in Anatomy in the Universities of Cambridge and Durham.

A HANDBOOK OF SURFACE ANATOMY AND LANDMARKS. Illustrated, post 8vo, 3s. 6d.

EDWARD WOAKES, M.D. LOND.
Senior Aural Surgeon and Lecturer on Aural Surgery at the London Hospital; Surgeon to the London Throat Hospital.

I.
ON DEAFNESS, GIDDINESS AND NOISES IN THE HEAD. Third Edition, with Illustrations, 8vo. [*In preparation.*

II.
POST-NASAL CATARRH, AND DISEASES OF THE NOSE CAUSING DEAFNESS. With Illustrations, cr. 8vo, 6s. 6d.

III.
NASAL POLYPUS: WITH NEURALGIA, HAY-FEVER, AND ASTHMA, IN RELATION TO ETHMOIDITIS. With Illustrations, cr. 8vo, 4s. 6d.

HENRY WOODS, B.A., F.G.S.

ELEMENTARY PALÆONTOLOGY—INVERTEBRATE. Crown 8vo, 6s.

[CAMBRIDGE NATURAL SCIENCE MANUALS.]

DAVID YOUNG, M.C., M.B., M.D.
Licentiate of the Royal College of Physicians, Edinburgh ; Licentiate of the Royal College of Surgeons, Edinburgh, etc.

ROME IN WINTER AND THE TUSCAN HILLS IN SUMMER. A Contribution to the Climate of Italy. Small 8vo, 6s.

HERMANN VON ZEISSL, M.D.
Late Professor at the Imperial Royal University of Vienna.

OUTLINES OF THE PATHOLOGY AND TREATMENT OF SYPHILIS AND ALLIED VENEREAL DISEASES. Second Edition, revised by M. von Zeissl, M.D., Privat-Docent for Diseases of the Skin and Syphilis at the Imperial Royal University of Vienna. Translated, with Notes, by H. Raphael, M.D., Attending Physician for Diseases of Genito-Urinary Organs and Syphilis, Bellevue Hospital, Out-Patient Department. Large 8vo, 18s.

OSWALD ZIEMSSEN, M.D.
Knight of the Iron Cross, and of the Prussian Order of the Crown.

THE TREATMENT OF CONSTITUTIONAL SYPHILIS. Post 8vo, 3s. 6d.

Lewis's Diet Charts.
Price 6s. 6d. per packet of 100 charts, by post 6s. 10½d.

A suggestive set of diet tables for the use of Physicians, for handing to patients after consultation, modified to suit individual requirements, for Albuminuria, Alcoholism, Anæmia and Debility, Constipation, Diabetes, Diarrhœa, Dyspepsia, Fevers, Gout, Nervous Diseases, Obesity, Phthisis, Rheumatism (chronic), with blank chart for other diseases.

Lewis's Four-Hour Temperature Charts.
25s. per 1000, 14s. per 500, 3s. 6d. per 100, 2s. per 50, 1s. per 20, carriage free.

This form has been drawn up to meet the requirements of a chart on which the temperature and other observations can be recorded at intervals of four hours. They will be found most convenient in hospital and private practice. Each chart will last a week.

Clinical Charts for Temperature Observations, etc.
Arranged by W. Rigden, M.R.C.S. 50s. per 1000, 28s. per 500, 15s. per 250, 7s. per 100, or 1s. per dozen.

Each Chart is arranged for four weeks, and is ruled at the back for making notes of Cases; they are convenient in size, and are suitable both for hospital and private practice.

Lewis's Clinical Chart, specially designed for use with the
Visiting List. This Temperature Chart is arranged for four weeks and measures 6 × 3 inches. 30s. per 1000, 16s. 6d. per 500, 3s. 6d. per 100, 1s. per 25, 6d. per 12.

Lewis's Nursing Chart.
25s. per 1000, 14s. per 500, 3s. 6d. per 100, 2s. per 50, or 1s. per 20.

These Charts afford a ready method of recording the progress of the case from day to day.
Boards to hold the Charts, price 1s.

Chart for Recording the Examination of Urine.
40s. per 1000; 25s. per 500; 15s. per 250; 7s. 6d. per 100; 1s. per 10.

These Charts are designed for the use of Medical Men, Analysts, and others making examinations of the urine of patients, and afford a very ready and convenient method of recording the results of the examination.

LEWIS'S PRACTICAL SERIES.

Under this title a Series of Monographs is published, embracing the various branches of Medicine and Surgery. The volumes are written by well-known Hospital Physicians and Surgeons, recognized as authorities in the subjects of which they treat. The works are of a THOROUGHLY PRACTICAL nature, calculated to meet the requirements of the practitioner and student, and to present the most recent information in a compact form.

DISEASES OF THE NOSE AND THROAT.
By F. de HAVILLAND HALL, M.D., F.R.C.P. Lond., Physician to Out-patients, and in charge of the Throat Department at the Westminster Hospital. With coloured plates and wood engravings, crown 8vo. 10s. 6d. [*Now ready.*

PUBLIC HEALTH LABORATORY WORK.
By HENRY R. KENWOOD, M.B., D.P.H., F.C.S., Instructor in the Hygienic Laboratory, University College, and Assistant to Professor Corfield in the Public Health Department, University College. With Illustrations, cr. 8vo, 10s. 6d.

MEDICAL MICROSCOPY: A GUIDE TO THE USE OF THE MICROSCOPE IN MEDICAL PRACTICE. By FRANK J. WETHERED, M.D., M.R.C.P., Demonstrator of Practical Medicine in the Middlesex Hospital Medical School, &c. With Illustrations, crown 8vo, 9s.

MEDICAL ELECTRICITY. A PRACTICAL HANDBOOK FOR STUDENTS AND PRACTITIONERS. By W. E. STEAVENSON, M D., and H. LEWIS JONES, M.A., M.D., M.R.C.P., Medical Officer in charge of the Electrical Department in St. Bartholomew's Hospital. With Illustrations, crown 8vo, 9s.

HYGIENE AND PUBLIC HEALTH.
By LOUIS C PARKES, M.D., D.P.H. LOND. UNIV., Fellow of the Sanitary Institute, and Member of the Board of Examiners. Third edition, with numerous Illustrations, cr. 8vo, 10s. 6d.

MANUAL OF OPHTHALMIC PRACTICE.
By C. HIGGENS, F.R.C.S., Ophthalmic Surgeon to Guy's Hospital; Lecturer on Ophthalmology at Guy's Hospital Medical School. Illustrations, cr. 8vo, 6s.

A PRACTICAL TEXTBOOK OF THE DISEASES OF WOMEN.
By A. H. N. LEWERS, M.D. Lond., M.R.C.P., Obstetric Physician to the London Hospital, etc. Fourth Edition, with Illustrations, crown 8vo, 10s. 6d.

ANÆSTHETICS THEIR USES AND ADMINISTRATION.
By DUDLEY W. BUXTON, M.D., B.S., M.R.C.P., Administrator of Anæsthetics and Lecturer in University College Hospital, etc. Second Edition, with Illustrations, crown 8vo, 5s.

TREATMENT OF DISEASE IN CHILDREN.
By ANGEL MONEY, M.D., F.R.C.P., late Assistant Physician to the Hospital for Sick Children, Great Ormond Street. Second edition, cr. 8vo, 10s. 6d.

ON FEVERS: THEIR HISTORY, ETIOLOGY, DIAGNOSIS, PROGNOSIS, AND TREATMENT. By ALEXANDER COLLIE, M.D. Aberd., M.R.C.P. Illustrated with Coloured Plates, crown 8vo, 8s. 6d.

HANDBOOK OF DISEASES OF THE EAR FOR THE USE OF STUDENTS AND PRACTITIONERS. By URBAN PRITCHARD, M.D. Edin., F.R.C.S. Eng., Professor of Aural Surgery at King's College, London. Second Edition, with Illustrations, crown 8vo, 5s.

A PRACTICAL TREATISE ON DISEASES OF THE KIDNEYS AND URINARY DERANGEMENTS. By C. H. RALFE, M.A., M.D. Cantab., F.R.C.P., Assistant Physician to the London Hospital; Examiner in Medicine to the University of Durham, etc., etc. Illustrations, cr. 8vo, 10s. 6d.

DENTAL SURGERY FOR MEDICAL PRACTITIONERS AND STUDENTS OF MEDICINE. By ASHLEY W. BARRETT, M.B. Lond., M.R.C.S., L.D.S., Dental Surgeon to, and Lecturer on Dental Surgery in the Medical School of, the London Hospital. Second edition, with Illustrations, cr. 8vo, 3s. 6d.

BODILY DEFORMITIES AND THEIR TREATMENT: A HANDBOOK OF PRACTICAL ORTHOPÆDICS. By H. A. REEVES, F.R.C.S. Edin., Senior Assistant Surgeon and Teacher of Practical Surgery at the London Hospital, etc. With numerous Illustrations, cr. 8vo, 8s 6d.

*** Further volumes will be announced in due course.

THE NEW SYDENHAM SOCIETY'S PUBLICATIONS.

President :—T. BRYANT, ESQ., F.R.C.S.
Honorary Secretary :—JONATHAN HUTCHINSON, ESQ., F.R.S.
Treasurer :—W. SEDGWICK SAUNDERS, M.D., F.S.A.
Secretary :—J. HUTCHINSON, JUN., ESQ., F.R.C.S.

Annual Subscription, One Guinea.

The Society issues translations of recent standard works by continental authors on subjects of general interest to the profession.

Amongst works recently issued are "Sir William Gull's Collected Works," "Laveran's Paludism," "Pozzi's Gynecology," "Flügge's Micro-Organisms," "Cohnheim's Pathology," "Henoch's Children," "Spiegelberg's Midwifery," "Hirsch's Historical and Geographical Pathology," "Ewald's Disorders of Digestion," works by Charcot, Duchenne, Begbie, Billroth, Graves, Koch, Hebra, Guttmann, etc.

The Society also has in hand an Atlas of Pathology with Coloured Plates, and a valuable and exhaustive "Lexicon of Medicine and the Allied Sciences."

The Annual Report, with full list of works published, and all further information will be sent on application.

PERIODICAL WORKS PUBLISHED BY H. K. LEWIS.

THE BRITISH JOURNAL OF DERMATOLOGY. Edited by William Anderson, H. G. Brooke, H. Radcliffe Crocker, T. Colcott Fox, Stephen Mackenzie, Malcolm Morris, J. F. Payne and J. J. Pringle. Published monthly, 1s. Annual Subscription, 12s. post free.

THE MEDICAL MAGAZINE. Edited by George J. Wilson, M.A., M.B. Published monthly, 2s. Annual subscription, 20s. post free.

THE NEW YORK MEDICAL JOURNAL. A Weekly Review of Medicine. Annual Subscription, Thirty Shillings, post free.

THE THERAPEUTIC GAZETTE. A Monthly Journal, devoted to the Science of Pharmacology, and to the introduction of New Therapeutic Agents. Edited by Dr. R. M. Smith. Annual Subscription, 10s., post free.

THE GLASGOW MEDICAL JOURNAL. Published Monthly. Annual Subscription, 20s., post free. Single numbers, 2s. each.

LIVERPOOL MEDICO-CHIRURGICAL JOURNAL, including the Proceedings of the Liverpool Medical Institution. Published twice yearly, 3s. 6d. each number.

TRANSACTIONS OF THE COLLEGE OF PHYSICIANS OF PHILADELPHIA. Volumes I. to VI., 8vo, 10s. 6d. each.

MIDDLESEX HOSPITAL, REPORTS OF THE MEDICAL, SURGICAL, AND Pathological Registrars for 1883 to 1892. Demy 8vo, 2s. 6d. *nett* each volume.

ARCHIVES OF PEDIATRICS. A Monthly Journal, devoted to the Diseases of Infants and Children. Edited by Dr. W. P. Watson. Annual Subscription, 12s. 6d., post free.

. MR. LEWIS is in constant communication with the leading publishing firms in America, and has transactions with them for the sale of his publications in that country. Advantageous arrangements are made in the interests of Authors for the publishing of their works in the United States.

Mr. Lewis's publications can be procured of all Booksellers in any part of the world.

London: Printed by H. K. Lewis, 136 Gower Street, W.C

www.ingramcontent.com/pod-product-compliance
Lightning Source LLC
Chambersburg PA
CBHW051743300426
44115CB00007B/680